T0100519

Grid and Cloud Database Management

Sandro Fiore • Giovanni Aloisio

Editors

Grid and Cloud Database Management

 Springer

Editors

Sandro Fiore, Ph.D.
Faculty of Engineering
Department of Innovation Engineering
University of Salento
Via per Monteroni
73100 Lecce, Italy

and

Euro Mediterranean Center
for Climate Change (CMCC)
Via Augusto Imperatore 16
73100 Lecce, Italy
sandro.fiore@unisalento.it

Prof. Giovanni Aloisio
Faculty of Engineering
Department of Innovation Engineering
University of Salento
Via per Monteroni
73100 Lecce, Italy

and

Euro Mediterranean Center
for Climate Change (CMCC)
Via Augusto Imperatore 16
73100 Lecce, Italy
giovanni.aloisio@unisalento.it

ISBN 978-3-642-20044-1   e-ISBN 978-3-642-20045-8
DOI 10.1007/978-3-642-20045-8
Springer Heidelberg Dordrecht London New York

Library of Congress Control Number: 2011929352

ACM Computing Classification (1998): C.2, H.2, H.3, J.2, J.3

Cover design: deblik, Berlin

Printed on acid-free paper

Springer is part of Springer Science+Business Media (www.springer.com)

Preface

Since the 1960s, database systems have been playing a relevant role in the information technology field. By the mid-1960s, several systems were also available for commercial purposes. Hierarchical and network database systems provided two different perspectives and data models to organize data collections. In 1970, E. Codd wrote a paper called *A Relational Model of Data for Large Shared Data Banks*, proposing a model relying on relational table structures. Relational databases became appealing for industries in the 1980s, and their wide adoption fostered new research and development activities toward advanced data models like object oriented or the extended relational. The online transaction processing (OLTP) support provided by the relational database systems was fundamental to make this data model successful. Even though the traditional operational systems were the best solution to manage transactions, new needs related to data analysis and decision support tasks led in the late 1980s to a new architectural model called data warehouse. It includes extraction transformation and loading (ETL) primitives and online analytical processing (OLAP) support to analyze data. From OLTP to OLAP, from transaction to analysis, from data to information, from the entity-relationship data model to a star/snowflake one, and from a customer-oriented perspective to a market-oriented one, data warehouses emerged as data repository architecture to perform data analysis and mining tasks. Relational, object-oriented, transactional, spatiotemporal, and multimedia data warehouses are some examples of database sources. Yet, the World Wide Web can be considered another fundamental and distributed data source (in the Web2.0 era it stores crucial information – from a market perspective – about user preferences, navigation, and access patterns).

Accessing and processing large amount of data distributed across several countries require a huge amount of computational power, storage, middleware services, specifications, and standards.

Since the 1990s, thanks to Ian Foster and Carl Kesselman, grid computing has emerged as a revolutionary paradigm to access and manage distributed, heterogeneous, and geographically spread resources, promising computer power as easy to access as an electric power grid. The term "resources" also includes the database, yet successful attempts of grid database management research efforts started only after 2000. Later on, around 2007, a new paradigm named *Cloud Computing* brought the promise of providing easy and inexpensive access to remote hardware and storage resources. Exploiting pay per use models, virtualization for resource provisioning, cloud computing has been rapidly accepted and used by researchers, scientists, and industries.

Grid and cloud computing are exciting paradigms and how they deal with database management is the key topic of this book. By exploring current and future developments in this area, the book tries to provide a thorough understanding of the principles and techniques involved in these fields.

The idea of writing this book dates back to a tutorial on Grid Database Management that was organized at the 4th International Conference on Grid and Pervasive Computing (GPC 2009) held in Geneva (4–8 May 2009). Following up an initial idea from Ralf Gerstner (Springer Senior Editor Computer Science), we decided to act as editors of the book.

We invited internationally recognized experts asking them to contribute on challenging topics related to grid and cloud database management. After two review steps, 16 chapters have been accepted for publication.

Ultimately, the book provides the reader with a collection of chapters dealing with *Open standards and specifications* (Sect. 1), *Research efforts on grid database management* (Sect. 2), *Cloud data management* (Sect. 3), and some *Scientific case studies* (Sect. 4). The presented topics are well balanced, complementary, and range from well-known research projects and real case studies to standards and specifications as well as to nonfunctional aspects such as security, performance, and scalability, showing up how they can be effectively addressed in grid- and cloud-based environments.

Section 1 discusses the open standards and specifications related to grid and cloud data management. In particular, Chap. 1 presents an overview of the WS-DAI family of specifications, the motivation for defining them, and their relationships with other OGF and non-OGF standards. Conversely, Chap. 2 outlines the OCCI specifications and demonstrates (by presenting three interesting use cases) how they can be used in data management-related setups.

Section 2 presents three relevant research efforts on grid-database management systems. Chapter 3 provides a complete overview on the Grid Relational Catalog (GRelC) Project, a grid database research effort started in 2001. The project's main features, its interoperability with gLite-based production grids, and a relevant showcase in the environmental domain are also presented. Chapter 4 provides a complete overview about the OGSA-DAI framework, the main components for the distributed data management via workflows, the distributed query processing, and the most relevant security and performance aspects. Chapter 5 gives a detailed overview of the architecture and implementation of DASCOSA-DB. A complete description of novel features, developed to support typical data-intensive applications running on a grid system, is also presented.

Section 3 provides a wide overview on several cloud data management topics. Some of them (from Chaps. 6 to 8) specifically focus only on database aspects, whereas the remaining ones (from Chaps. 9 to 12) are wider in scope and address more general cloud data management issues. In this second case, the way these concepts apply to the database world is clarified through some practical examples or comments provided by the authors. In particular, Chap. 6 proposes a new security technique to measure the trustiness of the cloud resources. Through the use of the metadata of resources and access policies, the technique builds the privilege chains and binds authorization policies to compute the trustiness of cloud database management. Chapter 7 presents a method to manage the data with dirty data and obtain the query results with quality assurance in the dirty data. A dirty database storage structure for cloud databases is presented along with a multilevel index structure for query processing on dirty data. Chapter 8 examines column-oriented databases in virtual environments and provides evidence that they can benefit from virtualization in cloud and grid computing scenarios. Chapter 9 introduces a Windows Azure case study demonstrating the advantages of cloud computing and how the generic resources offered by cloud providers can be integrated to produce a large dynamic data store. Chapter 10 presents CloudMiner, which offers a cloud of data services running on a cloud service provider infrastructure. An example related to database management exploiting OGSA-DAI is also discussed. Chapter 11 defines the requirements of e-Science provenance systems and presents a novel solution (addressing these requirements) named the Vienna e-Science Provenance System (VePS). Chapter 12 examines the state of the art of workload management for data-intensive computing in clouds. A taxonomy is presented for workload management of data-intensive computing in the cloud and the use of the taxonomy to classify and evaluate current workload management mechanisms.

Section 4 presents a set of scientific use cases connected with Genomic, Health, Disaster monitoring, and Earth Science. In particular, Chap. 13 explores the implementation of an algorithm, often used to analyze microarray data, on top of an intelligent runtime that abstracts away the hard parts of file tracking and scheduling in a distributed system. This novel formulation is compared with a traditional method of expressing data parallel computations in a distributed environment using explicit message passing. Chapter 14 describes the use of Grid technologies for satellite data processing and management within the international disaster monitoring projects carried out by the Space Research Institute NASU-NSAU, Ukraine (SRI NASU-NSAU). Chapter 15 presents the CDM ActiveStorage infrastructure, a scalable and inexpensive transparent data cube for interactive analysis and high-resolution mapping of environmental and remote sensing data. Finally, Chap. 16 presents a mechanism for distributed storage of multidimensional EEG time series obtained from epilepsy patients on a cloud computing infrastructure (Hadoop cluster) using a column-oriented database (HBase).

The bibliography of the book covers the essential reference material. The aim is to convey any useful information to the interested readers, including researchers actively involved in the research field, students (both undergraduate and graduate), system designers, and programmers.

The book may serve as both an introduction and a technical reference for grid and cloud database management topics. Our desire and hope is that it will prove useful while exploring the main subject, as well as the research and industries efforts involved, and that it will contribute to new advances in this scientific field.

Lecce   *Sandro Fiore*
February 2010   *Giovanni Aloisio*

Contents

Part I
Open Standards and Specifications

Chapter 1
Open Standards for Service-Based Database Access and Integration

Steven Lynden, Oscar Corcho, Isao Kojima, Mario Antonioletti, and Carlos Buil-Aranda

Abstract The Database Access and Integration Services (DAIS) Working Group, working within the Open Grid Forum (OGF), has developed a set of data access and integration standards for distributed environments. These standards provide a set of uniform web service-based interfaces for data access. A core specification, WS-DAI, exposes and, in part, manages data resources exposed by DAIS-based services. The WS-DAI document defines a core set of access patterns, messages and properties that form a collection of generic high-level data access interfaces. WS-DAI is then extended by other specifications that specialize access for specific types of data. For example, WS-DAIR extends the WS-DAI specification with interfaces targeting relational data. Similar extensions exist for RDF and XML data. This chapter presents an overview of the specifications, the motivation for defining them and their relationships with other OGF and non-OGF standards. Current implementations of the specifications are described in addition to some existing

S. Lynden
AIST, Information Technology Research Institute, National Institute of Advanced Industrial Science and Technology (AIST), Tsukuba 305-8568, Japan
e-mail: steven.lynden@aist.go.jp

O. Corcho · C. Buil-Aranda
Ontology Engineering Group, Departamento de Inteligencia Artificial, Facultad de Informática, Universidad Politécnica de Madrid, Boadilla del Monte, 28660, Madrid, Spain
e-mail: ocorcho@fi.upm.es; cbuil@fi.upm.es

I. Kojima (✉)
Information Technology Research Institute, National Institute of Advanced Industrial Science and Technology (AIST), Tsukuba 305-8568, Japan
e-mail: kojima@ni.aist.go.jp

M. Antonioletti
EPCC, The University of Edinburgh, JCMB, The Kings Buildings, Mayfield Road, Edinburgh EH9 3JZ, UK
e-mail: Mario.Antonioletti@ed.ac.uk

S. Fiore and G. Aloisio (eds.), *Grid and Cloud Database Management*,
DOI 10.1007/978-3-642-20045-8_1, © Springer-Verlag Berlin Heidelberg 2011

3

and potential applications to highlight how this work can benefit web service-based architectures used in Grid and Cloud computing.

1.1 Introduction and Background

Standards play a central role in achieving interoperability within distributed environments. By having a set of standardized interfaces to access and integrate geographically distributed data, possibly managed by different organizations that use different database systems, the work that has to be undertaken to manage and integrate this data becomes easier. Thus, providing standards to facilitate the access and integration of database systems on a large scale distributed scale is important.

The Open Grid Forum (OGF)[1] is a community-led standards body formed to promote the open standards required for applied distributed environments such as Grids and Clouds. The OGF is composed of a number of Working Groups that concentrate on producing documents that standardise particular aspects of distributed environments as OGF recommendations, which are complemented by informational documents that inform the community about interesting and useful aspects of distributed computing, experimental documents are more practically based and are required for the recommendation process and finally community documents inform and influence the community on practices anticipated to become common in the distributed computing community. A process has been established [1] that takes these documents through to publication at the OGF web site. An important aspect of the OGF recommendation process is that there must be at least two interoperable implementations of a proposed standard before it can achieve recommendation status. The interoperability testing is a mandatory step required to finalise the process and provide evidence of functional, interoperable implementations of a specification.

The Database Access and Integration Services (DAIS) Working Group was established relatively early within the lifetime of the OGF, which was at that time entitled the Global Grid Forum. The focus on Grids had up, to that point, predominantly been on the sharing of computational resources. DAIS was established to extend the focus to take data into account in the first instance to incorporate databases into Grids. The initial development of the DAIS work was guided by an early requirements capture for data in Grids in GFD.13 [2] as well as the early vision for Grids described by the Open Grid Services Architecture (OGSA) [3]. The first versions of the DAIS specification attempted to use this model only, however, much of the focus of the Grid community changed to the Web Services Resource Framework (WSRF) [4], and DAIS attempted to accommodate this new family of standards whilst still being able to use a non-WSRF solution – a requirement coming from the UK e-Science community [5] – the impact of which is clearly visible in the DAIS specification documents. The rest of this chapter describes the specifications in more detail.

---

[1] http://www.ogf.org.

1.2   The WS-DAI Family of Specification

1.2.1   Overview

The relationship between the WS-DAI (Web Services Database Access and Integration Services) family of specifications is schematically illustrated in Fig. 1.1. These provide a set of message patterns and properties for accessing various types of data. A core specification, the WS-DAI document [6], defines a generic set of interfaces and patterns which are extended by specifications dealing with particular data models: WS-DAIR for relational databases [7], WS-DAIX for XML databases [8] and WS-DAI-RDF(S) for Resource Description Framework (RDF) databases [9].

In WS-DAI, a database is referred to as a data resource. A data resource represents any system that can act as a source or sink of data. It has an abstract name which is represented by a URI and an address which shows the location of a resource. A data access service provides properties and interfaces for describing and accessing data resources. The address of a resource is a web service endpoint such as an EndPointReference (EPR) provided by the WS-Addressing [10] specification. A WSRF data resource provides compatibility with the WS-Resource (WSRF) [4] specifications. A consumer refers to the application or client that interacts with the interfaces provided by a data resource.

An important feature introduced by WS-DAI is the support for indirect access patterns. Typically, web services have a request-response access pattern – this is referred to as direct data access – where the consumer will receive the requested data in the response to a request, typically a query, made to a data access service. For example, passing an XPathQuery message to an XML data access service will result in a response message containing a set of XML fragments. An operation that directly inserts, updates or deletes data through a data access service also constitutes a direct data access. For example, passing an SQL insert statement to a data access service

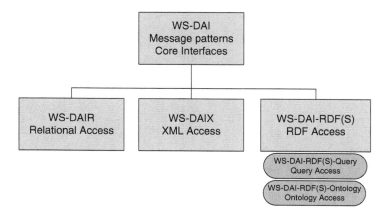

Fig. 1.1 The WS-DAI family of specifications

will result in a response message indicating how many tuples have been inserted. For indirect data access, a consumer will not receive the results in the response to the request made to a data access service. Instead, the request to access data is processed with the results being made available to the consumer indirectly as a new data resource, possibly through a different data service that supports a different set of interfaces. This is useful, for instance, to hold results at the service side minimising any unnecessary data movement. The type and behaviour of the new data resource are determined by the data access service and the configuration parameters passed in with the original request. This indirect access behaviour is different from the request-response style of behaviour found in typical web service interactions.

1.2.2   The Core Specification (WS-DAI)

The WS-DAI specification, also referred to as the core specification, groups interfaces into the following functional categories:

- Data description: Metadata about service and data resource capabilities.
- Data access: Direct access interfaces.
- Data factory: Indirect access interfaces.

It is important to note that data access and data factory operations wrap existing query languages to specify what data is to be retrieved, inserted or modified in the underlying data resources. The DAIS specifications do not define new query languages nor do they do any processing on the incoming queries nor do they provide a complete abstraction of the underlying data resource – for instance, you have to know that the data service you are interacting with wraps a relational database to send SQL queries to it. The benefit of DAIS is that it provides a set of operations that will function on an underlying data resource without requiring knowledge of the native connection mechanisms for that type of database. This makes it easier to build client interfaces that will use DAIS services to talk to different types of databases.

These interface groupings provide a framework for the data service interfaces and the properties that describe, or modify, the behaviour of these interfaces that can then be extended to define interfaces to access particular types of data, as is done by the WS-DAIR, WS-DAIX and WS-DAI-RDF(S) documents.

1.2.2.1   Data Description

Data Description provides the metadata that represents the characteristics of the database and the service that wraps it. The metadata are available as properties that can be read and sometimes modified. If WSRF is used, the WSRF mechanisms can be used to access and modify properties otherwise operations are available to do this for non-WSRF versions of WS-DAI. For instance, the message GetDataResource-PropertyDocument will retrieve metadata that includes the following information:

1. AbstractNames: A set of abstract names for the data resources that are available through that data services. Abstract names are unique and persistent name for a data resource represented as URIs.
2. ConcurrentAccess: A flag indicating whether the data service provides concurrent access or not.
3. DataSetMap: Can be used to retrieve XML Schema representing the data formats that the data service can return the results in.
4. LanguageMap: Shows the query languages that are supported by the data service. Note that DAIS does not require the service to validate the query language that is being used; improper languages will be detected by the underlying data resource.
5. Readable/Writable: A flag indicating whether the data service provides read and write capabilities to the resource. For instance, if the service were providing access to a digital archive it would clearly only have read-only access. This property is meant to describe the underlying characteristics of the data resource rather than authorization to access the resource.
6. TransactionInitiaton: Information about the transactional capabilities of the underlying data resource.

Using this information, a user can understand the database and service capabilities provided by that data service. The property set can be extended to accommodate particular properties pertaining to access to specific types of data, for example, relational, XML and RDF.

1.2.2.2 Data Access

Data access collects together messages that can directly access or modify the data represented by a data access service along with the properties that describe the behaviour of these access messages, as illustrated in Fig. 1.2, which depicts a use case where the WS-DAIR interfaces are used.

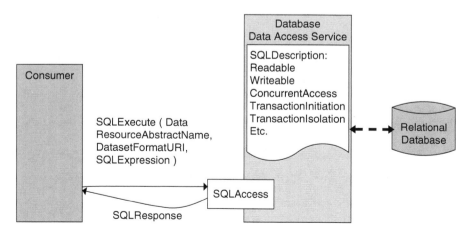

Fig. 1.2 Data access example

In this example, the data access service implements the SQLAccess messages and exposes the SQLDescription properties; more details about the interface and corresponding properties can be found in [7]. A consumer uses the SQLExecute message to submit an SQL expression. The associated response message will contain the results of the SQL execute request. When the SQL expression used is a SELECT statement, the SQL response will contain the data in a RowSet message serialized using an implementation-specific data format, for example the XML WebRowSet [11].

1.2.2.3 Data Factory

Factory messages create a new relationship between a data resource and a data access service. In this way, a data resource may be used to represent the results of a query or act as a place holder where data can be inserted. A data factory describes properties that dictate how a data access service must behave on receiving factory messages. The factory pattern may involve the creation of a new data resource and possibly the deployment of a web service to provide access to it (though existing web services can be re-used for this purpose – DAIS does not specify how this should be implemented). The WS-DAI specification only sets the patterns that should be used for extensions to particular types of data.

This ability to derive one data resource from another, or to provide alternative views of the same data resources, leads to a collection of notionally related data resources, as illustrated in Fig. 1.3, which again takes an example from the WS-DAIR specification. The database data access service in this example presents an SQLFactory interface. The SQLExecuteFactory operation is used to construct a new derived data resource from the SQL query contained in it. These results are then made available through an SQLResponseAccess interface which may be available through the original service or as part of a new data service. Access to the RowSet resulting from the SQL expression executed by the underlying data resource is made available through a suitable interface, assuming that the original expression contains a SELECT statement.

The RowSet could be stored as a table in a relational database or in a form decoupled from the database. DAIS does not specify how this should be implemented but the implementation does have a bearing on the properties of ChildSensitiveToParent and ParentSensitiveToChild which indicate whether changes in the child data affect the parent data or changes in the parent data affect the child data, respectively. The RowSet results are represented as a collection of rows via a data access service supporting the SQLResponseAccess collection of operations that allow the RowSet to be retrieved but does not provide facilities for submitting SQL expressions via the SQLAccess portType.

The Factory interfaces provide a means of minimising data movement when it is not required in addition to an indirect form of third party data delivery: consumer A creates a derived data resource available through some specified data service whose reference can be passed on to consumer B to access.

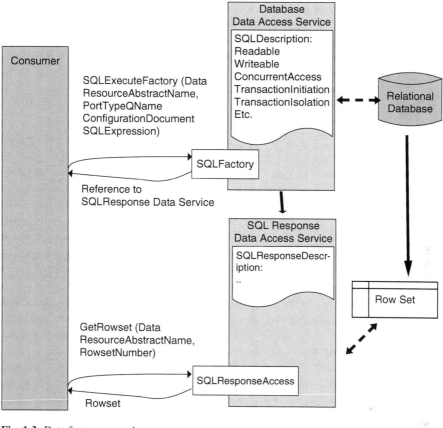

Fig. 1.3 Data factory example

The data resources derived by means of the Factory-type interfaces are referred to as data service managed resources as opposed to the externally managed resources which are database management systems exposed by the data services. Clearly, the creation of these derived data resources will consume resources, thus resulting in operations such as DestroyDataResource being provided. Soft state lifetime management of data resources is not supported by WS-DAI unless WSRF is used.

1.3   The Relational Extension (WS-DAIR)

Relational database management systems offer well-understood, widely used tools embodied by the ubiquitous SQL language for expressing database queries. As a natural result of this, the DAIS working group focused on producing the WS-DAIR extensions which defines properties and operations defined to deal with relational data. A brief overview of these extensions is given here, starting with the properties defined by WS-DAIR to extend the basic set of data resource properties defined by WS-DAI:

- SQLAccessDescription: Defines properties required to describe the capabilities of a relational resource in terms of its ability to provide access to data via the SQL query language.
- SQLResponseDescription: Defines a set of properties to describe the result of an interaction with a relational data resource using SQL. For example, the number of rows updated, the number of result sets returned and – or any error messages generated when the SQL expression was executed.
- SQLRowSetDescription: Defines properties describing the results returned by an SQL SELECT statement against a relational database, including the schema used to represent the query result and the number of rows that exist.

The following direct access interfaces are defined by WS-DAIR:

- SQLAccess: Provides operations for retrieving SQLAccessDescription properties (although for implementations that use WSRF should be able to employ the methods defined there as well) and executing SQL statements (via a SQLExecuteRequest message).
- SQLResponseAccess: Provides operations for processing the responses from SQL statements, for example, retrieval of SQLRowsets, SQL return values and output parameters.
- SQLRowSetAccess: Provides access to a set of rows through a GetTuples operation.
- SQLResponseFactory: Provides access to the results returned by an SQL statement. For example, the SQLRowsetFactory operation can be used to create a new data resource supporting the SQLRowset interface.

Example XML representations of an SQLExecuteRequest and a corresponding response message are shown in Fig. 1.4.

1.4   The XML Extension (WS-DAIX)

The growing popularity of XML databases and the availability of expressive query languages such as XQuery means that the provision of an extension to WS-DAI to cater for XML databases. Work on WS-DAIX was undertaken in addition to the WS-DAIR effort from the start. A key difference to the relational specification is that XML databases may support a number of different query languages that need to be catered for: XQuery, XUpdate and XPath, although XQuery can, in effect, encompass the capabilities of XUpdate and XPath. The following property sets are defined by WS-DAIX:

- XMLCollectionDescription: Provides properties describing an XML collection, such as the number of documents and the presence of an XML schema against which documents are validated.

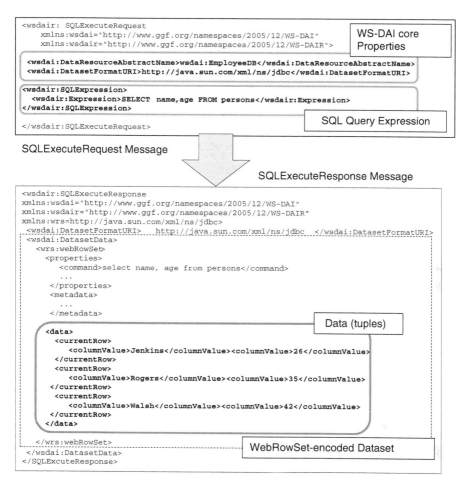

```
<wsdair: SQLExecuteRequest
    xmlns:wsdai="http://www.ggf.org/namespaces/2005/12/WS-DAI"
    xmlns:wsdair="http://www.ggf.org/namespaces/2005/12/WS-DAIR">
```
WS-DAI core
Properties

```
<wsdai:DataResourceAbstractName>wsdai:EmployeeDB</wsdai:DataResourceAbstractName>
<wsdai:DatasetFormatURI>http://java.sun.com/xml/ns/jdbc</wsdai:DatasetFormatURI>
```

```
<wsdair:SQLExpression>
  <wsdair:Expression>SELECT  name,age FROM persons</wsdair:Expression>
</wsdair:SQLExpression>
```

```
</wsdair:SQLExecuteRequest>
```
SQL Query Expression

SQLExecuteRequest Message

SQLExecuteResponse Message

```
<wsdair:SQLExecuteResponse
xmlns:wsdai="http://www.ggf.org/namespaces/2005/12/WS-DAI"
xmlns:wsdair="http://www.ggf.org/namespaces/2005/12/WS-DAIR"
xmlns:wrs=http://java.sun.com/xml/ns/jdbc>
  <wsdai:DatasetFormatURI>   http://java.sun.com/xml/ns/jdbc  </wsdai:DatasetFormatURI>
  <wsdai:DatasetData>
    <wrs:webRowSet>
      <properties>
        <command>select name, age from persons</command>
        ...
      </properties>
      <metadata>
        ...
      </metadata>
```

Data (tuples)

```
      <data>
        <currentRow>
          <columnValue>Jenkins</columnValue><columnValue>26</columnValue>
        </currentRow>
        <currentRow>
          <columnValue>Rogers</columnValue><columnValue>35</columnValue>
        </currentRow>
        <currentRow>
          <columnValue>Walsh</columnValue><columnValue>42</columnValue>
        </currentRow>
      </data>
```

```
    </wrs:webRowSet>
  </wsdai:DatasetData>
</SQLExecuteResponse>
```
WebRowSet-encoded Dataset

Fig. 1.4 An SQLExecuteRequest/response example (direct access)

- XMLSequenceDescription: Describes an XML sequence, usually created as the result of an XPath or XQuery expression. Specifically, a property to define the length of the sequence is provided. It should be noted that no extra properties are defined to describe data resources with XPath, XUpdate or XQuery capabilities as the WS-DAI-defined properties such as LanguageURI, DatasetFormatURI, etc. are adequate for this purpose.

The following direct data access interfaces are supported:

- XMLCollectionAccess: Provides access to an XML collection via operations supporting addition/removal or documents and sub-collections.
- XQueryAccess: Allows the evaluation of XQuery expressions across collections of XML documents represented by an XML resource.

- XUpdateAccess: Allows an XUpdate expression to be executed against an XML resource, returning the number of updated nodes.
- XPathAccess: Allows the evaluation of XPath expressions across collections of XML documents represented by an XML resource.
- XMLSequenceAccess: Provides access to an XML sequence created as a result of an XPath/XQuery query. The GetItems operation of this interface allows the client to obtain specific subsequences of the overall result.

The following indirect access interfaces are supported:

- XMLCollectionFactory: Provides access to collections and documents in collections.
- XPathFactory: Provides the XPathQueryFactory that allows new data resources (supporting the XMLSequenceAccess interface) to be created as the result of an XPath query.
- XQueryFactory: Provides the XQueryExecuteFactory operation to create new XMLSequenceAccess data resources as the result of an XQuery query.

1.5   The RDF Extension (WS-DAI-RDF(S))

The RDF is a World Wide Web Consortium (W3C) set of recommendations [12] focused on the representation and management of metadata. It includes two data models, RDF and RDF Schema, whose combination is known as RDF(S). The WS-DAI-RDF(S) extension to this domain provides data access mechanisms for RDF(S) data, divided into two types based on the style of access: declarative or programmatic. Hence, the following specifications are in the process of being defined within DAIS to access RDF(S) data.

1. WS-DAI RDF(S) Querying: This specification provides a query language interface to RDF data based on the W3C SPARQL query language [13] for RDF.
2. WS-DAI RDF(S) Ontology: This specification provides an API style of access based on ontology handling primitives conforming to the RDF(S) model. These primitives provide various operations including the possibility of performing updates to the ontology.

1.5.1   The WS-DAI RDF(S) Querying Specification

The objective of the querying specification is to provide an SPARQL interface to RDF data. The W3C has defined several related specifications based on SPARQL, including an XML-based query results format [14] and a protocol for accessing RDF resources [15]. The WS-DAI-RDF(S) Querying specification, the interaction patterns of which are illustrated in Fig. 1.5, is defined to be compatible with the

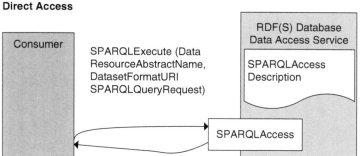

Direct Access

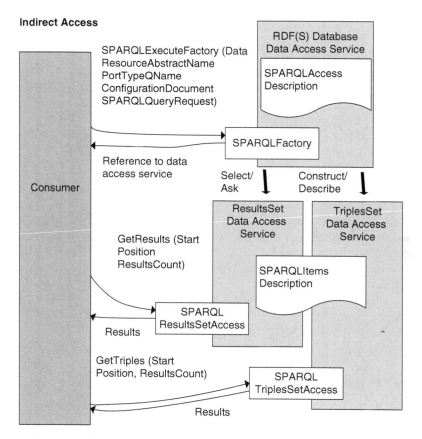

Fig. 1.5 Overview of WS-DAI RDF(S) querying specification

W3C standards (e.g. by supporting the SPARQL query language and the XML results format) while also benefitting from the WS-DAI approach. For example, indirect access is not supported by the W3C SPARQL protocol, meaning that when using the SPARQL protocol all query results are returned directly to the consumer accessing the service. In contrast, WS-DAI-RDF(S) allows the consumer to control the retrieval of query results, a feature that can be extremely useful in certain scenarios, such as when retrieving large result sets.

1.5.1.1   Indirect Access Using TriplesSetAccess and ResultsSetAccess

SPARQL has four query forms: CONSTRUCT, DESCRIBE, SELECT and ASK. The first and second forms return an RDF graph as a query result (CONSTRUCT returns an RDF graph constructed by substituting variables in query patterns, while DESCRIBE returns an RDF graph that describes the resources found). Other representations also exist but the important thing is that they are modeled as triples. For this purpose, the WS-DAI-RDF(S) specification introduces a TriplesSetAccess interface to provide access to the results.

In contrast to these two forms, the results of the other two forms are not RDF graphs: SELECT returns variables bound during the matching of an RDF graph against a basic graph pattern specified in the query; ASK returns a boolean value indicating whether there is a match for a query pattern. The WS-DAI-RDF(S) specification introduces a ResultsSetAccess interface to access the results of these query forms, based on the SPARQL Result Set XML Format specification.

1.5.2   The WS-DAI RDF(S) Ontology Specification

The object of the WS-DAI-RDF(S) Ontology access specification is to provide an integral access mechanism for RDF(S) sources that goes beyond the retrieval capabilities offered by the querying specification, whilst providing a simple but complete set of functionalities that abstract the most general necessities a consumer may have when accessing with RDF(S) data sources. To achieve this objective, the specification proposes a model-based access mechanism for accessing RDF(S) sources at the conceptual level, that is, an access mechanism that revolves around the concepts and semantics defined by the RDF(S) model. Thus, the specification details a set of ontology handling primitives for dealing with such models, hiding the syntactic aspects of RDF(S) and transparently exploiting its semantics.

1.5.2.1   Data Resources

The WS-DAI-RDF(S) Ontology specification differentiates several types of RDF(S) data resources, each of them provided to allow access to, and manage elements of RDF(S) sources at different levels of granularity. They can be divided into two groups:

- Placeholders for built-in RDF(S) classes (Resource, Class, Property, Statement, Container and List data resources): These data resources provide class-oriented views of an RDF(S) resource.

- Convenience abstractions (RepositoryCollection and Repository data resources), for RDF(S) sources that contain more than a resource.

1.5.2.2   Interfaces for Direct and Indirect Access

To interact with the data resources described above, several interfaces are provided in the WS-DAI-RDF(S) Ontology specification. The first group is for the direct access interfaces:

- RepositoryCollectionAccess: Provides access to the repositories of a collection.
- RepositoryAccess: Provides access to the repository content, offering functionality for managing the repository at RDF(S) resource level.
- ResourceAccess: Provides access to a particular RDF(S) resource, concentrating in those aspects common to every resource: property value management, resource description, etc.
- ClassAccess: Provides access to particular RDF(S) resources that are an RDF(S) class, focusing on the data that is specific to RDF(S) classes: class hierarchy traversal, instance retrieval, etc.
- PropertyAccess: Provides access to particular RDF(S) resources that are RDF(S) properties, focusing on the data that is specific to RDF(S) properties: range and domain management, property hierarchy traversal, etc.
- StatementAccess: Provides access to particular RDF(S) resources that are RDF(S) statements reified triples, not the triples themselves focusing on the management of the components that set up the reification.
- ListAccess and ListIteratorAccess: Provides access to particular RDF(S) resources that are RDF collections (List), focusing on the management of the members of a collection, as well as, the structure of the collection.
- ContainerAccess and ContainerIteratorAccess: Provides access to particular RDF(S) resources that are RDF(S) containers, focusing on the management of the members of the container, as well as the structure of the container, regardless the its specific type.
- AltAccess: Provides access to particular RDF(S) containers that are of the particular alt type.

There are also indirect access interfaces:

- RepositoryCollectionFactory: Provides access to the repositories in a collection.
- RepositoryFactory: Provides access to the repository content.
- ListFactory: Provides access to the contents of an RDF collection.
- ContainerFactory: Provides access to the contents of a container.

Finally, due to the large number of operations the aim will be to incrementally introduce the different levels of functionality described previously through three different profiles documents, schematically illustrated in Fig. 1.6. These will provide support for the different types of use case, of increasing complexity, with basic RDF support, RDF Schema support and, finally, full RDF support. It is envisaged that,

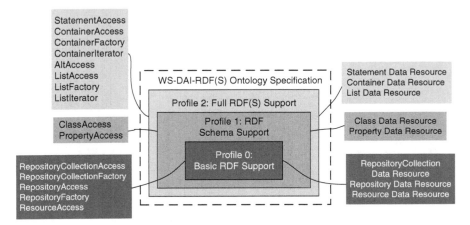

Fig. 1.6 Profile documents for the WS-DAI-RDF(S) ontology specification

like a Russian doll, implementation of a given level of profile will also require the previous levels to also be implemented.

1.6 Implementations

Implementations of the specifications are important for a number of reasons – first, they serve to debug and test the specifications during their development. Second, they provide examples to potential adopters of the specifications in use, allowing easier implementations to be constructed by developers. Third and most importantly, implementations are necessary to promote the specifications, allow them to become widely recognised, and foster adoption, a factor by which the success of specifications will ultimately by judged.

Several implementations of the DAIS specifications have been developed to serve as experimental platforms during the specification development process and following that, implementations have also been produced as part of research projects developing applications of the specifications. The following is a list of the implementations that have been made public to date.

1.6.1 WS-DAIR Implementations

- AMGA[2] is a metadata catalogue compliant with the EGEE grid environment. The implementation of the metadata catalogue provides various interfaces, including a SOAP WS-DAIR compliant implementation.

---

[2]http://amga.web.cern.ch/amga/.

- OGSA-DAI[3] is an open-source distributed data access and management system supporting Web service-based access to data. OGSA-DAI WS-DAIR, an implementation of the WS-DAIR interfaces using the OGSA-DAI middleware that can be obtained from the OGSA-DAI SourceForge site.[4]

1.6.2 WS-DAIX Implementations

- An implementation of WS-DAIX is also available from the OGSA-DAI source SourceForge site.

1.6.3 WS-DAI-RDF Implementations

- The EC funded ADMIRE (Advanced Data Mining and Integration for Eu-rope) [www.admire-project.eu/] project has developed an implementation of the WS-DAI-RDF Querying Specification.
- AIST's OGSA-DAI-RDF project has developed an implementation of the WS-DAI-RDF Querying Specification, which can be obtained from.[5]

The OGF process requires that two independent interoperable implementations exist before a proposed recommendation can become a full OGF recommendation. To date, two of the above implementations (WS-DAIR implementations from the OGSA-DAI and AMGA projects) have been utilised to validate the WS-DAI and WS-DAIR specifications as reported in [16]. A comparison of the functionality of these implementations is made in Table 1.1. The performance of the implementations is dependent on the underlying DBMS being utilised; however, the overhead incurred by the WS-DAI(R) Web service-based interfaces is similar for both implementations.[6]

1.7 Applications

The set of potential areas of application for the DAIS specifications is wide ranging and some research projects have already become early adopters of them. This section provides two examples of the application of the WS-DAI-RDF specification.

---

[3]http://ogsadai.org.uk.

[4]http://ogsa-dai.sourceforge.net/.

[5]http://dbgrid.org.

[6]A brief analysis of the performance of the AMGA WS-DAIR implementation and some comparisons with the OGSA-DAI implementation can be found under "Design and Implementation of WS-DAIR for AMGA" available at http://event.twgrid.org/isgc2009/program.htm.

Table 1.1 A comparison of the OGSA-DAI and AMGA implementations of WS-DAI and WS-DAIR

	OGSA-DAI	AMGA
	Apache Axis	gSoap
Infrastructure	Java 1.4	C++
SOAP binding	rpc/literal	Document/literal
Underlying DBMS	Any JDBC-enabled relational database	Supports MySQL, SQLLite PostgreSQL, Oracle
Supported languages	SQL (dependent on underlying DBMS)	SQL-92 AMGA Metadata Language
Stored procedures	Yes (if supported by DBMS)	No
Datasets	WebRowSet comma-separated values	WebRowSet
Security features	None	SSL, GSI, VOMS permission, ACL
Un-supported features	ServiceBusyFault GenericQuery	ServiceBusyFault SQL CommunicationArea (for fault messages)

One of them is the ADMIRE registry, which uses the WS-DAI-RDF specifications to provide support in a data mining and integration (DMI) context. The second presents a scenario in which the specifications can be applied to distributed SPARQL query processing. Other applications that make use of the other specifications have already been pointed out above, such as the AMGA and OGSA-DAI projects. These provide additional examples where the WS-DAI and WS-DAIR specifications have been implemented for the convenience and benefit of their users.

1.7.1 ADMIRE

The ADMIRE[7] registry allows a range of DMI components, called processing elements (PE), to be registered and discovered, together with the set of types, in the context of their inputs and outputs, that can be handled by those processing elements. The descriptions used in the registrations contain the data types of the input and output parameters for each PE and any restrictions associated with these, such as: the relationships between the inputs and outputs, termination conditions, are error conditions and all these information are available at the registry in an RDF format.

In ADMIRE, users create these PEs and register their descriptions in a registry by means of a register operation as defined in the DISPEL language [17]. Users can

---

[7]EU FP7 ICT 215024 www.admire-project.eu.

then retrieve PE descriptions by using SPARQL. By adding a web service layer, the registry may be accessed by different users at different times in different contexts (binding the states to the users). The WS-DAI-RDF(S) specification thus provides a convenient way of providing standardised access to this RDF-based data repository, and this is what has been achieved by ADMIRE.

1.7.2   Distributed Query Processing

The WS-DAI-RDF(S) specifications allow data integration applications to be constructed on top of the consistent interfaces provided by WS-DAI-RDF(S) data resources. When integrating data from distributed data sources, it is necessary to deal with syntactic heterogeneities that may be present between the interfaces used to interact with data resources. Furthermore, data retrieval mechanisms must support delivery mechanisms that allow clients some form of control over the rate at which data is delivered, especially when scalability is desired.

This is important for grid-based distributed databases, where data is federated and accessible via service-based interfaces. The standardised interfaces provided by the WS-DAI-RDF(S) specifications mean that many heterogeneities present amongst the individual data sources are resolved when performing these tasks. Data integration may be performed by multiple computational resources, and the WS-DAI indirect data access pattern can be used to execute sub-queries which result in the creation of a new data resource for each set of query results. The various data integration tasks (e.g., joins, unions) that need to be performed can then delegated to appropriate nodes in a set of computational resources, which are given references to the created data resources that need to be accessed to perform their allocated tasks. This therefore allows parallel and distributed query processing to take place following the approach used by the OGSA-DQP [18] distributed query processor, which uses OGSA-DAI data resources. The DAIS specification's operations allow similar applications to be developed accessing data resource using open standards.

1.8   Conclusions

This chapter has given an overview of the WS-DAI family of specifications that have been the focus of the DAIS Working Group of the OGF. A core specification provides a framework which can then be extended to deal with specific types of data. This process has already been realised for relational, XML and RDF data, and some initial proposals have been also made for other types of databases. Generally, the DAIS approach provides a core specification and a flexible framework that allows extensions if further requirements are specified of the core specification, which may in turn impact on the other extension specified.

This chapter's review of the interfaces provided by the specifications has focused, in particular, on the novel use of indirect data access to provide a means of minimising data movement, allowing derived data resources to be deployed and exposed at the server side.

These specifications provide a means of abstracting out some of the variability in the data resources used in distributed environments and presenting uniform interfaces to specific types of data – for now: relational, XML and RDF data – to clients. The use of web services to do this ensures a certain degree of programming language neutrality and portability across different computer systems. For these reasons, it is expected that the adoption of these specifications will facilitate the management and integration of data across the distributed environments presented by Grids and Clouds.

Acknowledgements The authors thank all those people who have participated in the process of developing and ratifying the DAIS specification documents and OGF for hosting the process.

References

1. Catlett, C., de Laat, C., Martin, D., Newby, G., Skow, D.: Open Grid Forum Document Process and Requirements [Obsoletes GFD.1] GFD.152. Open Grid Forum, 2009. http://www.ogf.org/documents/GFD.152.pdf
2. Atkinson, M.P., Dialani, V., Guy, L., Narang, I., Paton, N.W., Pearson, P., Storey, T., Watson P.: Grid Database Access and Integration: Requirements and Functionalities. GFD-I-13. Open Grid Forum, 2003. http://www.ogf.org/documents/GFD.13.pdf
3. Foster, I., Kishimoto, H., Savva, A., Berry, D., Djaoui, A., Grimshaw, A., Horn, B., Maciel, F., Subramaniam, R., Treadwell, J., Von Reich, J.: The Open Grid Services Architecture, Version 1.0. OGF GFD-I.030. Open Grid Forum, 2005. http://www.ogf.org/documents/GFD.30.pdf
4. Web Service Resource Framework (WSRF) Specifications. OASIS. http://www.oasis-open.org/committees/tc_home.php?wg_abbrev=wsrf. Accessed 9 Oct 2010
5. Atkinson, M., De Roure, D., Dunlop, A., Fox, G., Henderson, P., Hey, T., Paton, N., Newhouse, S., Parastatidis, S., Trefethen, A., Watson, P., Webber, J.: Web Service Grids: An Evolutionary Approach. Technical report, UK e-Science Institute. http://www.nesc.ac.uk/technical_papers/UKeS-2004-05.pdf. Accessed 9 Oct 2010
6. Antonioletti, M., Atkinson, M., Laws, S., Malaika, S., Paton, N.W., Pearson, D., Riccardi, G.: Web Services Data Access and Integration (WS-DAI) Specification Version 1.0. OGF GFD.74. Open Grid Forum, 2006. http://www.ogf.org/documents/GFD.74.pdf
7. Antonioletti, M., Collins, B., Krause, A., Laws, S., Magowan, J., Malaika, S., Paton, N.W.: Web Services Data Access and Integration The Relational Realisation (WS-DAIR) Specification Version 1.0. OGF GFD.76. Open Grid Forum, 2006. http://www.ogf.org/documents/GFD.76.pdf
8. Antonioletti, M., Hastings, S., Krause, A., Langella, S., Laws, S., Malaika, S., Paton, N.W.: Web Services Data Access and Integration. The XML Realisation (WS-DAIX) Specification Version 1.0. OGF GFD.75. Open Grid Forum, 2006. http://www.ogf.org/documents/GFD.75.pdf
9. Antonioletti, M., Aranda, C.B., Corcho, O., Esteban-Gutirrez, M., Gmez-Prez, A., Kojima, I., Lynden, S., Pahlevi. S.M.: WS-DAI RDF(S) Realization: Introduction, Motivational Use Cases and Terminologies GFD-I 163. Open Grid Forum, 2009. http://www.ogf.org/documents/GFD.163.pdf

10. Gudgin, M., Hadley, M., Rogers, T.: Web Services Addressing 1.0 – Core. W3C Recommendation. World Wide Web Consortium, 2006. http://www.w3.org/TR/ws-addr-core
11. Bruce, J.: JSR-000114 JDBC RowSet Implementations, 07 April 2004. http://jcp.org/aboutJava/communityprocess/final/jsr114
12. The Resource Description Framework (RDF) Specifications, (last visited on 10/10/10). http://www.w3.org/standards/techs/rdf#w3c_all
13. Prud'hommeaux, E., Seaborne, A.: SPARQL Query Language for RDF. W3C Recommendation. World Wide Web Consortium, 2008. http://www.w3.org/TR/rdf-sparql-query
14. Beckett, D., Broekstra, J.: SPARQL Query Results XML Format – W3C Recommendation. World Wide Web Consortium, 15 January 2008. http://www.w3.org/TR/rdf-sparql-XMLres
15. Grant Clark, K., Feigenbaum, L., Torres, E.: SPARQL Protocol for RDF. W3C Recommendation. World Wide Web Consortium, 15 January 2008. http://www.w3.org/TR/rdf-sparql-protocol
16. Lynden, S., Antonioletti, M., Jackson, M., Ahn, S.: WS-DAI and WS-DAIR Implementations – Experimental Document GFD.160. Open Grid Forum, 2009. http://www.ogf.org/documents/GFD.160.pdf
17. Atkinson, M., Brezany, P., Krause, A., van Hemert, J., Janciak, I., Yaikhom, G.: DISPEL: Grammar and Concrete Syntax, version 1.0. Deliverable report D1.7, the ADMIRE Project, February 2010. http://www.admire-project.eu/docs/ADMIRE-D1.7-research-prototypes.pdf
18. Dobrzelecki, B., Krause, A., Hume, A., Grant, A., Antonioletti, M., Alemu, T., Atkinson, M., Jackson, M., Theocharopoulos, E.: Integrating distributed data sources with OGSADAI DQP and Views. Phil. Trans. Roy. Soc. A **368**, 4133–4145 (2010)

Chapter 2
Open Cloud Computing Interface in Data Management-Related Setups

Andrew Edmonds, Thijs Metsch, and Alexander Papaspyrou

Abstract The Cloud community is a vivid group of people who drive the ideas of Cloud computing into different fields of Information Technology. This demands for standards to ensure interoperability and avoid vendor lock-in. Since such standards need to satisfy many requirements, use cases, and applications, they need to be extremely flexible and adaptive. The Open Cloud Computing Interface (OCCI) family of specifications aims to achieve this goal: originally developed for the deployment of infrastructure Clouds, it can also be used in different service and deployment models. This article will outline the OCCI specifications and demonstrate how they can be used in data management-related setups. Not only can OCCI be easily integrated but it can also be used to deploy data-centric applications (which are secured by SLAs), support data-awareness in scheduling, as well as directly interface with data management tools in a PaaS-based manner. To demonstrate this, three use cases are discussed in this article.

2.1 Introduction

Next to traditional HPC and Grid computing, Cloud computing has become a new driver for the global IT market. The overall idea is to deliver a service to the customer. Instead of traditionally boxing and shipping of software products,

A. Edmonds (✉)
Intel Ireland Branch, Collinstown Industrial Park, Leixlip, County Kildare, Ireland
e-mail: andrewx.edmonds@intel.com

T. Metsch
Platform Computing GmbH, Europaring 60, 40878 Ratingen, Germany
e-mail: tmetsch@platform.com

A. Papaspyrou
Technische Universität Dortmund, Institut für Roboterforschung, 44221 Dortmund, Germany
e-mail: alexander.papaspyrou@tu-dortmund.de

S. Fiore and G. Aloisio (eds.), *Grid and Cloud Database Management*,
DOI 10.1007/978-3-642-20045-8_2, © Springer-Verlag Berlin Heidelberg 2011

software is now delivered as a service to the customer directly. This change in use of computing services changes the IT landscape drastically – not only will data centers most probably transform into service providers but also the way service providers and customers interact will change.

One example is billing in all businesses where a Pay-per-Use model can be easily established. The next major change in this area will be the management of data: starting with the idea of moving compute resources to the data (data-aware scheduling) as an obvious step also the way how data is treated in the Cloud (manipulation of data – NoSQL vs. Relational Databases vs. Virtual Disc Images) will evolve. Countless other opportunities such as signing, tracing changes and movement of data are still ahead of us.

Since many customers move into the cloud the deployment of their data and the applications becomes very important to them. Still, most Cloud computing providers currently focus on providing *Infrastructure-as-a-Service* (IaaS)[1] but this might change as the industry moves its focus into the idea of providing *Platform-as-a-Service* (PaaS) where services are constructed on a higher (non-OS, but rich API) level to provide services surrounding the data.

Still, the underlying technology is evolving: standards are being developed and technologies emerge (like virtualisation). As such, there is a demand for ensuring clean interfaces and protocols which are easy to use and can be used for multiple kinds of service offerings to prevent a vendor lock-in.

In the context of these developments, the Open Cloud Computing Interface (OCCI) working group works towards forming such a standard. The OCCI family of specifications can be used for IaaS and PaaS offerings. In this paper, it is demonstrated how OCCI can be used in data-centric setups for IaaS and PaaS offerings. To this end, a setup is described in which Virtual Machines (containing Databases etc.) can be deployed in a Cloud environment while ensuring certain Service Level Agreements (SLAs). Another use case demoes the ability of OCCI for moving compute resource towards large datasets. The last scenario works (in contrast to the former two) towards a PaaS scenario: it shows a Key-Value store implementation over OCCI.

The purpose of these use cases is to show the need for an interoperable Cloud interface/protocol which can be used in all layers of the Cloud stack. Furthermore, it demonstrates that OCCI provides flexible usage models for a very heterogeneous field of scenarios in the broader field of data management in the Cloud.

The rest of the paper is organised as follows: in Sect. 2.2, the OCCI family of specification is introduced. Next, three use cases for the application of OCCI are exemplified in Sects. 2.3–2.5. Finally, the paper concludes with a summary of achievements and shows future work.

---

[1] Usually in the form of virtual machines.

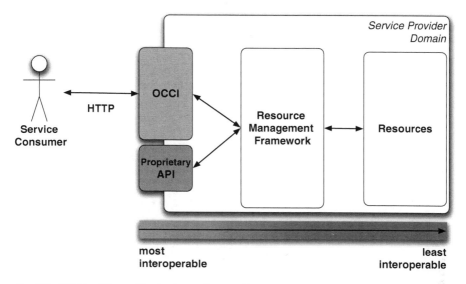

Fig. 2.1 OCCI and its position in the service provider context

2.2   Open Cloud Computing Interface

OCCI is an effort driven by a working group in the standards track of the Open Grid Forum.[2] It strives to create an open, interoperable protocol and API for the Cloud.

The group started with a clear focus on provisioning IaaS but later extended the focus to include other layers in the Cloud stack as well. The following diagram (Fig. 2.1) shows where OCCI fits in the service provider context.

The OCCI protocol can be used for integration, ensuring interoperability and portability between service providers. Proprietary APIs can be used alongside OCCI in the case that other features than those of OCCI are maintained.

The specification strives to be very easy, flexible and extensible. Therefore, it is broken into different modules. It starts with a module describing the core models. Another module describes how this model can be mapped and rendered using a HTTP/REST approach. The third module describes the infrastructure entities and how they related to the core model.

2.2.1   Motivation for Standards

Main driver for standards in the past has been interoperability. This is still a fundamental part of what standards want to achieve. Still there are nuances in the term interoperability which are important and need to be looked upon separately:

---

[2]http://www.ogf.org/.

Interoperability. Describes how two services can inter-operate on the fly. This demands a standardised API and protocol (e.g. live migrating a virtual machine from one host to another, which are in different management domains).

Integration. Describes how a service provider can bring together different technologies and interconnect them within his domain (e.g. integrate a virtual machine management tool with an identity management system).

Portability. This is mostly about the porting between service providers. In comparison with interoperability, there is no direct connection between the service provider. This demands that there are standardised data formats which providers can understand (e.g. porting a virtual machine from one hypervisor to another).

Innovation. Standards have always been started when a field in the IT community gains popularity, is widely adopted and begins on a path of commoditisation. Next to interoperability, standards can be a driver for innovation as well as widely adopted innovations can demand standards.

Reusability. This can be seen on two levels. First the reuse of (legacy) codes through basic standardised APIs and the reuse of the standard itself in different fields.

2.2.2 The Core Model

The core meta-model [10] for OCCI imposes a general means of handling general resources, providing semantics for defining the type of a given entity, describing interdependencies in between different entities, and defining operating characteristics on them. Although the meta-model aims to ease the implementation burden by setting a common ground for other OCCI-related specifications, it can be used as a standalone component in other contexts (e.g. Resource Oriented Architectures (ROAs)) as well.

The UML class diagram shown (Fig. 2.2) gives an overview of the OCCI core meta-model. At its heart lies the `Resource` type. Any resource exposed through OCCI is a `Resource` or sub-type thereof. A resource can be for example a virtual machine, a job in a job submission system, a user, etc. The `Resource` type contains a number of common attributes that domain-specific `Resource` types inherit. The `Resource` type is complemented by the `Link` type which associates one `Resource` instance with another. The `Link` type also contains a number of common attributes that domain-specific `Link` types inherit.

`Entity` is an abstract type which both `Resource` and `Link` inherit. Each sub-type of `Entity` is identified by a unique `Kind` instance. The `Kind` type comprise the classification system built into the OCCI model. `Kind` is a specialisation of `Category` and introduces additional capabilities in terms of `Action` types.

2.2.2.1 Classification and Identification

The OCCI model provides a built-in classification system allowing for safe extension towards domain-specific usage. This system is like a "type system" but with the possibility of being easily exposed over a text-based protocol.

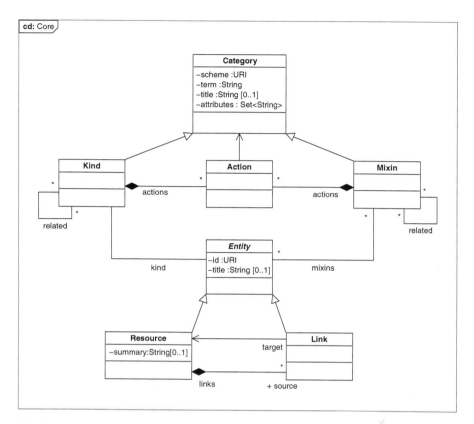

Fig. 2.2 UML class diagram of the OCCI model. The diagram provides an overview of the OCCI model but is not a standalone definition thereof

The classification system can be summarised with the following key features:

- Each OCCI base type and extension thereof is assigned a unique identifier, a structural `Kind`, which allows for dynamic discovery of available types.
- The relationship of structural `Kinds` is part of the system and thus the inheritance model is also discoverable.
- The classification system allows non-structural `Kinds` to be assigned to resource instances adding new capabilities using a mix-in-like model.
- Tagging of resource instances is supported through mix-in of non-structural `Kinds` which have no additional capabilities defined.
- A collection of associated resources is implicitly defined for each structural and non-structural `Kind`. That is all resource instances associated with a particular `Kind` instance form a collection.

2.2.2.2 Categorisation

The Category type comprises the basis of the identification mechanism used by the OCCI classification system. Instances of the Category type are only used to identify Action types. All other uses of Category properties are managed through its sub-type Kind.

A Category is uniquely identified by concatenating the categorisation scheme with the category term, for example *http://example.com/category/scheme#term*. This is done to enable discovery of Category definitions in text-based renderings such as HTTP. Sub-types of Category such as Kind inherit this property.

2.2.2.3 Kind Relationships

The OCCI base types Resource and Link extend Entity. This together with any further sub-typing implies a hierarchy of related structural Kind instances. The Kind relationships thus mirror the type inheritance structure of the OCCI model and any extension thereof.

In an example where a domain-specific "Custom Compute Resource" is a sub-type, the OCCI infrastructure type Compute, which in turn is a sub-type of the Resource type, four related structural Kinds would be involved.

One or more Entity instances associated with the same Kind, automatically form a collection, and each Kind identifies a collection consisting of all Entity instances of it. For example, an instance of the Resource type will always be associated with the structural Kind (*http://scheme.ogf.org/occi/core#resource*) and thus part of the collection implied by the Kind.

Collections are, by definition of the core model, navigable and support the following operations:

- Retrieve the whole collection.
- Retrieve a specific item in a collection.
- Retrieve a subset of a collection.

2.2.2.4 Discovery

In addition to that, Kinds and Category instances a particular service provider support can be discovered. By examining these instances a client is enabled to deduce the following information:

- The Entity sub-types available from a service provider, including domain-specific extensions.
- The attributes associated with each Entity sub-type.
- The invocable operations, that is Actions, defined for each Entity sub-type.
- Additional mix-ins or tags, that is non-structural Kinds, applicable to Entity sub-type instances.

Overall, the OCCI core meta-model provides a solid foundation for the remote management of resources offered in an *as-a-Service* manner, allowing for the development of interoperable tools for common tasks including deployment, automatic scaling and monitoring. The explicit split-out of it allows the leverage of the developed models, protocols, and APIs in manners not anticipated and to foster modularity and extensibility for future usage paradigms.

2.2.3   RESTful HTTP Rendering of the OCCI Model

The OCCI Core model which is described in the previous Sect. 2.2.2 is free of any rendering and forms the base of OCCI. Based upon this model, OCCI describes a serialisation rendering. This rendering – or serialisation format – is passed on the wire between client and service, see [11].

OCCI has a default rendering which is text based and uses the HTTP protocol and implements a ROA, see [14]. In this architecture, a system is modelled as a set of related resources. ROA's use Representation State Transfer (REST), see [6], to cater for client and service interactions. In these interactions, clients request to perform operations on the state of an individual or set of resources managed by the service.

HTTP is commonly used in most ROA systems. It provides means to uniquely identify resources through URIs as well as operating upon them with a set of general-purpose operations called verbs. These HTTP verbs map loosely to the resource-related operations of *create* (POST), *retrieve* (GET), *update* (POST, PUT) and *delete* (DELETE).

2.2.3.1   Rendering of Resources

Each Resource in the OCCI core model will be rendered as a unique URI (for example *http://example.com/foo*). Each resource can be identified uniquely by an URI and has at least one Category assigned, which defines the type and the operations that can be performed. This means that from this standpoint a resource can be almost anything like a Database entry, a Virtual Machine, an Image, etc.

Resources can be linked and actions can be performed upon them. Resource of the same type (as in have the same Category assigned) can be found under a certain path relative to the root of the service provider (e.g. all storage devices will appear under the path */storage* – still the path name "storage" is freely defined by the Service Provider and can do discovered through the Query interface).

Since Categories cannot only be used to define the type of the resource, but also to tag or group resources, resource can show up under multiple paths. The following URL hierarchy demonstrates this feature:

```
/compute/123
/storage/discABC
/database/tableXYZ
/nosql/entry_1
/my-linux-vms/123      // links to /compute/123
/my-datasets/discABC   // links to /storage/discABC
/my-datasets/tableXYZ  // links to /database/tableXYZ
/my-datasets/entry_1   // links to /nosql/entry_1
```

This very flexible system allows that the OCCI model can be used for several use cases including for Data Management operations.

2.2.3.2 Discovery of Capabilities Through a Query Interface

One of the main features of OCCI is that clients can discover the capabilities of the service provider through a standardised query interface. This is important since OCCI is designed for extensibility. To query the capabilities of a Service provider implementing OCCI, the Client needs to do a HTTP GET on the URI */-/*.

This Category management URL allows the client to get a listing of all categories supported by the provider. Should the provider allow and support client-created categories, then this URL endpoint must support the creation of user categories as well.

2.2.3.3 Linking and Performing Actions on Resources

Each of these resources can be linked with other resources. Links again are RESTful resources and have a source and a target attribute. Each link resource is bound to a category identifying it as a link.

Next to linking, some type specific actions can be performed. The set of possible actions is defined by the Category of the resource. Actions are triggered by adding a fragment to the URI of the resource indicating which action should be triggered (e.g. *http://example.com/foo;action=shutdown*). Parameters of the action are described in the HTTP message.

2.2.3.4 Use of HTTP Features

The HTTP rendering of OCCI makes use of many HTTP features. This includes for example HTTP headers for Versioning and all Authentication features. OCCI does not explicitly define those but makes use of those features.

Next to these basic features, OCCI also makes use of the *Content-Type* definitions. At a minimum, all information for OCCI resources is transferred in the HTTP

Body. This is defined as the basic *text/plain* content type. Other content types also exist. For example, the information can also be rendered in the HTTP Header or as HTML (e.g. for browsing the OCCI interface using a Web Browser) by supplying the appropriate content-type header as specified in the specification.

Rendering of data is done through simple key value associations. Also, more structural data representations such as JSON of RDFa can easily be added to OCCI.

2.2.4 OCCI for Virtual Machine (Infrastructure) Provisioning

Having described the core model and a way of rendering it on the wire, a concrete compliment to the core model is now explored [12].

The infrastructure specification extends the core model at two key points:

1. To represent various infrastructure-related resources, it extends Resource using inheritance.
2. To represent concrete relationships between infrastructural resources, it extends Link using inheritance.

To represent the main elemental resources found in infrastructure-type services, OCCI has three specialisations of Resource:

1. Compute: Information processing resources.
2. Network: Interconnection resources.
3. Storage: Information recording resources.

Complimenting these, to allow linkage are:

- NetworkInterface: Represents an L2 client device (e.g. network adapter).
- StorageLink: Represents a link from a Resource to a target Storage Resource.

The relations of these infrastructure resources are shown in the UML diagram (Fig. 2.3).

When modeling elements, it was found that OCCI needed to support not only generic cases but also specific cases. This issue was exemplified by Network. It might be immediately attractive to model all functionalities within this Resource, including aspects of IP configuration, however, then the model would force certain technology choices upon implementers. To avoid this, the working group chose to utilise the OCCI mix-in capabilities to avoid such a situation. Where an implementer wishes to offer TCP/IP functionality on top of the Network resource, they can do so by implementing the IPNetworking mix-in. The IPNetworking mix-in allows to supplement the Network Resource with the necessary TCP/IP features. Should an implementer wish to offer another type of L3/L4 technology for example AppleTalk or IPX, then they only need implement a custom network mix-in.

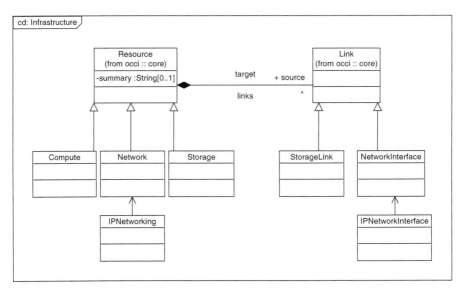

Fig. 2.3 Extended OCCI core model showing infrastructure elements

It was the infrastructure model, along with the OCCI core and HTTP rendering specifications, that aided a successful collaboration that investigated the integration of two large European Union FP7 research projects, SLA@SOI[3] and RESERVOIR.[4] The proposed integration was detailed in a subsequent technical paper [13].

2.2.5 Related Standards and Specifications

A guiding principle in OCCI is to make use of existing standards and specifications where appropriate.

OCCI and the Storage Networking Industry Association's (SNIA)[5] Cloud Data Management Interface (CDMI) working groups have collaborated together so that both specifications are interoperable with each other. It states that

> "The SNIA Cloud Data Management Interface (CDMI) is the functional interface that applications will use to create, retrieve, update and delete data elements from the cloud. As part of this interface the client will be able to discover the capabilities of the cloud storage offering and use this interface to manage containers and the data that is placed in them. In addition, meta-data can be set on containers and their contained data elements through this interface" [16].

---

[3]http://www.sla-at-soi.eu/.

[4]http://www.reservoir-fp7.eu/.

[5]http://www.snia.org/.

OCCI and the Distributed Management Task Force's (DMTF)[6] Open Virtualization Format (OVF), see [4], can be easily integrated through the use of the resource-type Link. Where a provider wishes to supply an OVF representation of a client's resource instance(s), they can do so by associating the instance(s) with a mirror representation, only the serialisation format is OVF.

Other than this, the OCCI working group is closely working together with other groups inside of the Open Grid Forum. The Distributed Computing Infrastucture Federation (DCI-fed[7]) working group focuses on the creation of models and APIs for setting up distributed federated computing environments. Other than this, the OCCI working group uses Standards like those developed by the Distributed Resource Management Application API (DRMAA[8]) working group for common Job operations on Clusters via the OCCI protocol.

2.3 SLA Assured Provisioning of Database Services Using OCCI

In today's service marketplaces including cloud-based ones, there exist basic limitations in service offerings. Typically, the customer has little say in what is offered by a service provider and is left with a "take it or leave it" situation. Not only is the customer faced with such a dilemma, with little possibility of negotiation but if they do accept the service offering there is little in the way of service transparency and so detections of service violations are impossible unless that customer implements custom violation detection systems. The SLA@SOI project seeks to address these challenges by providing three major benefits:

Predictability and Dependability: The quality characteristics of service can be predicted and enforced at run-time.

Transparent SLA Management: SLAs defining the exact conditions under which services are provided/consumed can be transparently managed across the whole business and IT stack.

Automation: The whole process of negotiating SLAs and provisioning, delivery and monitoring of services will be automated allowing for highly dynamic and scalable service consumption.

In this section, a use-case that combines the OCCI model and API with an SLA management framework to provide an SLA assured database service is described. In today's service marketplace, there exists a number of service providers who offer database services, for example, the Amazon Relational Database Service,[9]

---

[6]http://www.dmtf.org.

[7]http://forge.gridforum.org/sf/projects/dcifed-wg.

[8]http://www.drmaa.org.

[9]http://aws.amazon.com/rds.

Microsoft SQL Azure[10] and Longjump Platform as a Service.[11] Other than offering a basic, non-negotiable, non-machine readable SLA, these service providers do not offer certain guarantees that particular consumers will require. A case in point is where a third party service provider wishes to process personal and identifying information. Many law jurisdictions will require that user-supplied data and the processed resultant data remain within the protection of that jurisdiction, which may mean that the physical location of that data must always remain in the country or region where that jurisdiction has powers to protect. If that data at any one time falls outside of those defined physical locations due to actions taken by the service provider that the third party uses, then regardless of knowing or not knowing about such actions, the third party can be liable under the relevant laws set out by the jurisdiction. In the use case presented here, an SLA management framework provides the means to:

1. Customise a service offering.
2. Negotiate on that service offering to the satisfaction of the third party and their legal responsibilities.
3. Be notified when terms of the agreed service offering deviate and have deviations logged as an audit trail.

The use case is realised by the third party provisioning the offered database service using the OCCI API through the facilities of the SLA manager. OCCI provides the standard and interoperable means of provisioning the required database service and the SLA manager provides the means as outlined above. That database service is realised as a pre-built virtual machine with all the requisite database software installed and configured, which once provisioned is accessible by the consumer. The service provider offers means to monitor the agreed terms in the SLA and, in particular for this use case, allows for the physical location of the virtual machine to be monitored. This allows the SLA management framework to monitor constantly the physical location of the virtual machine and in the case that the virtual machine is migrated to an inappropriate physical location the third party will instantly receive notification of that event and logs will provide an audit trail.

2.3.1 SLA@SOI SLA Management Framework

SLA@SOI defines a holistic view for the management of SLAs and implements an SLA management framework that can be easily integrated into a service-oriented infrastructure (SOI), see [17]. The main innovative features of the project are:

• An automated e-contracting framework
• Systematic grounding of SLAs from the business level down to the infrastructure

---

[10]http://www.microsoft.com/en-us/sqlazure.

[11]http://www.longjump.com.

- Exploitation of virtualisation technologies at infrastructure level for SLA enforcement
- Advanced engineering methodologies for creation of predictable and manageable services

The accompanying diagram (Fig. 2.4) illustrates the anticipated SLA management activities throughout the business/IT stack.

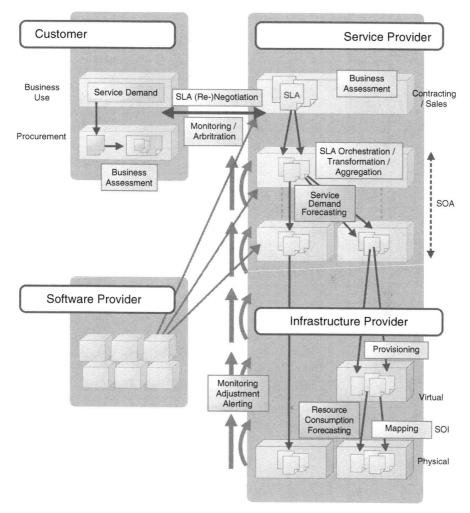

Fig. 2.4 SLA@SOI overview

2.3.2   SLA@SOI and OCCI

In this use case, there are two main components that are required for realisation. The first and most fundamental is a Service Manager that offers a database service.

The Service Manager is the entity that is responsible for providing the client's service. Relevant to this use case is that the Service Manager provides database services as preconfigured virtual machines and that the location of those virtual machines can be monitored. The SLA@SOI framework makes no assumption on this Service Manager only that it has an interface:

1. That can create, retrieve, update and delete its managed services.
2. Through which service instance metrics can be listed and retrieved.

The second entity required is the SLA@SOI framework's SLA Manager. This is a set of both generic and domain-specific components. What is generic relates to the management of SLA templates (what a provider offers) and SLA instances (what a provider runs on their clients behalf and guarantees). The domain-specific components are those that interact with the particular service manager that provides the client services. Further details of the SLA@SOI framework and its architecture can be found, see [18].

The SLA Manager offers to clients one or more SLA Templates, which is expressed using the SLA@SOI SLA model. Through either a UI or API, the client can select, customise, negotiate and provision an SLA-guaranteed service. In the use case scenario, this would entail the third party specifying what physical location (e.g. region, country) is required for their regulatory compliance.

Once the SLA Manager is acting on the client's behalf, it first negotiates with the service manager using the OCCI query interface. The OCCI query interface allows for the various Resource types to be queried for and interrogated and in particular to this use case, the locations that a provider can provision their virtual machines. As an extension to the query interface, SLA@SOI will also allow for per-user quotas to be queried. Using this extension, an SLA Manager can tell whether a client's request will be fully satisfied or not by the current service provider. Having established that the client's quota is sufficient, the next step can either take two paths. The first is that the provisioning of the requested service is done automatically or, second, the provisioning must be explicitly executed by the client. Where a provisioning request is executed in one or the other manner, the next responsibility of the SLA Manager will be to call the provisioning functionality of the Service Manager (relationship and interaction is shown in Fig. 2.5). This again is looked after by the OCCI API and an OCCI request from the SLA manager's domain specific components is sent to the Service Manager. As soon as the provisioning request is successful, the SLA manager then begins to monitor the provisioned service, including the location of the service's virtual machine. It does this by monitoring the various terms of the agreed SLA (e.g. QoS metrics).

For the SLA Manager, the major advantage of choosing to implement OCCI as a means to talk with Service Managers is that in the case where a particular

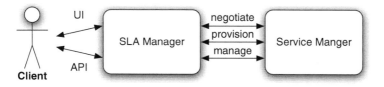

Fig. 2.5 Relationship between SLA manager and service manager

Service Manager cannot satisfy the provisioning of the requested service resources due to insufficient client quotas or unsuitable virtual machine deployment locations, the SLA Manager can, with the necessary logic implemented, look up the next registered service provider and seek to have the remaining service resources provisioned there, without any need for Service Manager protocol or API changes. Such functionality makes an excellent case for SLA-mediated cloud brokerage use cases.

2.4 On-Demand Data-Aware Provisioning of Services

A different application area for OCCI appears in the context of traditional community-based Distributed Computing Infrastructures (DCIs): modern research more and more relies on cross-institutional, cross-project data processing. In many communities, scientists quickly state the requirement to enable exchange of information beyond traditional boundaries such as project collaborations or long-term Virtual Organisations. Rather than that, a more flexible, more agile approach is expected.

This development poses a major challenge not only for the management of data itself (i.e. ensuring authentication and authorisation, planning distribution and replication, and tracking provenance), but also for the management of its computation: workload needs to run close to the data in most cases (since data is usually large), but the compute infrastructure available in the direct proximity of the data may not necessarily provide the correct environment. That is, applications to process the data might be missing, the operating system does not match the application requirements, or – on a higher level – certain services needed for data analysis and manipulation have not been deployed on-site.

Beside, many communities run their own, proprietary workload management software, tailored to the specific needs of their users. As such, it is usually not an option to require a central system, often referred to as a *meta scheduler*[12] for all users of all communities. Rather than that, additional technology needs to be incorporated, which allows dynamic federation of planning domains depending on the current demand.

---

[12]Mostly found in the context of traditional Grid Computing.

The D-Grid Scheduler Interoperability project (DGSI) in the context of the German D-Grid Initiative[13] aims to provide a solution both issues through the development of a standards-based protocol between Meta schedulers. DGSI approaches interoperability DCIs from two sides, namely *Activity Delegation* (taking care of the handover of workload from one domain to the other), and *Resource Delegation* (taking care of the leasing of resources from one domain to another). Assuming the DGSI protocols and services in place, the notion of delegation can help to avoid traditional data management techniques such as decoupled copying, prefetching, and replication at all.

2.4.1 The Climate Community Use Case

The effects of climate change are one of the major challenges of mankind: stakeholders of many areas strive for strategies to deal with the consequences of pollution and man-made changes to the environment. The basis of all decision making are models of climate processes and the understanding of interplay of the enormous amount of parameters in them. Since the beginning of industrialisation in the nineteenth century, Earth System Science, one of the data sciences, investigates these processes, their chemical formation in the diverse subsystems such as oceans, atmosphere or biosphere, and their long-term influence on climate.

From those investigations, researchers nowadays possess very detailed insight into climate development. This rests on the permanent acquisition, cataloging, and processing of very large (Peta scale) volumes of experimental and model data, as well as the continuous re-evaluation of scientific results using refined models.

Current information technology provides potent means to accelerate these processes of data evaluation and simulation. High Performance Computing (HPC) infrastructure, high speed networks, and modern storage architectures support archival, preprocessing, selection, and transportation of large data amounts as well as the computation of highly demanding simulations (e.g. short term weather forecast or storm track analysis).

For the latter scientific analysis, researchers filter and examine geopotential heights to track and predict the movement of low-pressure areas over time with regard to a given climate model. This is essential as storms and cyclones typically cross such areas [3]. This analysis and simulation is based on long-term acquired global climate data.

Usually, scientists are only interested in a restricted area for a Stormtrack analysis and have to reduce the amount of available base data to the region of interest. Besides a complex combination of several steps, this resorts to either access to a specific amount of climate data (Fig. 2.6a) or execute simple visualisation workflows on selected and preprocessed data (Fig. 2.6b).

---

[13]http://www.d-grid.de.

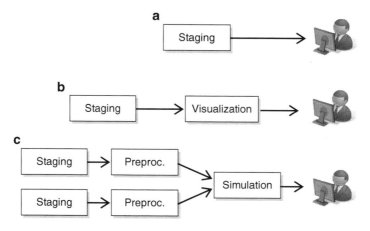

Fig. 2.6 Three simple workflow examples from C3Grid: (**a**) selective data download; (**b**) simple simulation on input data; (**c**) simulation on a set of preprocessed input data

Today, they generally have two possibilities to retrieve the desired data: they either get access to the storage and download the full amount of data (i.e. full replication is performed) or use proprietary programs to reduce the amount of data at the storage site and download the desired data set afterwards. In the first case, the required local storage may simply not be available to the single scientist, or the providing institution may not be willing to provide an external party with access to the full archive due to strategical considerations. In the second case, the user needs to cooperate directly with the data provider, basically via two mechanisms:

1. Having to use tools that are installed, but potentially not known to him,[14] or
2. Having to roll out the software on her own, either doing this as part of the batch processing job or in cooperation with the resource provider.

Obviously, the former is not acceptable from a user's perspective. The latter in turn requires extensive manual intervention and additionally necessitates the acquisition of user rights to retrieve or even locally process the requested data. As most of the climate data is stored in a distributed way, the procedure often has to be repeated for several data sites. Furthermore, it leaves intelligent, automatic load balancing totally to the user, which is generally not desirable. In addition to that, this traditional approach of application deployment massively hinders cross-community collaboration, if they rely on different infrastructure technologies: if the user takes care by himself to deploy the application as part of his computational workload, the number of resources compatible will usually be very restricted.

---

[14]With the exception of widely accepted and distributed tools such as the Climate Data Operators [15].

2.4.2 An Approach for Dynamic, Cross-Community Resource Allocation

Most DCI environments share the ability to efficiently distribute user workload to the resources available within their community. This issue, usually generalised under the term *Meta Scheduling*, is already very diverse within a community: both submitted jobs and available resources differ considerably, to the extent that coordination has to handle specialised knowledge about usage scenarios and infrastructure. This leads to very different, community-specific approaches for the development of Grid scheduling services [9].

The DGSI provides a standards-based interoperability layer for scheduling and resource management services in DCI ecosystems. By allowing the users of a community to distribute the workload among resources within the management domain of another community while keeping the individual, specialised scheduling solutions being run by the communities, it offers new perspectives for community collaboration, resource federation, and efficient utilisation. The general architecture is depicted in Fig. 2.7.

2.4.2.1 Delegation Models

The DGSI protocols foresee two scenarios to be considered: the delegation of activities and the delegation of resources:

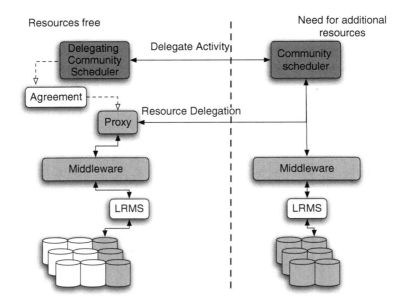

Fig. 2.7 Overall delegation architecture using the DGSI protocols. Meta schedulers from different domains (architectural, organisational, and technological) cooperate using activity and resource delegation

Delegation of activities. By means of DGSI, one meta-scheduler is able to delegate activities, that is single or parallel jobs or workflows, to another meta-scheduler from a different management domain (i.e. another community). By use of WS-Agreement [1], JSDL [2], and OGSA-BES [8], DGSI provides a standardised way of handing over workload to the other domain: a set of jobs that cannot be executed in the local scheduling domain can be channeled to another one (assuming the resource requirements of the jobs match the provided environment) to minimise waiting time induced by a high load on the originating side of the delegation. Via the mechanisms of SLA negotiation and agreement (as provided by the WS-Agreement protocol), it is ensured that both requirements and fulfilment can be negotiated in a reliable manner.

While the initial use case for activity delegation assumes an environment that requires cross-domain load balancing for workload to amplify user experience in a federated DCI environment, it is obvious that, with respect to data management, the very same mechanisms enable Meta Scheduling systems to easily move the workload close to the data: even if the data is assumed in a different community domain, proximity-based approaches for data-aware scheduling systems are easy to implement over the federated nature of the DGSI protocols.

Delegation of resources. To complement the handover of workload between DCIs in a more "as-a-Service"-related manner, the DGSI protocols also support the delegation of resources from one domain to another. This allows one meta-scheduler to effectively "lease" resources from another one over a given period of time and use them in the same way as managed resources within the own domain. Again, by use of WS-Agreement, GLUE, and middleware provisioning, a standardised means for requesting, negotiating, agreeing, monitoring, and provisioning those resources is available: after successfully agreeing on the "lease" contract, the scheduler that requests resources can effectively incorporate them into his planning algorithms for management over the time of lease.

The original use case was tailored to the specific needs of cross-community collaboration: the provisioning mechanisms were merely used to dynamically provide a management endpoint (i.e. a specific Grid middleware) that the requesting scheduler is able to cope with. For example, an environment that is generally managed by UNICORE [5] can provide a resource lease to a scheduler that manages its resources through Globus Toolkit [7] just by provisioning a Globus GRAM endpoint for the leased resources while – at the same time – ensuring the fulfillment of the negotiated SLAs through injected monitoring and enforcement mechanisms. From the perspective of data management, especially in the context of proximity-aware deployment of applications close to their data, much more can be done: by leveraging the provisioning interface to the deployment of the user's application rather than the middleware only, the user is provided with a unified view on the lower-level infrastructure and thus can run his application on a much larger resource space than given in traditional approaches. On the other hand, the meta-scheduler enjoys much more freedom in deploying the application close to the data, without having to give up its planning mandate (as in activity delegation).

2.4.3 The Role of OCCI for a Data-Aware Delegation Scenarios

The OCCI family of specifications, especially the infrastructure rendering, is the key enabler for introducing data-awareness into the different delegation scenarios.

Figure 2.8 depicts the role of OCCI in the overall process.

Enabling activity delegation. While OCCI is not strictly necessary for the activity delegation scenario, it makes the dynamic provisioning of a delegation channel (in case of the initial usage scenario of DGSI a service such as OGSA-BES) much easier. That is, the meta-scheduler that accepts workload delegation can dynamically

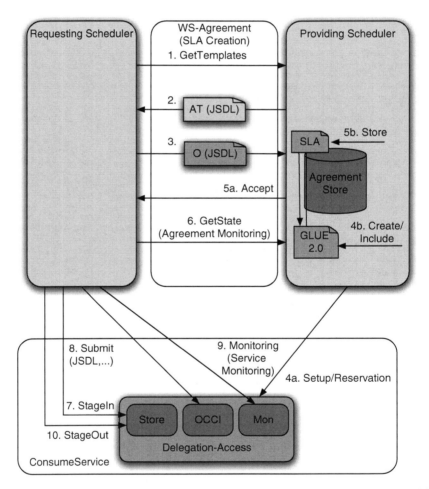

Fig. 2.8 Ten steps for negotiating a delegation within the DGSI protocol stack. While activity delegation only requires six steps (up to the *GetState* call for agreement monitoring), resource delegation runs to completion of the tenth step. Note, however, that – from step six and forth – each step is specific to the concrete usage of the delegated resource (e.g. a single job submission)

create instances of a submission and monitoring interface on its own infrastructure without having to provide own resources for the purpose. OCCI ensures in this context a unified view on underlying resources that can run demand-based, lifecycle-managed middleware services to the DCI ecosystem.

Enabling resource delegation. Here, OCCI is a strong requirement for allowing the user to deploy her own applications in the context of a virtualised environment over a unified interface. With OCCI in place, description, status management, and provisioning of virtual machines can be not only unified within the community itself, thus even there providing strong benefit, but rather beyond the boundaries of domains, allowing easy deployment of applications on the resources of a different community.

As such, OCCI fulfills two major requirements to enable this technology: interoperability and portability of the applications, and dynamic provisioning of infrastructure. The packaging paradigm of Virtual Machines additionally allows easy movement and infrastructure-agnostic capacity planning with data requirements in mind. Speaking of such, OCCI is the enabling technology for making data aware, proximity-based scheduling and resource management happen in federated DCI environments.

2.5 Use of OCCI for a Simple Key-Value Store

The previous Sects. 2.3 and 2.4 showed the usage of OCCI in virtualised environments (but data-centric setups). This last use case shows how the exact same standard can be used to give a database application a RESTful standard OCCI compliant interface.

A very simple use case is taken to demonstrate the abilities of OCCI as a data management front-end interface. Many NoSQL databases such as CouchDB[15] are deployed with a built-in RESTful interface. With the proliferation of NoSQL databases and their various RESTful APIs, there is a perceivable need for a standardised interface through which a client could discover the abilities and functionalities of the service provider (and in this use case the NoSQL Database).

Clients can then decide which service provider to use. This is essentially important since Cloud computing is all about delivering services experience to the customer. The customer should decide which service to use based on the experience, the functionalities and the price the service provider offers.

The discovery interface described in the OCCI section of this paper describes these self-discovery features. Section 2.3 on SLA@SOI describes how the OCCI core model can be extended for provisioning virtual machines.

---

[15]http://couchdb.apache.org/.

In this use case, the Core model is only extended by the one class which derives from `Resource`. It is called a *Key-Value resource* and has two attributes: `key` and `value`. A simple flip actions is defined. When the client queries the discovery interface, it will see the `Category` definition of this resource type:

```
> GET /-/ HTTP/1.1
> User-Agent: curl/7.21.0 (x86_64-pc-linux-gnu) [...]
> Host: localhost:8888
> Accept: */*
> Cookie: [...]
>
< HTTP/1.1 200 OK
< Content-Length: 517
< Etag: "89c0aeace4f7209b57d38cb0c4877bb9b22ad7a4"
< Content-Type: text/plain
< Server: pyocci OCCI/1.1
<
Category: keyvalue;
          scheme=http://example.com/occi/keyvalue;
          title=A Resource which holds a Key and a Value;
          location=/keyvalues/;
          rel=http://schemas.ogf.org/occi/core#resource;
          attributes=key value;
          actions=flip
Category: flip;
          scheme=http://example.com/occi/keyvalue;
          title=Flips the key and the value;
          attributes=foo bar
Category: keyvaluelink;
          scheme=http://example.com/occi/keyvalue;
          title=A link between two Key Value Resources;
          location=/keyvalues/links/;
          rel=http://schemas.ogf.org/occi/core#link;
          attributes=source target
```

The `GET` on the path */-/* indicates that one wants to discover what the service provider offers. It returns a `Category` definition showing the scheme of the category and which attributes it supports. As there will be no actions, this is all the `Category` features.

Now Key-Value resources can easily be created using this `Category` and retrieved through the OCCI interface:

```
> POST / HTTP/1.1
> User-Agent: curl/7.21.0 (x86_64-pc-linux-gnu) [...]
> Host: localhost:8888
> Cookie: [...]
> Content-Type: text/occi
> Category: keyvalue;
            scheme=http://example.com/occi/keyvalue;
```

```
> X-OCCI-Attribute: key=foo, value=bar
>
< HTTP/1.1 200 OK
< Content-Length: 2
< Content-Type: text/html; charset=UTF-8
< Location: /users/foo/keyvalues/dba17696-[...]
< Server: pyocci OCCI/1.1
<
```

The request indicates the type (via `Category`), a Key-Value resource, that the new resource should be. Also, two attributes are delivered alongside providing values to the key and value attributes of this new resource instance. The service will return a location of the new resource. This location can be used to retrieve the resource instance:

```
> GET /users/foo/keyvalues/dba17696-[...] HTTP/1.1
> User-Agent: curl/7.21.0 (x86_64-pc-linux-gnu) [...]
> Host: localhost:8888
> Accept: */*
> Cookie: [...]
>
< HTTP/1.1 200 OK
< Content-Length: 191
< Etag: "6bad49cb7785101006593a9fe79d5b54a4a19516"
< Content-Type: text/plain
< Server: pyocci OCCI/1.1
<
Category: keyvalue;
          scheme=http://example.com/occi/keyvalue
Link: </users/foo/keyvalues/dba17696-[...]?action=flip>
X-OCCI-Attribute: value=bar
X-OCCI-Attribute: key=foo
```

The response tells what type the REST resource is (via the `Category` header). It also returns us the two attributes which where defined during the creation of the resource.

Updating the attributes can be done using the HTTP `PUT` verb and it provides a new set of attributes. Deletion of the resource can be done through the HTTP `DELETE` verb.

Next to these HTTP basic renderings, the implementation[16] used for this example can also render OCCI using HTML by specifying the *text/html* content-type. This allows the user on client side to use the browser to discover the Query interface and the resources using a web browser (Fig. 2.9).

---

[16]http://pyssf.sf.net/.

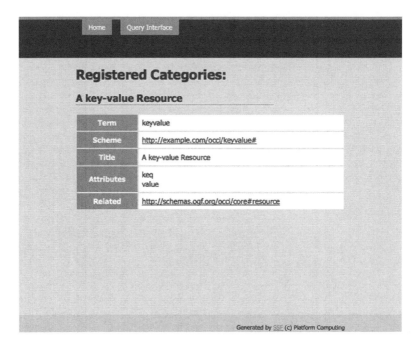

Fig. 2.9 Screen-shot of an HTML rendering of the OCCI query interface

This is a very generic interface and shows that OCCI can be used for provisioning infrastructure as well as PaaS based offerings.

Regardless of the type (OCCI `Kind`), a REST-Resource (represented through an URI) represents the interface will not change. This even means that the GUI (Fig. 2.9) is also generic and it would look and work the same for different types of `Resources`. The implementation of the OCCI interface demoed here can therefore be used for Infrastructure provisioning or other Platform offerings (like Job submission for Clusters).

2.6 Conclusions

With the last use case, the authors of this paper want to demonstrate the flexibility and extensibility of the OCCI interface. OCCI can be used using different setups especially the discovery functionalities and the extensible Core model support this. This demonstrates that OCCI can be used in IaaS and PaaS setups which relate to Data management.

This paper strives not to give a complete overview of all possible setups regarding OCCI and Data management. Still it demoed how some setups can be implemented using OCCI. Most often OCCI plays the role of ensuring and safe-guarding Interoperability as described in Sect. 2.2.1.

Several implementations of the OCCI specification exist notworthly does in research projects and currently ongoing working in commercial applications. Most notable is the currently in development effort which tries to incorporate the OCCI standard in the OpenStack[17] Cloud framework. Research Projects like the previously noted RESERVOIR and SLA@SOI (see Sect. 2.2.4) have adopted OCCI as well.

Next to this interoperability aspect, it is important to state that OCCI does not try to replace existing proprietary interfaces. It is defined for interoperability means as described before. Service Providers can still use their proprietary API/Interface to deliver higher-level functionalities, which is very specific to their offerings.

This idea of brokerage could either be realised in an automated fashion or with the user's interaction. Still OCCI makes this idea possible. Without an interoperable interface, a Cloud Broker of querying different cloud providers using one client would be impossible. Indeed, there is a current trend to enable interoperability through the use of facade/proxy service intermediaries. This is but a temporary solution as this approach leads to additional overhead in terms of inefficiencies, additional maintenance, configuration and management. This is something that OCCI seeks to remove and solve by doing so.

Next to driving adoption, the OCCI working group will focus on standardized interfaces for advanced reservation, monitoring and billing techniques. Also seman-tic enabled renderings will be added to the specification. Currently, the group is looking into JSON, XML or RDF/RDFa renderings.

What this paper demonstrated is that OCCI can be used on many layers of the Cloud Stack (IaaS and PaaS) and is possibly one of the small but important contributions to realise Cloud offerings. Even when narrowing the field to Data Management in Cloud and Grids, the OCCI interface can and must play a roll as an enabler.

Acknowledgements The authors acknowledge the contributions of all members of the OCCI working group. This work is partially supported by the German Ministry of Education and Research under project grant #01IG09009, and is partially supported by the European Community Seventh Framework Programme (FP7/2001-2013) under grant agreement no.216556.

References

1. Andrieux, A., Czajkowski, K., Dan, A., Keahey, K., Ludwig, H., Nakata, T., Pruyne, J., Rofrano, J., Tuecke, S., Xu, M.: Web Services Agreement Specification (WS-Agreement). In: Standards Track, no. GFD-R.107 in The Open Grid Forum Document Series, Grid Resource Allocation Agreement Protocol (GRAAP) Working Group, Muncie (IN) (2007)
2. Anjomshoaa, A., Brisard, F., Drescher, M., Fellows, D., Ly, A., McGough, S., Pulsipher, D., Savva, A.: Job Submission Description Language (JSDL) Specification, Version 1.0. In: Standards Track, no. GFD-R.56 in The Open Grid Forum Document Series, Job Submission Description Language (JSDL) Working Group, Muncie (IN) (2005)

---

[17]http://www.openstack.org.

3. Blackmon, M.L.: A climatological spectral study of the 500 mb geopotential height of the northern hemisphere. J. Atmos. Sci. **33**, 1607–1623 (1976)
4. Crosby, S., Doyle, R., Gering, M., Gionfriddo, M., Hand, S., Hapner, M., Hiltgen, D., Johanssen, M., Leung, J., Machida, F., Maier, A., Mellor, E., Parchem, J., Pardikar, S., Schmidt, S.J., Warfield, A., Weitzel, M.D., Wilson, J.: Open virtualization format specification. In: Grarup, S., Lamers, L.J., Schmidt, R.W. (eds.) Standards and Technology, no. DSP0243 in DMTF Specifications, Distributed Management Task Force (2009)
5. Erwin, D.W., Snelling, D.F.: UNICORE: A grid computing environment. In: Sakellariou, R., Gurd, J., Freeman, L., Keane, J. (eds.) Proceedings of the 7th International Euro-Par Conference, Lecture Notes in Computer Science (LNCS), vol. 2150, pp. 825–834. Springer, Heidelberg (2001)
6. Fielding, R.T.: Architectural styles and the design of network-based software architectures. PhD thesis, University of California, Irvine (2000)
7. Foster, I., Kesselman, C.: Globus: A toolkit-based grid architecture. In: The grid: Blueprint for a future computing infrastructure, pp. 259–278, 1st edn. Morgan Kaufman, San Mateo (1998)
8. Foster, I., Grimshaw, A., Lane, P., Lee, W., Morgan, M., Newhouse, S., Pickles, S., Pulsipher, D., Smith, C., Theimer, M.: OGSA Basic Execution Service Version 1.0. In: Standards Track, no. GFD-R.108 in The Open Grid Forum Document Series, Open Grid Services Architecture Basic Execution Services (OGSA-BES) Working Group, Muncie (IN) (2006)
9. Grimme, C., Papaspyrou, A.: Cooperative negotiation and scheduling of scientific workflows in the collaborative climate community data and processing grid. Future Generat. Comput. Syst. **25**, 301–307 (2009)
10. Metsch, T., Edmonds. A., et al.: Open Cloud Computing Interface – Core and Models. In: Standards Track, no. GFD-R in The Open Grid Forum Document Series, Open Cloud Computing Interface (OCCI) Working Group, Muncie (IN) (2010)
11. Metsch, T., Edmonds, A., et al.: Open Cloud Computing Interface – HTTP Rendering. In: Standards Track, no. GFD-R in The Open Grid Forum Document Series, Open Cloud Computing Interface (OCCI) Working Group (2010)
12. Metsch, T., Edmonds, A., et al.: Open Cloud Computing Interface – Infrastructure. In: Standards Track, no. GFD-R in The Open Grid Forum Document Series, Open Cloud Computing Interface (OCCI) Working Group, Muncie (IN) (2010)
13. Metsch, T., Edmonds, A., Bayon, V.: Using cloud standards for interoperability of cloud frameworks. Tech. rep., SLA@SOI project (FP7 ICT-2007.1.2-216556. http://sla-at-soi.eu/wp-content/uploads/2010/04/RESERVOIR-SLA@SOI-interop-techReport.pdf(2010)
14. Richardson, L., Ruby, S.: RESTful Web Services. O'Reilly Media, Sebastopol (CA) (2007)
15. Schulzweida, U., Kornblueh, L.: CDO User's Guide, Climate Data Operators, Version 1.0.6 (2006)
16. Slik, D., Siefer, M., Hibbard, E., Schwarzer, C., Yoder, A., Bairavasundaram, L.N., Baker, S., Carlson, M., Nguyen, H., Ramos, R.: Cloud data management interface. In: SNIA Technical Position Series, 1st edn. Storage Network Industry Association, San Francisco (2010)
17. Theilmann, W., Yahyapour, R., Butler, J.: Multi-level SLA management for service-oriented infrastructures. In: Mähonen, P., Pohl, K., Priol, T. (eds.) Towards a Service-Based Internet, Lecture Notes in Computer Science (LNCS), vol. 5377, pp, 324–335. Springer, Heidelberg (2008)
18. Theilmann, W., Happe, J., Ellahi, T., Torelli, F., Kearney, K., Lambea, J., Fuentes, B., Vuk, M., Guinea, S., Edmonds, A., Nolan, M., Brosch, F., Kotsokalis, K.: Deliverable D.A1a: Framework Architecture (full lifecycle). Tech. rep., SLA@SOI project (FP7 ICT-2007.1.2-216556, http://sla-at-soi.eu/wp-content/uploads/2009/07/D.A1a-M26-FrameworkArchitecture.pdf(2010)

Part II
Research Efforts on Grid Database Management

Chapter 3
The GRelC Project: From 2001 to 2011, 10 Years Working on Grid-DBMSs

Sandro Fiore, Alessandro Negro, and Giovanni Aloisio

Abstract This chapter provides a complete overview on the Grid Relational Catalog (GRelC) Project, a grid database research effort started in 2001 at the University of Salento. The project's main features, its interoperability with gLite-based production grids, and a relevant show-case in the environmental domain are presented.

3.1 Introduction

The management of large volume of data is a big challenge for several scientific domains such as Bioinformatics, Earth Science, High Energy Physics, and Astronomy. Computational and Data grids [1] provide the proper foundations to store, access, and analyze such a huge amount of data taking advantage of the large number of available distributed computational cores.

Over the last two decades, several data grid research efforts, such as the European DataGrid [2], the Storage Resource Broker [3], and the Globus GridFTP [4] tried to address several issues concerning file management in a grid environment, distributed file systems, grid storage services, data replication, efficient transfer protocols, etc. Yet, it was only in the last 10 years that the interest in grid-database systems distributed query processing, grid-database replication, concurrency management in a grid environment has significantly increased. A research effort started in 2001

S. Fiore (✉) · G. Aloisio
Euro-Mediterranean Centre for Climate Change (CMCC), via Augusto Imperatore 16, 73013 Lecce, Italy
and
University of Salento, via per Monteroni, 73100, Lecce, Italy
e-mail: sandro.fiore@unisalento.it; giovanni.aloisio@unisalento.it

A. Negro
University of Salento, via per Monteroni, 73100, Lecce, Italy
e-mail: alessandro.negro@unisalento.it

S. Fiore and G. Aloisio (eds.), *Grid and Cloud Database Management*,
DOI 10.1007/978-3-642-20045-8_3, © Springer-Verlag Berlin Heidelberg 2011

at the University of Salento and falling in this second category is the GRelC project [5], which is the main topic of this chapter.

The outline is as follows: Section 3.2 presents the 10-year history of the GRelC project. Section 3.3 provides a complete overview about the most relevant work in this area. Section 3.4 presents the grid database management system vision from the GRelC project perspective. Section 3.5 summarizes the key points related to the GRelC service, whereas Sect. 3.6 highlights the most important gLite-based features of this middleware. Section 3.7 presents a GRelC service showcase in the climate change domain. Section 3.8 concludes this chapter highlighting some future work.

3.2   The GRelC Project: A Decade of Research Efforts on Grid-DBMS

The GRelC project started in 2001 as a research effort at the University of Salento with a Ph.D. thesis. The initial goal was both simple and ambitious: *to provide a set of data grid services to transparently, securely and efficiently manage relational databases in a grid environment.* Such a piece of software was completely missing in the existing middleware (Globus [6]). The first use case was a relational information system named dynamic grid catalog (DGC) [7]. At that time, the most relevant grid information system was the MDS of Globus and it exploited a hierarchical data model. Conversely, the DGC service exploited a relational data model in the back-end. The same MDS information, but with a relational schema, was managed by the DGC service. A super-peer model was also proposed in 2003 to implement distributed scenarios. Preliminary performance results were successful and proved that such an approach could be suitable to manage relational information systems and more in general relational back-ends.

The GRelC service was an evolution of the DGC one. While the DGC service managed a relational database with a specific schema (for information system purposes), the GRelC service was able to manage whatever database regardless of its schema. New scenarios related to data access, integration, analysis, federation of grid-enabled databases were then possible. The scope and the number of potential use cases became wider making the GRelC service as general as the existing middleware services (e.g., GRAM, GridFTP, MDS).

Before its release, many tests were carried out to check the performance, the security framework (in terms of both authentication and authorization), the server robustness, etc. The EU GridLab project was an important test-case for several GRelC libraries and internal components of this service [8].

Until 2004, the GRelC releases exploited a client–server architecture, a proprietary communication protocol and the Grid Security Infrastructure [9]. Moreover, an integration service (named GRelC Gather Service [10]) was also developed to integrate relational databases geographically spread and adopting the same schema. A relevant use case was a health grid information system [11]. Supported DBMS were basically PostgreSQL and MySQL.

In the same period, there was a community migration toward open grid service architecture (OGSA) [12] and WebService (WS)-based implementations. The GRelC service was completely re-engineered to address interoperability through a WS-based approach. Instead of moving toward OGSI (which seemed to be too heavy), the GRelC service was implemented as a web service WS-I compliant and GSI enabled that is a very light implementation.

For a couple of years (until 2006), a lot of features were added to the GRelC service.

Most of them addressed:

– Performance (in particular grid-enabled queries in a geographical environment), through new and advanced query delivery mechanisms relying on compression, streaming, chunking, and prefetching. National and international performance tests [13, 14] were carried out to prove the efficiency of grid-enabled queries with regard to the well-known direct database connection approaches (ODBC/JDBC).
– *Database schema:* Dump, restore, and automatic registration of existing databases were successfully implemented to further extend the available set of functionalities.
– DBMS support, extending the back-end libraries to access to Oracle and IBM DB2 systems [15].

In 2006, the GRelC team started two important activities: the GRelC Portal [16] and a gLite-based release (see Sect. 3.6). The GRelC Portal was the first general-purpose web application able to access to grid-databases via Internet, and it was really effective to provide a transparent and ubiquitous access to data.

On the contrary, the gLite-based release (2007) was a crucial step to meet the EGEE community, their use cases and needs. This community provided new important requirements, particularly in the Earth Science context (EGEE NA4). That release was also available for training and dissemination purposes through the GILDA t-infrastructure [17].

From 2007, the GRelC service has also been adopted at the *Euro Mediterranean Center for Climate Change (CMCC)* [18] to address and solve challenging data and metadata issues in the Environmental domain. New releases were customized for CMCC users according to new requirements and needs coming from this scientific domain (the XML support to access to ISO19139-compliant metadata documents is a relevant example).

From 2008, the GRelC software has been included into the Italian grid release (gLite-based) and distributed into the Worker nodes and User Interfaces components across the Italian country. This way, several performance tests based on the gLite middleware were also carried out to stress the system and prove its stability. In the same year, GRelC was included into the EGEE RESPECT Program [19] due to its compatibility with the gLite middleware and its added value with regard to new database-oriented functionalities that were not available in the gLite release at that time.

At the end of 2008, the GRelC service was adopted into the Climate-G testbed (see Sect. 3.7) to provide distributed data access to climate change datasets. This testbed currently represents the most valuable use case for this software.

From 2009 to 2010, new GRelC releases (server and portal) addressing stability, management, and monitoring were made available to the user community.

In 2011, the GRelC team will face new challenges. The most relevant one will be related to the EGI Database of Databases (a global registry hosting the list of DB resources available in the EGI context). The registry will complement the EGI Application Database allowing scientists to know more about existing DBs, their location, main purpose, available data, etc. This will help the co-operation and interaction among research groups, promoting a more effective publishing and sharing of grid-enabled data sources.

3.3 Related Work

Over the last decade, several projects have addressed the main goal of *managing databases in a grid environment*. This section presents the most relevant ones, along with the main differences with the GRelC project.

The Spitfire Project [20] was part of the Work Package 2 of the European Data Grid Project and provided a means to access relational databases from the grid. Both GRelC and Spitfire started in 2001 and can be considered as the first efforts in this area.

The Open Grid Services Architecture Data Access and Integration (2002) [21] is a project concerning the development of middleware to assist with access and integration of data from separate data sources via the grid. OGSA-DAI is also strongly connected with OGF [22] standardization bodies and activities, and it is/was exploited in several international projects and in different scientific domains. With regard to GRelC, which is able to handle data integration among databases with the same schema, OGSA-DAI provides a more general distributed query processor (DQP) [23]. A technical difference relates to the programming language (C for GRelC, Java for OGSA-DAI). Finally, while GRelC has a stronger support on the client side (GRelC Portal, command line interface for end-users and administrators and the XGRelC GUI) and a better integration with the gLite middleware (support for BDII is also available), OGSAI-DAI provides WS-DAI* [24] compliant interfaces and dataflow support (through the *activity* concept).

The Grid Miner Project [25] focuses its effort on data mining and on-line analytical processing (OLAP), two complementary technologies able to provide a highly efficient and powerful data analysis and knowledge discovery solution on the Grid when applied in conjunction with each other. With regard to Grid Miner, both OGSA-DAI and GRelC focus more on low-level access and integration layers and services.

The Mobius Project [26] aims at developing an array of tools and middleware components to coherently share and manage data and metadata in a Grid and/or distributed computing environment.

AMGA [27] is the gLite Metadata Catalogue, designed to meet the requirements of the EGEE applications. The main features are connected with a hierarchical organization of metadata (organized internally in a tree-like structure), dynamic schema creation, and replication of metadata collections. VOMS [28] support is also available. Even though AMGA is today able to access relational databases, its initial focus was mainly on metadata. With regard to GRelC, AMGA does not provide support for XML resources, but it already implements the OGF WS-DAIR specification.

3.4   Grid Database Management System: The GRelC Perspective

The GRelC service represents a (partial) implementation of a grid database management system (GDMS). In our vision, a GDMS is an architectural stack consisting of several layers (*fabric, data access, management, and collective*) addressing in particular access, integration, management, monitoring, harvesting, and replication (see Fig. 3.1). From 2004 [29] to 2011 [30], this stack has been refined by considering and including all of the needed building blocks for this kind of system. Yet, it is important to point out that the final picture has not been revolutionized after approximately 10 years, due to its initial far-sighted design.

A complete overview of the four layers is presented as follows to have a complete understanding of our GDMS vision.

The *Fabric* layer represents the typical fabric layer that can be found in most of the grid-oriented architectures. Obviously, it refers to the database resources,

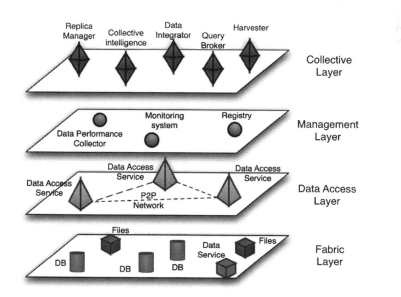

Fig. 3.1 The GDMS stack

and it is characterized by a high level of heterogeneity with regard to the DBMS servers, the data models, the supported data formats, the available APIs, the security frameworks, the supported platforms, etc.

The *Data Access* refers to the grid database access interface. According to the hourglass model, such an interface must provide a uniform and grid-enabled entry point to the underlying and heterogeneous database resources. From a security point of view, a data access service must basically provide support for the Grid Security Infrastructure, the de-facto standard for security in a grid environment.

The *Management* layer relates to a user-transparent level devoted to monitoring, management, and control functionalities. At this level, several metrics can be collected to check the status of the underlying data access services. ECA rules mechanisms can also be implemented at this layer along with global, advanced, and automatic detection tools or completely/semiautomatic diagnosis tools.

The *Collective* layer consists of a set of services to carry out data integration, data harvesting, data replication, query distribution, etc. Collective services base their decisions on statistics collected by and available from the *Management* layer. There is also a direct interaction between these services and the underlying layers (*Management* and *Data Access*) to carry out different activities.

In the context of the GRelC Project, the *Data Access* layer has been completely implemented. However, the *Management* and *Collective* layers have been only partially addressed.

3.5   The GRelC Service in a Nutshell

The GRelC service architecture and infrastructure have been already discussed in detail in several works [31, 32]. A brief description of this service is presented as follows focusing on few, simple but relevant concepts. This section actually provides a concise and complete summary (*the GRelC service in a nutshell*) by highlighting the *ten* most important concepts related to this service.

1. The GRelC service is a WS-I compliant, GSI, and VOMS enabled web service. These are three fundamental aspects related to the access interface (WS-based), the security infrastructure (GSI), and the authorization support (VOMS).
2. This software provides a complete command line interface able to support the user in terms of access and administration needs.
3. The same can be done through the GRelC Portal, which provides a web-based ubiquitous, transparent, and seamless access to the GRelC service.
4. As regards interoperability, the GRelC service runs both on gLite- and Globus-based environments. Since the software only depends on the GSI as far as grid middleware is concerned, it could basically run on any GSI-based middleware.
5. Support is provided for relational (MySQL, PostgreSQL, SQLite, IBM DB2 and Oracle) and XML-based (Xindice, eXist, XML flat-files) back-ends.

6. The service does not change the SQL query submitted by the user (SQL tunnelling), allowing the users to take advantage of proprietary SQL extensions available in the back-end system. Anyway, it performs some checks on it to avoid query injection and to prevent malicious users from carrying out disruptive actions on the target data sources.
7. The application programming interfaces (APIs) and the SDK are available for C and Java developers.
8. The available grid-enabled queries exploit compression, chunking and streaming to enhance performance. Performance comparisons are available in [13] and [14].
9. The asynchronous query support is available to implement distributed and gLite-based scenarios. Additional details can be found in [33].
10. A complete authorization framework and a rich set of policies support different authorization schemas and scenarios as stated in [34].

3.6 A Crucial Step: Moving Toward EGEE and gLite

The GRelC service was ported on the gLite platform in 2007 for preliminary tests; yet, the GRelC roadmap on how to move toward EGEE and gLite was a bit more articulated [35].

The first release was GSI enabled but not VOMS compliant, allowing just the implementation of local authorization schemas. The VOMS support was added to the GRelC service in few weeks, to fully address the security framework compatibility with the other gLite components.

The second action was at the architectural level. The EGEE middleware consisted of several components to manage data and computational resources, schedule jobs on the grid, manage authorization at VO level, etc. In particular, the EGEE farm model relied only on few components: a storage element to manage files and a computing element and worker nodes to run computational jobs. Owing to the fact that a grid-database support was completely missing in that model, an *extended farm model* was proposed to include database oriented support and functionalities (see Fig. 3.2). *The proposed farm model extension improved the user/VO support adding to the EGEE farm new capabilities concerning the grid-DB management, without changing or limiting those already existing.*

To have the GRelC service completely integrated into the farm structure, a BDII support was also added. This means that the GRelC service was able to publish on the BDII (like the computing element or the storage element) information about the available database resources. This was the first research effort in this area (the provided extensions were completely proprietary). The opportunity to publish data on the information systems enabled gLite broker components to discover the grid-database resources available on the grid.

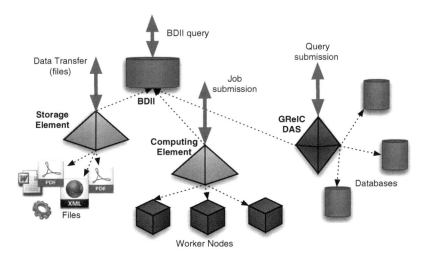

Fig. 3.2 The EGEE extended farm model

Finally, the asynchronous support [31] was added to the GRelC service to enable it to act, with regard to the database queries, as the computing element dealing with the computational jobs.

All these activities were carried out in about a year, allowing the GRelC service to be included into the EGEE RESPECT Program in mid-2008.

An important activity in the EGEE context was related to training. Thanks to the GILDA team, a training environment for the GRelC service was quickly set up, enabling EGEE users to carry out training activities, submit queries, and exploit the GRelC command line interface and portal. In the EGEE context, GILDA acted as a crucial component of the project's t-infrastructure (training infrastructure) program, helping to pass on knowledge and experience, as well as computing resources, to the scientific community and Industry.

3.7 An International and Multidisciplinary Use Case: Climate-G

This section presents the most relevant use case exploiting the GRelC service, namely the Climate-G testbed [30]. It is an interdisciplinary research effort devoted to the Environmental domain and involving both computer science and climate change researchers and scientists. It acts as a virtual laboratory, across Europe and USA, addressing data and metadata management issues at a very large scale.

The main goal of Climate-G is to allow scientists to carry out geographical and cross-institutional data discovery, access, visualization and sharing of climate data.

Such a testbed has been conceived in the context of the EGEE Earth Science Cluster Community. The main partners are: Centro Euro-Mediterraneo per i Cambiamenti Climatici (CMCC, Italy), Institut Pierre-Simon Laplace (IPSL/CNRS, France), Fraunhofer Institut für Algorithmen und Wissenschaftliches Rechnen (SCAI, Germany), National Center for Atmospheric Research (NCAR, USA) and Rensselaer Polytechnic Institute (RPI, USA), University of Reading (Reading, UK), University of Cantabria (UC, Spain), and University of Salento (UniSalento, Italy).

Distributed data and metadata management (hundreds of Petabytes of climate datasets) represents the key challenge related to the GRelC service that will be presented in this section. Data distribution comes from the need of sharing data among centers without moving it to a central repository, whereas metadata distribution is strongly needed to address local autonomy, scalability, and fault tolerance.

Each site participating in the Climate-G testbed hosts an OPeNDAP/THREDDS [36] server (domain-oriented service) to manage some climate change datasets and makes them available to the users. Furthermore, some of the sites host a GRelC server (grid-oriented service) to manage the metadata experiments about the available datasets. The co-existence of grid and domain-related services is an important user requirements addressed by the testbed.

Figure 3.3 depicts the Climate-G network of metadata services. A grid-enabled harvester (basically, a GRelC client) is also part of the infrastructure to collect and gather the relevant metadata from each GRelC service.

While the OPeNDAP services provide access, subsetting and download functionalities, the GRelC services enable search and discovery functionalities, making distributed data effectively accessible and shareable by the scientific community.

In the Climate-G testbed, metadata are stored both in relational and in XML databases, and they are available through the same grid-enabled GRelC interface. Although the relational databases (even including the harvester DB) contain just key information about the available experiments, the XML databases store the full experiments descriptions (ISO 19115/19139 and INSPIRE [37] compliant).

The entry point of the testbed is the Climate-G Portal. It exploits the GRelC client Java package to implement the search and discovery functionalities, as well as the XML metadata access, the web-based proxy creation and the access to the list of datasets, experiments and projects. It is worth mentioning that monitoring facilities are also available in the Climate-G Portal. They give the administrators full control of the underlying metadata systems with real-time monitoring capabilities, reports, and statistics about the involved resources.

This testbed represents the most valuable showcase for GRelC. In June 2009, during the first year review of the EGEE-III Project, a demonstration of the Climate-G testbed was presented to a European Commission appointed panel. The work was evaluated by the EGEE NA4 Steering Committee, EGEE Activity Management Board and the European Commission *"as indicative of the excellent scientific work being done on the grid and of the advancement of grid services/tools"*.

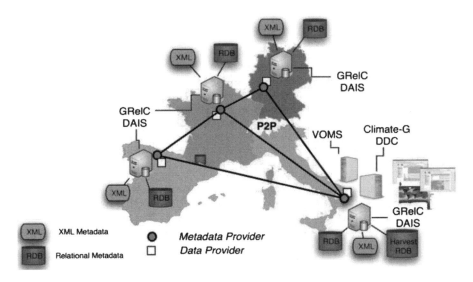

Fig. 3.3 The grid metadata system in the Climate-G testbed

3.8 Conclusions and Future Work

This chapter provided an overview of the GRelC project, its history, the grid database management vision, the main features of the GRelC service, the path toward EGEE and gLite, and finally a real showcase (the Climate-G testbed) related to the Earth Science and the Environmental domains. Due to the huge amount of scientific work that has been done in the context of this research project, it was not easy to single out the most relevant milestones achieved in approximately 10 years of history. The authors tried to highlight the key points of this research effort, leaving out many technicalities, implementation aspects, and low level details.

Future work will address the major issues at the management and collective layers, to complete and improve what has been done so far. The OGF WS-DAI* specifications will also be implemented and tested to address interoperability and to enable new challenging scenarios involving different, yet interoperable, grid database access and integration services.

References

1. Foster, I., Kesselman, C.: The Grid: Blueprint for a New Computing Infrastructure. Morgan Kaufmann, CA (1998)
2. Cameron, D., et al.: Replica management in the European DataGrid Project. J. Grid Comput. **2**(4), 341–351 (2004). doi:10.1007/s10723-004-5745-x
3. Rajasekar, A., Wan, M., Moore, R., Schroeder, W., Kremenek, W., Jagatheesan, A., Cowart, C., Zhu, B., Chen, S.-Y., Olschanowsky, R.: Storage resource broker – managing distributed data in a grid. Comp. Soc. India J. Special Issue on SAN **33**(4), 42–54 (2003)

4. Bresnahan, J., Link, M., Khanna, G., Imani, Z., Kettimuthu, R., Foster, I.: Globus GridFTP: What's new in 2007. In: Proceedings of the First International Conference on Networks for Grid Applications (GridNets 2007), pp. 1–5 (2007)
5. Aloisio, G., Cafaro, M., Fiore, S., Mirto, M.: The grid relational catalog project, advances in parallel computing. In: Grandinetti, L. (ed.) Grid Computing: The New Frontiers of High Performance Computing, pp. 129–155. Elsevier, Amsterdam (2005)
6. Foster, I.: Globus toolkit version 4: Software for service-oriented systems. In: IFIP International Conference on Network and Parallel Computing, LNCS, vol. 3779, pp. 2–13. Springer, Heidelberg (2006)
7. Aloisio, G., Cafaro, M., Blasi, E., Epicoco, I., Fiore, S., Mirto, M.: Dynamic grid catalog information service. In: Proceedings of the First European Across Grids Conference, 13–14 Feb 2003. Santiago de Compostela (Spain), LNCS, vol. 2970, pp. 198–205. Springer, Heidelberg (2003)
8. Aloisio, G., Cafaro, M., Epicoco, I., Fiore, S., Lezzi, D., Mirto, M., Mocavero, S.: Resource and service discovery in the iGrid information service. In: Gervasi, O., et al. (eds.) ICCSA 2005: International Conference – Grid Computing and Peer-to-Peer Systems, Lecture Notes in Computing Science, Singapore, 9–12 May 2005, vol. 3482, pp. 1–9. ISBN 3-540-25862-0
9. Foster, I., Kesselmann, C., Tsudik, G., Tuecke, S.: A security architecture for computational grids. In: Proceedings of 5th ACM Conference on Computer and Communications Security Conference, pp. 83–92 (1998)
10. Aloisio, G., Cafaro, M., Fiore, S., Mirto, M., Vadacca, S.: GRelC data gather service: A step towards P2P production grids. In: Proceedings of 22nd ACM Symposium on Applied Computing (SAC 2007), pp. 561–565
11. Mirto, M., Aloisio, G., Cafaro, M., Fiore, S.: A gather service in a health grid environment. In: CD-Rom of Medicon and Health Telematics 2004, IFMBE Proceedings, vol. 6, 31 July–05 August, Island of Ischia, Naples, Italy
12. Foster, I., Kesselman, C., Nick, J., Tuecke, S.: The physiology of the grid: An open grid services architecture for distributed system integration. www.globus.org/research/papers/ogsa.pdf
13. Aloisio, G., Cafaro, M., Fiore, S., Mirto, M.: Advanced delivery mechanisms in the GRelC project. In: ACM Proceeding of 2nd International Workshop on Middleware for Grid Computing (MGC 2004), 18 Oct 2004, Toronto, ON, Canada, pp. 69–74
14. Fiore, S., Negro, A., Vadacca, S., Cafaro, M., Aloisio, G., Barbera, R., Giorgio, E.: Advances in the GRelC data access service. In: Proceedings of ISPA 2008, 10–12 Dec 2008, Sydney, Australia, pp. 849–854
15. Mirto, M., Epicoco, I., Fiore, S., et al.: The LIBI grid platform for bioinformatics. In: Cannataro, M. (ed.) Handbook of Research on Computational Grid Technologies for Life Sciences, Biomedicine and Healthcare, pp. 577–613. ISBN: 978-1-60566-374-6, May 2009. Published under Medical Information Science Reference, IGI Global. University Magna Graecia of Catanzaro, Italy. http://www.igi-global.com/reference/details.asp?id=34292
16. Fiore, S., Negro, A., Vadacca, S., Verdesca, E., Leone, A., Aloisio, G.: The GRelC portal: A seamless and ubiquitous way to manage grid databases. In: Proceedings of PDCAT 2008 – 01–04 Dec 2008, Dunedin, New Zealand, pp. 413–418
17. Andronico, G., Ardizzone, V., Barbera, R., Catania, R., Carrieri, A., Falzone, A., Giorgio, E., La Rocca, G., Monforte, S., Pappalardo, M., Passaro, G., Platania, G.: GILDA: The grid INFN virtual laboratory for dissemination activities. TRIDENTCOM, pp. 304–305 (2005)
18. Fiore, S., Vadacca, S., Negro, A., Aloisio, G.: Data issues at the Euro-mediterranean Centre for Climate Change. J. Earth Sci. Inform. 2(1–2), 23–35 (2009). doi:10.1007/s12145-009-0023-x
19. EGEE RESPECT Program. http://technical.eu-egee.org/index.php?id=290
20. Hoschek, W., McCance, G.: Grid enabled relational database middleware. Informational Document Global Grid Forum, Frascati, Italy, 7–10 October 2001

21. Antonioletti, M., Atkinson, M.P., Baxter, R., Borley, A., Chue Hong, N.P., Collins, B., Hardman, N., Hume, A., Knox, A., Jackson, M., Krause, A., Laws, S., Magowan, J., Paton, N.W., Pearson, D., Sugden, T., Watson, P., Westhead, M.: The design and implementation of grid database services in OGSA-DAI. Concurrency Comput. Pract. Ex. **17**(2–4), 357–376 (2005)

22. Antonioletti, M., Krause, A., Paton, N.W.: An outline of the global grid forum data access and integration service specifications, VLDB DMG 2005. Lect. Notes Comput. Sci. **3836**, 71–84 (2005)

23. Dobrzelecki, B., Krause, A., Hume, A., Grant, A., Antonioletti, M., Alemu, Y., Atkinson, M., Jackson, M., Theocharopoulos, E.: Integrating distributed data sources with OGSA–DAI DQP and Views. Phil. Trans. R. Soc. A **368**(1926), 4133–4145 (2010). doi: http://dx.doi.org/10.1098/rsta.2010.0166

24. Antonioletti, M., Krause, A., Paton, N.W., Eisenberg, A., Laws, S., Malaika, S., Melton, J., Pearson, D.: The WS-DAI family of specifications for web service data access and integration. ACM SIGMOD Rec. **35**(1), 48–55 (2006)

25. Brezany, P., Janciak, I., Min Tjoa, A.: Data mining on the grid: Perspective from the GridMiner experience. In: 5th Cracow Grid Workshop, Poland, 21–23 Nov 2005

26. Hastings, S., Langella, S., Oster, S., Saltz, J.: Distributed data management and integration framework: The Mobius project. In: Proceedings of the Global Grid Forum 11 (GGF11) Semantic Grid Applications Workshop, pp. 20–38 June 2004

27. Santos, N., Koblitz, B., Distributed metadata with the AMGA metadata catalog. In: The Proceedings of the Workshop on Next-Generation Distributed Data Management HPDC-15, Paris, France, June 2006

28. Alfieri, R., Cecchini, R., Ciaschini, V., dell'Agnello, L., Frohner, A., Gianoli, A., Lorentey, K., Spataro, F.: VOMS, an Authorization System for Virtual Organizations. LNCS, vol. 2970, pp 33–40. Springer, Heidelberg (2004)

29. Aloisio, G., Cafaro, M., Fiore, S., Mirto, M.: The GRelC project: Towards GRID-DBMS. In: Proceedings of Parallel and Distributed Computing and Networks (PDCN) – IASTED, Innsbruck, Austria, pp. 1–6, 17–19 February 2004

30. Fiore, S., Negro, A., Aloisio, G.: The data access layer in the GRelC system architecture. Future Generat. Comput. Syst. **27**(3), 334–340 (2011)

31. Fiore, S., Negro, A., Vadacca, S., Cafaro, M., Aloisio, G., Barbera, R., Giorgio, E.: An architectural overview of the GRelC data access service. In: Udoh, E., Wang, F. (ed.) Handbook of Research on Grid Technologies and Utility Computing: Concepts for Managing Large-Scale Applications, pp. 98–108. IGI Global, PA (2009)

32. Fiore, S., Cafaro, M., Mirto, M., Vadacca, S., Negro, A., Aloisio, G.: The GRelC project: State of the art and future directions. In: Grandinetti, L. (ed.) High Performance Computing and Grids in Action, vol. 16, pp. 331–344. IOS Press, VA (2008)

33. Fiore, S., Cafaro, M., Vadacca, S., Negro, A., Verdesca, E., Mirto, M., Aloisio, G.: Asynchronous query mechanisms within the GRelC data access service. In: Proceedings of the IASTED International Conference, Parallel and Distributed Computing and Networks, pp. 49–54. Innsbruck, Austria, 12–14 Feb 2008

34. Fiore, S., Negro, A., Aloisio, G.: Data virtualization in grid environments through the GRelC data access and integration service. In: Proceedings 4th International Conference for Internet Technology and Secured Transactions, IEEE, (ICITST 2009), pp. 817–822. London, UK, 9–12 Nov 2009

35. Fiore, S., Cafaro, M., Negro, A., Vadacca, S., Aloisio, G., Barbera, R., Giorgio, E.: GRelC DAS: A Grid-DB access service for gLite based production grids. In: IEEE Proceedings of the Fourth International Workshop on Emerging Technologies for Next-generation GRID (ETNGRID 2007), pp. 261–266. Paris, France, 18–20 June 2007

36. Cornillon, P., Gallagher, J., Sgouros, T.: OPeNDAP: accessing data in a distributed, heterogeneous environment. Data Sci. J. 2, 164–174 (2003)

37. INSPIRE Directive: Directive 2007/2/EC of the European Parliament and of the Council of 14 March 2007 establishing an Infrastructure for Spatial Information in the European Community (INSPIRE)

Chapter 4
Distributed Data Management with OGSA–DAI

**Michael J. Jackson, Mario Antonioletti, Bartosz Dobrzelecki,
and Neil Chue Hong**

Abstract OGSA–DAI provides a framework for sharing and managing distributed data. OGSA–DAI is highly customizable and can be used to manage, share and process distributed data (e.g. relational, XML, files and RDF triples). It does this by executing workflows that can encapsulate complex distributed data management scenarios in which data from one or more sources can be accessed, updated, combined and transformed. Moreover, the data processing capabilities provided by OGSA–DAI are further augmented by a powerful distributed query processor and relational views component that allow distributed data sources to be viewed and queried as if they were a single resource. OGSA–DAI allows researchers and business users to move away from logistical and technical concerns such as data locations, data models, data transfers and optimization strategies for data integration and instead focus on application-specific data analysis and processing.

4.1 Introduction

The Open Grid Services Architecture–Data Access and Integration Services (OGSA–DAI) framework has, since its inception in 2002, been designed to serve as a solution for complex distributed data management challenges in academia, industry and commerce. OGSA–DAI provides an environment for the execution of complex distributed data management scenarios in which data from multiple sources and of multiple types (e.g. relational, XML, files, RDF triple stores, web services) can be accessed, updated, combined, filtered, transformed and delivered.

M.J. Jackson (✉) · M. Antonioletti · B. Dobrzelecki · N.C. Hong
EPCC, The University of Edinburgh, James Clark Maxwell Building, The King's Buildings,
Mayfield Road, Edinburgh EH9 3JZ, UK
e-mail: Mike.Jackson@ed.ac.uk; Mario.Antonioletti@ed.ac.uk;
bartosz.dobrzelecki@googlemail.com; N.P.ChueHong@ed.ac.uk

S. Fiore and G. Aloisio (eds.), *Grid and Cloud Database Management*,
DOI 10.1007/978-3-642-20045-8_4, © Springer-Verlag Berlin Heidelberg 2011

Instead of being tailored as a solution to a specific distributed data management problem, OGSA–DAI has been designed to be extensible. It allows customizations to be made for individual application-specific requirements, whether this is in terms of the data resources supported, the data processing operations executed or the way in which the framework is accessed or exposed. Data streaming is fundamental to OGSA–DAI enabling the processing of large data sets and the implicit exploitation of any parallelism available on the machines on which it runs.

OGSA–DAI includes a distributed query processor for relational data sources [1], which has its origin in the OGSA–DQP distributed query processor [2, 3] developed by the Universities of Manchester and Newcastle. This distributed query processor allows complex queries involving distributed data sources to be expressed declaratively.

Features such as these have facilitated OGSA–DAI's adoption in the solution of distributed data management challenges in a range of projects, organizations and communities in different domains including: astronomy [4], earth sciences [5], geo-spatial information systems [6, 7], chemistry [8], biochemistry [9], medicine [10–12], social sciences [13], transportation [14], environment [15], classics [16] and many more [17].

OGSA–DAI is free and available as a 100% Java open source product released under an Apache 2.0 license. In October 2009, OGSA–DAI moved to being an open source project employing open development processes, hosted at SourceForge [18] and currently has project members from the University of Edinburgh and the Universidad Politécnica de Madrid as well as other individual world-wide contributors.

4.1.1 Overview

The remainder of this chapter motivates and illustrates the use of OGSA–DAI through a worked example. In Sect. 4.2, the need for distributed data management is illustrated via an example from health informatics. Section 4.3 introduces the OGSA–DAI framework, its main components and how OGSA–DAI represents and executes distributed data management scenarios via workflows. OGSA–DAI's distributed query processor is described in Sect. 4.4 and its SQL views components in Sect. 4.5. How OGSA–DAI supports interoperability with computational or data analysis tools is shown in Sect. 4.6. Performance issues are discussed in Sect. 4.7. Section 4.8 reviews related work. The chapter concludes with a discussion of the strengths and weaknesses of OGSA–DAI and examines possible future directions.

4.2 A Distributed Data Use Case: Health Informatics

Every time a patient visits a health centre or hospital, or a doctor performs a home visit, information is recorded in the patient's medical records. This includes personal details, for example name, address, medical details, symptoms. This data,

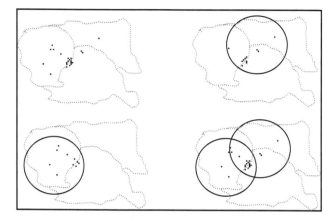

Fig. 4.1 Three regions are shown with two surgeries (not shown) whose catchment areas may span the regions. Illness incidents are shown by *dots*. (**a**) (*top left*) Shows a global view where a cluster of patients with symptoms between regions may give cause for concern; (**b**) (*top right*) and (**c**) (*bottom left*) show the data available at each surgery; (**d**) (*bottom right*) shows the combined data from both surgeries reconstituting the global view

if analysed, can provide important information, for example, whether an outbreak of a contagious disease, such as swine flu, is imminent. One way to extract this information would be to aggregate the patient numbers across surgeries within a given region, as depicted in Fig. 4.1a. If the number of patients exhibiting particular symptoms exceeds some critical threshold this might then give cause for concern and set off an alarm. This would not be difficult to implement but it does assume that all the patient data is readily available and can be easily accessed and analysed. In reality, access to the data will typically not be so straightforward.

For instance, the region of interest may be covered by two surgeries with overlapping catchment areas. Visualizing patients with symptoms in one surgery might yield Fig. 4.1b, where the cluster does not exceed a critical threshold of 10. For the other surgery, shown in Fig. 4.1c, there is, likewise, no cluster greater than the threshold because the cluster is within an area where the catchment areas overlap. Only when the data from both surgeries is combined is the fact that there is a cluster of patients with symptoms that exceed the threshold revealed, see Fig. 4.1d.

So, if the data is held across multiple sources, or databases, to identify clusters of patients, there are a number of activities that need to be done:

- Need to obtain the numbers of patients exhibiting the symptoms of concern together with the patient's post codes from each surgery. For example, an SQL query could be executed if the data is held in a relational database. An example query, in SQL, might then be `SELECT COUNT(*) AS count, postcode FROM patients WHERE symptom = "FLU" GROUP BY postcode;`
- Combine, or union, this data together.
- Determine the final total counts of occurrences for each post code.

A dedicated application could be written to do this, which would have to handle geographically distributed data, stored in different database products, which use different data formats and use different authentication mechanisms. Keeping the clients operational with all the surgery databases would be a nontrivial task.

To address this issue, a proxy server can be introduced. The server can be used to manage the connections with the databases. If a surgery then changed its database, only the server would need updating. A client only needs to connect to the server and so is protected from changes that take place behind its interfaces. As a consequence, clients can be more lightweight.

To further increase efficiency and flexibility, the server could also manage execution of the data processing activities on a client's behalf. A client then needs to do is to tell the server what activities it wants to run. To do this, three activities can be envisaged: one to query the first surgery's database, one to query the second and one to combine and summarize the results. Furthermore, if the data could be streamed through the server, and the server execute these activities concurrently, for example, combining and summarizing data as additional data is still being retrieved from the databases, then this could yield a reduced load on the server and, potentially, a faster response time for the client.

Furthermore, if the databases could be made to appear as a single database to a client, then the activities the client needs to request would be simpler: a client would only need to request the execution of a single query activity. Given the expressive power of query languages, the client could specify how the data is to be combined and summarized through a single query rather than having a separate activity to do this. In other words, it would be easier if the client could specify the whole task through a single query. For example:

```
SELECT SUM(count) AS total, postcode
FROM
((SELECT * FROM HealthCentreOne.patients
  UNION ALL
 (SELECT * FROM HealthCentreTwo.patients))
WHERE symptom="FLU" GROUP BY postcode ORDER BY total;
```

The server can take care of determining which queries need to be sent to which database and what additional activities need to be run to execute the client's query. This is called distributed query processing (DQP). With DQP, the activities that the client needs to tell the server to execute become much simpler: just run the SQL query above. The processing complexity is transferred to the server. DQP also opens up the possibility of optimizing queries to ensure that as much work is driven down to the databases as possible and so spares the server from pulling in unnecessary data. For example, it could transform the user's query above into:

```
SELECT SUM(count) AS total, postcode
FROM
((SELECT COUNT(*) AS count, postcode
```

```
FROM HealthCentreOne.patients
WHERE symptom="FLU" GROUP BY postcode)
UNION ALL
(SELECT COUNT(*) AS count, postcode
FROM HealthCentreTwo.patients
WHERE symptom="FLU" GROUP BY postcode))
GROUP BY postcode ORDER BY total;
```

Note that the WHERE clause has been pushed to the databases themselves. Going one stage further, it would be useful if it could make the patient data appear to be as if in one single table, for example allpatients, and then the user could just submit a very simple query like:

```
SELECT SUM(count) AS total, postcode
FROM allpatients
WHERE symptom="FLU" GROUP BY postcode ORDER BY total;
```

The server can then replace allpatients with the following and run the rewritten query using DQP:

```
((SELECT * FROM HealthCentreOne.patients
UNION ALL
(SELECT * FROM HealthCentreTwo.patients))
```

Finally, if a client needs to visualise the data it can transform it into a suitable format, for example, a JPG image file or a document written in the geographical markup language KML [19]. As this is just a data transformation, it is again desirable to let the server handle this. Also, instead of delivering the visualization data to the client, the server can hold the image until the client is ready for it. The server could return a URL telling the client from where it can retrieve the data. This would let the client do other things while the server is processing and allow the server to cache the result for other clients. The results can then just be obtained from the same URL. In such a scenario, the set of activities that a client requests the server execute becomes:

- Run a query to get the total counts of occurrences of patients with the symptoms of concern distributed by post code.
- Map the post codes to latitudes and longitudes.
- Convert the counts, latitudes and longitudes to a visualisation format, for exampe a JPG image or KML document.
- Cache the format on the server and return the URL to the client.

How such a server might work in practice is examined in the next section.

4.3 OGSA–DAI

OGSA–DAI provides a framework for the execution of distributed data management scenarios such as the one presented in the previous section. OGSA–DAI's primary design goals are to:

- Support the execution of collections of discrete data-related operations including data access, updates transformation, integration and delivery.
- Allow concurrent processing of different parts of a stream of data and so provide implicit parallelism and reduce the processing memory footprint required.
- Not restrict users to specific data source types or data-related operations.

In the remainder of this section, the main components of the framework and how the framework operates are described.

4.3.1 Data Representation

In OGSA–DAI, data is represented by Java objects. These objects can include elementary types such as strings, integers, floats, booleans and arrays. For binary and character data, byte and character arrays as well as binary and character large objects (blobs and clobs) are often used. The framework does not mandate or preclude the use of any Java object – application-specific objects can also be used within OGSA–DAI. A number of special objects are also provided which developers may find useful. These are:

- A MetadataWrapper object which can be used as a generic container for metadata about the data being passed within OGSA–DAI. If present, this typically occurs at start of the data stream which it describes. The contents of a MetadataWrapper are determined by the resources or activities that created it and the activities that consume it. The framework itself makes no assumptions as to the content of this wrapper.
- A Tuple object can be used to represent a row of relational data. This consists of the values for each field in a row. In what follows, these are denoted as: ("Joe Bloggs", "flu", 35, "EH9 5Z").
- A TupleMetaData object represents metadata about Tuples such as column names and their SQL and Java types. This is an example of a concrete metadata object that can be wrapped in a MetadataWrapper.
- ListBegin and ListEnd objects can be used to delimit data that is related in some way. For example, a group of tuples corresponding to the results of an SQL query or a group of character arrays that, when combined, form a valid XML document. Lists allow data to be broken into smaller chunks and contribute to the framework by allowing different operations on different parts of the data to be executed concurrently as the data is streamed.

4.3.2 Resources

Resources manage the state and behaviour of OGSA–DAI. All interactions with the framework are done via a resource. There are six types of resource in OGSA–DAI which can be partitioned into three groups:

- *Data resources.* An OGSA–DAI *data resource* is how a database, or other source of data, is represented in OGSA–DAI. It manages the data transfer between a source of data and OGSA–DAI. A source of data can be: a relational database, an XML database, a file system, an RDF triple store or a SPARQL endpoint. OGSA–DAI data resources are an extensibility point so new implementations can be developed to access other sources of data, for example web services, Microsoft Excel spreadsheets or application-specific sources of data.
- *Data cache resources.* These may be divided into two types:
  - A *data source* allows data to be cached locally for later retrieval by a client. Data sources support data pull operations; data is pulled by the client from an OGSA–DAI server.
  - A *data sink* allows a client to push data into a cache local to an OGSA–DAI server to be retrieved later by the framework. Data sinks support data push operations; data is pushed by a client to the OGSA–DAI server.
- Request execution and management resources of which there are three types:
  - A *data request execution resource*, or *DRER*, executes workflows submitted by clients.
  - *Session resources*, or sessions, allow state information to be held between workflow executions. State information can be stored during the execution of a workflow and then retrieved during execution of a subsequent one.
  - *Request resources* allow clients to monitor the execution of workflows and manage their lifetime.

4.3.3 Activities

Activities are the fundamental building blocks of an OGSA–DAI workflow. Each activity implements a data-resource related operation or acts on the data being streamed through the framework. An activity, depending on its implementation, can: read data in from a data resource, stream in data from another activity, manipulate the data in some way, create a new OGSA–DAI resource, output data to another activity or update a data resource.

OGSA–DAI 4.1 currently ships with more than 80 activities that implement a range of operations including: querying or updating relational or XML databases, listing directories and getting files; transforming data in various ways, for example to WebRowSet XML or comma-separated values; project tuples' columns or do an XSL transform. It includes activities to deliver data by: FTP, e-mail, HTTP or GridFTP [20]. It also includes activities related to OGSA–DAI itself, for example for creating data sources and sinks. Activities are an extensibility point and developers can write application-specific data-related operations to use within the OGSA–DAI framework.

Each activity has zero or more inputs, which are either mandatory or optional (if no value is given for an optional input, then the activity uses an internally defined default value). For example, the SQLQuery activity has one mandatory input: an SQL query; DeliverToFTP has three mandatory inputs: the FTP server URL, a file name and the data to be delivered and one optional input – a boolean to indicate the use of passive mode (defaults to false when no value is given). Inputs are typed and the expected types are declared to the framework by the activities. The framework validates the outputs and connects these to inputs on behalf of the activity at runtime. More specific checks on the nature of the input data are the responsibility of the activity implementations themselves. Equally, activities also have zero or more outputs. For example, SQLQuery has one output: a list of tuples; DeliverToFTP has no outputs.

Activities may target one or zero OGSA–DAI resources. An activity declares whether it needs a resource and what its type should be to the framework. For example, SQLQuery expects a relational data resource, DeliverToFTP expects no resource and WriteToDataSource expects a data source. Finally, activities also declare to the framework any other information they expect. For example, the URLs of any web services through which the framework is accessed, the current security configuration or a factory component with which to create new resources.

4.3.4 Workflows

A *workflow* is a specification of one or more activities connected together to form a directed graph. Each mandatory activity input must either be connected to the output of another activity or have a value provided by an *input literal*. An input literal is a container for values provided by the client when they submit a workflow, for example an SQL query, an FTP server URL, a column index or an output filename. Activities then use these instead of the output from another activity. Activities themselves have no knowledge of other activities or input literals – the connections are managed by the framework. All activity outputs must be connected to another input. Figure 4.2 depicts an example workflow to query a relational database, project onto a column and convert the results to comma-separated values.

4.3.5 How Workflows Are Executed

A workflow is executed as follows:

- Instances of each activity in the workflow are created.
- Input and output pipes are created. These are buffers which represent the connections between inputs and outputs. The default buffer size is configurable.

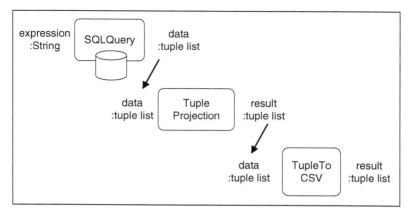

Fig. 4.2 An example workflow consisting of three activities: input and output names as well as types are shown

- Where applicable, each activity is given its target resource and any other information they have requested from the framework. The framework checks whether the type of resource matches that expected by the activity and raises an error if this is not the case.
- The workflow is validated to ensure that all activity outputs are connected and all mandatory inputs are connected either to an activity output or an input literal. An error is raised if this validation fails.
- All activities start executing concurrently. Each activity:

  - Attempts to pull data from its inputs.
  - Checks the data is of the type expected. Typically, any activity can accept objects of the same Java class or sub-classes thereof.
  - Blocks until some data is extracted either from an input literal or from the output of another activity.
  - Executes its data-related operation.
  - Pushes data onto its outputs if it has any.

- Activities terminate when:

  - They encounter an error which may be: a generic error, for example the input data is not of the expected type; or an activity-specific error, for example a query is syntactically incorrect; a database connection disappears; a parameter value is illegal or an internal problem arises.
  - They receive no more data from their inputs.

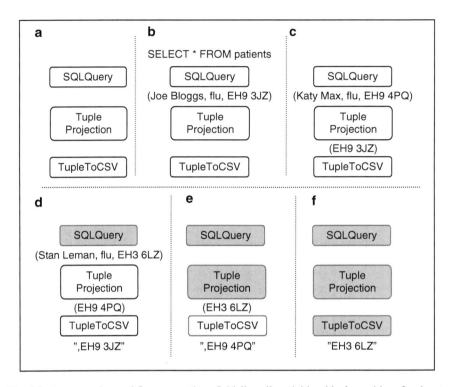

Fig. 4.3 An example workflow execution. Initially, all activities block, waiting for inputs (**a**). SQLQuery receives an input literal with an SQL query, this unblocks it, and it runs the query and outputs the first row of data as a tuple (**b**). This unblocks TupleProjection which projects on to the postcode column and outputs a new tuple. Meanwhile SQLQuery outputs the next row as a tuple (**c**). TupleProjection's output unblocks TupleToCSV which converts the tuple to comma-separated values and outputs these as a character array. SQLQuery has no more inputs so sets its state to completed (**d**). Other activities continue execution to completion until they too have no inputs (**e**, **f**) and set their states to: completed. The workflow completes when all activities have completed

As an example, consider the following data:

Name	Illness	Postcode
Joe Bloggs	Flu	EH9 3JZ
Katy Max	Flu	EH9 4PQ
Stan Leman	Flu	EH3 6LZ

Now, imagine the workflow of Fig. 4.2 is executed where SQLQuery queries the table, TupleProjection projects each row onto the postcode column and TupleToCSV converts the projected columns into a list of comma-separated values. The execution might proceed as shown in Fig. 4.3.

4.3.5.1   Using List Markers to Logically Group Outputs

Since activities process inputs one after the other, there is a risk that the outputs corresponding to one input are confused with the outputs arising from the next input. For example, if SQLQuery received two SQL query statements as an input, it would output the tuples for the first query then the tuples for the second. How can subsequent activities tell which results were from the first query and which from the second? This is where the ListBegin and ListEnd blocks are used. By grouping the tuples from a query's execution between these list blocks, the correspondence between inputs and outputs is preserved. For each query input, the activity outputs an OGSA–DAI list – a sequence of data blocks delimited by ListBegin and ListEnd. By using list delimiters, rather than a single list object, the ability for different activities to work on different parts of the data stream is preserved.

4.3.5.2   Concurrent Execution

All activities execute concurrently by which we mean that their operation is threaded, but in theory explicit parallelism would be possible. This can lead to more efficient processing as activities can work on different parts of the data stream concurrently as shown in the previous example. Streaming potentially gives a reduced memory footprint on the server, as the entire data set does not have to be read in prior to processing starting as well as the reduced execution time due to the implicit parallelism. This allows efficient processing of large data volumes to take place though it does require one to have care with activity implementations and how they are composed into workflows to reduce or avoid the risk of deadlocks.

4.3.6   Clients and Requests

OGSA–DAI is typically deployed as a client–server architecture. A server hosts the OGSA–DAI framework. Clients submit workflows to this framework. The submission of a workflow by a client is termed a *request*. When the framework receives a request, it creates a request resource, which holds a *request status*. The framework can execute a number of requests concurrently and queue a number more, the exact numbers being configurable.

The request status contains information about the status of each activity in the workflow (i.e. it has started execution, it has completed, it encountered an error and what it was). It also contains the status of the workflow as a whole. A request status can also contain data. A DeliverToRequestStatus activity can be used to stream data into the request status. How the request resource and request status are used by the client depends on the type of request. This is selected by the client when they submit their request and specifies how they want the workflow to be executed. There are two types of request:

- *Synchronous requests*: If the request is synchronous, then the framework will not return a request status to the client until the workflow has completed execution or it has encountered an error. A client has to block and wait for the workflow to complete.
- *Asynchronous requests*: if the request is asynchronous, then the framework returns a request status immediately on receiving the workflow together with a name for the new request resource. The client can then monitor the status of the request by querying this request resource. This allows the client to do other things and only get the data when it is ready, which is useful if the workflow will take a long time to complete (e.g. if it is going to process a million rows of data). It also allows a client to submit a workflow and then another to get the request status.

Asynchronous requests in conjunction with data sources or data sinks, support scalable data movement. If a synchronous request is sent to get million rows of data, then all this data would be returned to the client in the request status which may cause problems (web services can experience problems when returning large amounts of data in one go, for example). In OGSA–DAI, a client can submit a request to create a data source and then submit an asynchronous request to query a database for the one million rows and populate the data source with these. The client can then use the data source to get the data back in smaller chunks, for example a 1,000 or a 100 rows at a time. Conversely, if the client wishes to push large amounts of data to OGSA–DAI, they can request creation of a data sink on the server and then push data to it in chunks. However, in any such scenario there is a trade-off – the more data that is to be transported in chunks implies a greater number of client–server interactions.

4.3.7 Accessing the OGSA–DAI Framework

By default, the OGSA–DAI framework is exposed and accessed through web services. There are six classes of OGSA–DAI web service corresponding to the six resource types previously described.

- The *data request execution service* (DRES) accepts workflows from clients.
- The *request management service* (RMS) allows access to the request status of a request resource and management of its lifetime.
- The *data source service* allows a client to pull data from a data source on a server.
- The *data sink service* allows a client to push data to a data sink on a server.
- The *data resource information service* allows a client to find information about a data resource, for example the underlying product type and version of a database product.
- The *session management service* allows a client to manage a session's lifetime.

Each service allows information about each of the associated resources to be accessed, for example the activities that can be executed in conjunction with that resource. The web services conform to the Web Services Resource Framework (WSRF) [21, 22] OASIS standards providing access to underlying OGSA–DAI resources.

OGSA–DAI services can support the transport of strings, numbers, booleans and binary data (byte arrays and character arrays) between clients and servers. Activities on the server can be used to convert data into one of these formats (e.g. XML data is usually converted into an OGSA–DAI list of character arrays).

4.3.7.1 Data Delivery and Web Services

Delivering data, especially binary data, through web services, over SOAP/HTTP, will be slower than direct methods, such as FTP or GridFTP, especially for large data volumes. Data sources and data sinks provide one way to offset this, transporting data in smaller chunks. However, OGSA–DAI is distributed with a range of activities that support non-web service-based delivery methods to obtain more efficient data movement. For example, data could be delivered:

- To an e-mail address.
- From a URL via an HTTP GET directed at an OGSA–DAI servlet.
- To or from FTP servers.
- To or from GridFTP servers. GridFTP is specifically designed for the efficient movement of large amounts of data.

Other delivery mechanisms can be implemented and supported, via application-specific activity implementations.

4.3.7.2 Security

Rather than enforce a specific security protocol or infrastructure, the OGSA–DAI framework treats this as a quality of the presentation layer through which the framework is exposed. The framework only has the notion of a security context which serves as a generic container for any security-related information which it passes around. Specific components, for example data resources or activities can query this security context for the information they need. For example, OGSA–DAI's Globus Toolkit version uses Globus Toolkit security [23] at the presentation layer, adding a client's credentials to the security context when a request arrives at the presentation layer. To exploit these credentials, for example to map them to a database username and password, an OGSA–DAI data resource can use a *login provider* component to map the security context to the appropriate database credentials. OGSA–DAI provides an example login provider that maps security contexts containing Globus Toolkit credentials to usernames and passwords, though, other implementations (e.g. from VOMS [24] attributes) are also possible.

4.4   Distributed Query Processing

OGSA–DAI's distributed query processor (DQP) [1] is a set of OGSA–DAI data resources, activities and other components that runs within the OGSA–DAI framework to support distributed query processing across relational data sources. The framework has no special awareness of DQP; DQP has been developed using the framework's standard extensibility points. However, from the client's perspective, the DQP extension allows creation of a single database view of many distributed databases exposed as OGSA–DAI resources, which can then be orchestrated declaratively.

The *DQP resource*, a type of OGSA–DAI data resource, is a central component of DQP. From the client's perspective, it is a read-only relational resource that is able to answer declarative queries expressed in SQL. It exposes a global schema, which is constructed by concatenating individual schemas obtained from all of the federated relational resources. A federation of resources can be created statically using a server-side configuration document or dynamically using a factory activity. This not only allows OGSA–DAI deployers to expose predefined federations, which can be tightly controlled in terms of access rights but also affords the possibility of creating highly dynamic ad-hoc federations that can, for example, take into account up-to-date information about the state of a distributed system. This configuration document is a simple list of OGSA–DAI services and resource names.

The DQP resource encapsulates a *query parser, query plan optimizer* and *execution coordinator*. *Query plans* are evaluated by the OGSA–DAI framework using OGSA–DAI's relational, transformation and delivery activities and a set of activities implementing various relational operators and control structures.

A client interacts with a DQP resource by submitting a workflow that contains an SQLQuery activity, which takes an SQL expression as its input, targeted at this resource. This query is translated into an abstract syntax tree and converted to a *logical query plan* (LQP). An LQP represents the internal representation of the client's query and has the form of a tree of relational operators. The initial LQP is validated to check whether the client's query can be satisfied using information about the schemas of the federated resources. Once validated, the query plan goes through a chain of optimizers, which modify the plan using heuristic- and cost-based rules. The most important heuristic rule used in the optimization phase makes sure that as much processing as possible is pushed down to the underlying resources. This not only lets underlying databases make use of their indexes but also limits the amount of data transferred by filtering tuples at each local source. If a query plan scans distributed resources, it is split into several partitions, each assigned to an evaluation node (an OGSA–DAI service and associated DRER).

The optimized and partitioned query plan is translated into a set of OGSA–DAI workflows. Some relational operators like SELECT or PROJECT have one-to-one mapping to activities, some map to complex sub-workflows. Data is transferred between partitions via data sinks or data sources (depending on the choice of either push or pull data transfer modes). A special *Coordinator* component creates all the required data sinks/sources and submits the workflows for execution. The query results are pipelined to through to the client's original workflow as soon as they start arriving.

To illustrate how queries are translated into query plans and then into workflows, let us consider the health use case and assume that one health centre (h1) exposes a database with two tables: patient records – patients(zip, reason) – and locations – locations(zip, lat, long) – and that another health centre (h2) exposes a table with patient records only – patients(postcode, disease). Our goal is to count all occurrences of flu at each postcode and link these numbers with spatial information. The global schema exposed by a DQP resource federation over the above resources will have three tables: h1_patients, h1_locations and h2_patients. A possible query may be as follows:

```
SELECT lat, long, COUNT(*) FROM
(SELECT lat, long, zip
FROM h1_locations l, h1_patients p
WHERE p.reason LIKE '%FLU%' AND l.zip = p.zip)
UNION ALL
(SELECT lat, long, postcode AS zip
FROM h1_locations l, h2_patients p
WHERE p.disease LIKE '%FLU' AND l.zip = p.postcode) t
GROUP BY zip, lat, long
```

Figure 4.4 presents a possible query plan generated for the above query. This plan is translated into a set of OGSA–DAI workflows and submitted for execution.

4.4.1 DQP and Extensibility Points

DQP is highly extensible. It is possible to provide alternative implementations of optimizers and use them in conjunction with the default optimizers or replace the default optimization chain altogether. The mapping between relational operators and their equivalent executable workflows is also configurable. This allows provision of alternative physical implementations for relational operators. It is possible to assign several implementations for a given operator. An optimizer can use annotations to tag which implementation should be used at translation time. *User defined functions* (UDFs) can also be defined; both scalar and aggregate UDFs are supported. Other extensibility points include cardinality estimation module and pluggable pre- and post-execution code. An example extension that exploits this high flexibility federates SPARQL queries over multiple endpoints [25].

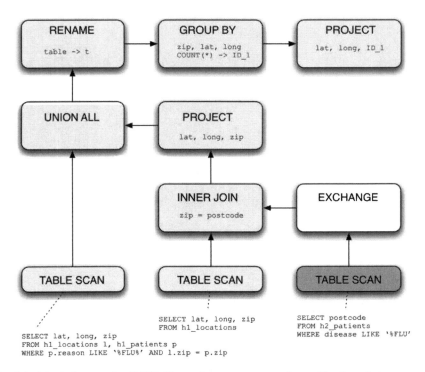

Fig. 4.4 A logical query plan (LQP). The exchange operator marks partition boundaries

4.4.2 DQP and Non-relational Resources

Although the resources over which DQP can federate queries must expose a relational interface, the underlying data source does not have to be relational. It is possible to develop an OGSA–DAI data resource that provides a relational interface, or wrapper, for non-relational resources. This is often used to model web services as relational resources, so they can be used as members within a DQP federation.

4.5 Relational Views

A database view is a well-known relational tool. In essence, it is a named, virtual table composed of the result set of a stored query. An OGSA–DAI *views resource* [1] allows for a relational view to be created on top of any resource exposed by OGSA–DAI and is able to execute an SQL query. A view resource looks like a read-only relational resource providing table schema and executing queries. A view in OGSA–DAI can be defined without requiring write access to an underlying database, meaning that views can be defined over remotely located read-only databases.

There are many uses of views. A view can be used to join and simplify multiple tables into a single virtual table. This is often applied to simplify the writing of queries from the client's perspective. In our health use case, a view could be used to encapsulate the union of patient records tables. Another use of views is to limit the exposure of tables to the world outside the organization. For example, it may be acceptable to expose postcode and disease information but keep private a patient's name. Views can also be used to smooth out differences between table schemas. These mappings exploit the expressiveness of the SQL language and can range from simple column renaming to complex, value-replacing joins. In our use case, we could define a view that renames column names in one of the patient's tables, so they use identical schema.

Note that OGSA–DAI's views and DQP components are distinct and complementary. They both expose a relational interface and can be mixed to build complex data integration scenarios. It is possible to build federation of views as well as define a view on top of a DQP resource.

4.6   Interoperability

It is important for OGSA–DAI be able to work with other applications, or middleware, within distributed computing, grid and cloud contexts. This facilitates OGSA–DAI's adoption into existing infrastructures and allows OGSA–DAI to be used with other products to construct solutions to complex distributed computing problems. Although OGSA–DAI's workflows form a de facto standard, various activities have been undertaken to make OGSA–DAI accessible via other means. For example, the open grid forum (OGF) database access and integration services (DAIS) family of specifications [26–28] provide a common set of interfaces for accessing data. These interfaces provide operations with tighter semantics than that of OGSA–DAI workflows. These interfaces allow OGSA–DAI to be used within service orchestration workflow environments such as Taverna [29] or ActiveBPEL [30] while providing access to distributed databases underneath. Another example is the implementation of a JDBC driver to access OGSA–DAI [31]. This allows legacy applications that access databases via JDBC to access OGSA–DAI without any change. The application gains as it can be used to access any federation of databases exposed through DQP, for example. OGSA–DAI may also be used as a basis for implementing other standards. For instance, the SEE-GEO [7] project implemented OGC-compliant web services [32] based on OGSA–DAI to integrate census data with geographical data relating to the boundaries of UK regions.

Interoperability can be facilitated either through the use of mediators or through alternative presentation layers. A mediator component sits between an application and OGSA–DAI. The mediator converts requests from the application into invocations of OGSA–DAI. For example, submitting workflows and responses from OGSA–DAI back into responses to be consumed by other application. For example, the JDBC driver for OGSA–DAI translates Java JDBC API calls, typically SQL queries and requests for database meta-data, into OGSA–DAI workflows and the results, query results and meta-data, back into JDBC API responses. A similar design can be envisaged for an ODBC database API [33].

Another approach is to use an alternative OGSA–DAI presentation layer. Although OGSA–DAI ships with a web services presentation layer, OGSA–DAI has been designed so that the underlying framework does not rely upon web services and so it can be replaced. The SEE-GEO project replaced OGSA–DAI services with OGC-compliant geo-linking services. These services map requests to OGSA–DAI workflows, where the operation invoked together with its arguments determines the workflow that is selected, its input literals and the data resources it targets. When the workflow is completed, the request status is parsed and the appropriate response from the operation constructed. Another example is provided by the OGSA–DAI implementations of the OGF DAIS specifications for relational and XML web services. Again, this involved replacing the OGSA–DAI web services with those compliant with the DAIS specifications and mapping between the operations of these specifications and suitable workflows.

Using servlets, allows OGSA–DAI to expose data that can be accessed via HTTP rather than a web service. This, too, can facilitate interoperability. For example, the Google Map API [34] allows a Google Map rendered in an internet browser to be overlaid with features specified in a KML document. The KML document is specified by a URL. An OGSA–DAI server can query data and convert it into a KML document and then return the URL of a servlet that provides access to this document. The URL can be passed to the Google Map API for rendering. The use of servlets and URLs that provide access to data when dereferenced also allows OGSA–DAI to be used in the context of web service orchestration engines such as Taverna. These typically pass round references to data. The use of a URL means that such engines do not have to be extended with OGSA–DAI-specific components.

4.7 Performance

OGSA–DAI components mediate between data producers and data consumers. As a consequence, performance is an important consideration – does the benefit of being able to execute complex distributed data management operations mitigate the increased time incurred when accessing and updating data. Throughout the lifetime of OGSA–DAI, both the OGSA–DAI development team and other researchers have undertaken performance evaluations to understand a number of issues including:

- How OGSA–DAI performs compared to using a direct JDBC connection to access relational data [1, 35–37].
- The pros and cons of alternative delivery mechanisms and presentation layers [1, 35, 38, 39].
- How an OGSA–DAI server behaves when confronted with concurrent access by a large number of clients or when manipulating large volumes of data [40].

- Which OGSA–DAI components give rise to bottlenecks and how these could be resolved [35, 38, 39].

These evaluations have yielded a number of findings:

- OGSA–DAI uses JDBC so will inevitably be slower than JDBC. However, OGSA–DAI's overhead increases more sharply than JDBC as the number of rows in a result increases.
- WebRowSet XML should only be used as a delivery format for small data sets. Unless a client specifically needs data as XML, alternative representations, for example comma-separated values or custom binary formats, yield both faster delivery times and increases in the volume of data a server can return to a client within a single web service invocation.
- Web services are not optimal for delivering large amounts of data, especially binary data. Alternative delivery methods (e.g. SOAP attachments, FTP or GridFTP) can yield improved performance in terms of both delivery time and data volume.
- For a specific deployment, and depending upon the specific qualities of an OGSA–DAI server's host machine, there is a critical number of concurrent clients that the server can manage, beyond which OGSA–DAI's performance will degrade markedly both in terms of CPU consumption and memory usage.
- Using Globus Toolkit security adds an overhead but this is generally constant and forms a negligible part of the round-trip time for an OGSA–DAI invocation when large amounts of data are being processed.

Ultimately, OGSA–DAI users need to decide whether the additional overhead introduced by OGSA–DAI is acceptable based upon their application-specific requirements. However, OGSA–DAI has been designed to allow users as much flexibility as possible to improve performance along a number of dimensions, for example by providing activities that support non-web service-based delivery, by allowing the number of concurrent requests that a DRER can execute to be configured or by allowing its web services presentation layer to be replaced by alternatives, for example by REST endpoints or a direct Java connection.

4.8 Related Work

A number of tools have been developed to solve problems in distributed data management. These tools differ in subtle or significant ways depending on the specific distributed data management challenges they have been designed to solve. AMGA [41] is a metadata catalogue primarily targeted at dealing with metadata within grid environments. In particular, the environment that has been developed for processing outputs from the Large Hadron Collider. AMGA implements its own query language which is similar to SQL although, from version 1.9, there is some SQL-92 support. AMGA has a web services front-end, which conforms to the OGF WS-DAIR standard [26], as well as a proprietary client API. Although AMGA can serve as a generic wrapper for databases, it is primarily intended to serve as a metadata catalogue.

The Grid Relational Catalog (GRelC) [42,43], such as OGSA–DAI, can provide a web services wrapper for relational, XML and file-based resources. This allows a level of abstraction to be made from the specifics of how connections are made to each of these types of resource but not from the underlying type of data resource being accessed, that is the type of queries that are composed by users. GRelC also supports data integration through the use of their GRelC Data Gather Service, which allows an SQL query to be propagated over various other GRelC-mediated services and the results to be merged back at the original submission node [44]. This is similar to, though less sophisticated than, OGSA–DAI's distributed query processor. GRelC, like AMGA, is a component of the gLite middleware [45].

The Integrated Rule-Oriented Data System (iRODS) [46], the open source successor to SRB [47], provides a virtualization layer that federates and replicates many different types of data but primarily in the file space domain. They have a service mode of operation where micro-services can be composed together to form rules which effectively act as server-side workflows, which are triggered by an event, for example a file is placed in an iRODS repository, which is then replicated to other servers by a rule. iRODS, like SRB, has been widely adopted in a number of communities and, although it can access databases, its strength lies in file access so its application space overlaps with OGSA–DAI.

There have been various products developed to support distributed query processing (DQP) capabilities, an example was demonstrated by SkyQuery [48]. SkyQuery implements a mediator-wrapper architecture for integration of astronomy data archives. Similar data federation functionality is delivered by the MOBIUS Project [49], a distributed query processor where individual data resources are exposed as XML services able to answer XPath queries. Mediation based on XML processing is also used by the XAware [50] data integration system. Commercial products providing similar data federation functionality include, among others, the IBM WebSphere Information Integrator [51] and the Virtuoso Virtual Database [52].

Many of the DQP solutions mentioned above adopt a centralised query processing approach. This means that what can be computed remotely is constrained by the capabilities of the data source. For example, for XML-based solutions, this is limited to what can be expressed in XPath or XQuery. In contrast, OGSA-DAI not only delegates as much processing as possible to the underlying data source but also it can perform arbitrary data processing operations on a remote OGSA-DAI server. This can be exploited in queries with user-defined functions or extended through bespoke relational operators. Fully distributed query evaluation provides a means for improved utilization of resources by exploiting parallelism implicit in many data processing workflows. There is current work to extend the DQP capabilities, already available to the relational domain, to the RDF domain [25].

4.9   Conclusions and Future Directions

This chapter has given an overview of OGSA–DAI and its approach to distributed data management. Workflows provide a simple yet powerful way of representing complex scenarios that involve data-related operations and which can involve multiple distributed heterogeneous data sources. OGSA–DAI provides an abstraction layer between a client and the data of interest. At the simplest level, OGSA–DAI provides an abstraction of underlying data resources, removing native access connection complexities, hiding the physical location of the data and providing a common interface to access many different types of data. Various forms of security policies can be enforced within this layer controlling who can access the data, what they can access and how they can manipulate it.

A complete data abstraction is not possible in that clients remain aware of the type of data resource, for example relational, XML, exposed by OGSA–DAI. Also, it will be noted that an additional layer means one more set of components between data producers and data consumers with the consequent implications upon performance.

However, the level of abstraction is sufficient for OGSA–DAI to be used to build powerful higher level capabilities, and the performance implications are offset by the enhanced distributed data management capabilities that are delivered in compensation. Exemplars of this are the DQP and SQL views components. Ultimately, of course, the benefit of OGSA–DAI is very much determined on a case-by-case basis dependent on specific problems to be addressed and the requirements of any sought solution. However, as the introduction demonstrated, OGSA–DAI has been, and continues to be used to deliver innovative data management solutions in a whole range of areas.

Acknowledgements We acknowledge our past and present collaborators including the National eScience Centre at The University of Edinburgh, the eScience Centre of the North West of England at The University of Manchester, the North East of England eScience Centre at The University of Newcastle, IBM UK and Oracle UK. The project has been funded by the UK Department of Trade and Industry (under the Grid Core Programme I and II), the UK Engineering and Physical Sciences Research Council (under OGSA–DAI: an OMII-UK node, EP/D043956/1), by the European Union as part of the Framework Program 6 project BEinGRID (reference 034702) and by The University of Edinburgh.

References

1. Dobrzelecki, B., Krause, A., Hume, A., Grant, A., Antonioletti, M., Alemu, Y., Atkinson, M., Jackson, M., Theocharopoulos, E.: Integrating distributed data sources with OGSA–DAI DQP and Views. Phil. Trans. R. Soc. A **368**(1926), 4133–4145 (2010). doi: http://dx.doi.org/10.1098/rsta.2010.0166
2. Lynden, S., Pahlevi, S., Kojima, I.: Service-based data integration using OGSA-DQP and OGSA-WebDB. In: Proceedings of the 2008 9th IEEE/ACM International Conference on Grid Computing, 29 Sept–01 Oct 2008. IEEE Computer Society, Washington, DC, pp. 160–167 (2008). doi:10.1109/GRID.2008.4662795

3. Lynden, S., Mukherjee, A., Hume, A., Fernandes, A., Paton, N., Sakellariou, R., Watson, P.: The design and implementation of OGSA-DQP: a service-based distributed query processor. Future Gen. Comput. Syst. **25**, 224–236 (2009). doi:10.1016/j.future.2008.08.003

4. Xiang, H.X.: Experiences running OGSA-DQP queries against a heterogeneous distributed scientific database. In: Proceedings of the 2009 15th International Conference on Parallel and Distributed Systems, pp. 706–710 (2009)

5. Tanimura, Y., Yamamoto, N., Tanaka, Y., Iwao, K., Kojima, I., Nakamura, R., Tsuchida, S., Sekiguchi, S.: Evaluation of large-scale storage systems for satellite data in GEO GRID. In: The International Archives of the Photogrammetry, Remote Sensing and Spatial Information Sciences, **XXXVII**, Part B4. pp. 1567–1574. Beijing (2008)

6. Groeper, R., Kunz, C., Grimm, C.: Connecting OGC web services and the Grid using Globus Toolkit 4 and OGSA–DAI. Conference on Grid Computing. In: 10th IEEE/ACM International, pp. 66–73, October 2009. doi: 10.1109/GRID.2009.5353080

7. Higgins, C., Koutroumpas, M., Sinnott, R.O., Watt, J., Doherty, T., Hume, A.C., Turner, A.G., Rawnsley, D.: Spatial data e-infrastructure, GLAS-PPE/2009-23 Preprint, 2009. http://ppewww.physics.gla.ac.uk/preprints/2009/23/2009--23.pdf. Accessed 5 Aug 2010

8. Koehler, M., Ruckenbauer, M., Janciak, I., Benkner, S., Lischka, H., Gansterer, W.N.: Supporting molecular modeling workflows within a grid services cloud. Computational Science and Its Applications – ICCSA 2010, Lecture Notes in Computer Science, 2010, vol. 6019/2010, pp. 13–28. doi: 10.1007/978–3–642–12189–0_2

9. Swain, M., Silva, C.G., Loureiro-Ferreira, N., Ostropytskyy, V., Brito, J., Riche, O., Stahl, F., Dubitzky, W., Brito R.M.M.: P-found: Grid-enabling distributed repositories of protein folding and unfolding simulations for data mining. Future Gen. Comp. Syst. **26**(3), 424–433 (2010)

10. Brochhausen, M., Weiler, G., Martín, L., Cocos, C., Stenzhorn, H., Graf, N., Dörr, M., Tsiknakis, M., Smith, B.: Applications of the ACGT Master Ontology on Cancer. On the Move to Meaningful Internet Systems: OTM 2008 Workshops Lecture Notes in Computer Science, vol. 5333/2010, pp. 1046–1055 (2010). doi: 10.1007/978-3-540-88875-8_132

11. Garcia Ruiz, M., Garcia Chaves, A., Ruiz Ibañez, C., Gutierrez Mazo, J.M., Ramirez Giraldo, J.C., Pelaez Echavarria, A., Valencia Diaz, E., Pelaez Restrepo, G., Montoya Munera, E.N., Garcia Loaiza, B., Gomez Gonzalez, S.: mantisGRID: a grid platform for DICOM medical images management in Colombia and Latin America. J. Digit. Imaging **24**(2), 271–283 (2010). doi:10.1007/s10278-009-9265-x

12. Espino, J., Hall, K., Washington, D., White, P., Grant, A., Hume, A., Antonioletti, M., Krause, A., Jackson, M., Heinbaugh, W., Fu-Chiang, T.: Open-source collaboration in practice between the real-time outbreak and disease surveillance laboratory. Public Health Information Network (PHIN) Conference 2008, pp. 24–28, Atlanta. The National Center for Public Health Informatics Research Lab, The University of Edinburgh and Tarrant County Public Health (2008)

13. Tan, K.L.L., Gayle, V., Lambert, P.S., Sinnott, R.O., Turner, K.J.: GEODE – Sharing occupational data through the grid. In: Proceedings of the UK e-Science All Hands Meeting 2006 (2006)

14. Graham, P.J., Sloan, T.M., Carter, A.C., Gregory, I.: FirstDIG: Data investigations using OGSA–DAI. In: Proceedings of the UK e-Science All Hands Meeting 2004 (2004)

15. MESSAGE (Mobile Environmental Sensing System Across Grid Environments) Project. http://bioinf.ncl.ac.uk/message/ (2010). Accessed 5 Aug 2010

16. Antonioletti, M., Blanke, T., Bodard, G., Hedges, M., Hume, A., Jackson, M., Rajbhandari, S.: Building bridges between islands of data – an investigation into distributed data management in the humanities. In: 5th IEEE International Conference on e-Science, pp 33–39, Oxford, 7–9 Dec 2009. (2009). ISBN 978-0-7695-3877-8

17. OGSA–DAI open source project publications. http://sourceforge.net/apps/trac/OGSA--DAI/wiki/Publications (2010). Accessed 7 Oct 2010

18. OGSA–DAI open source project. http://sourceforge.net/projects/OGSA--DAI (2010). Accessed 7 Oct 2010

19. KML. http://code.google.com/apis/kml/documentation/ (2010). Accessed 5 Aug 2010

20. Globus Project GridFTP. http://globus.org/toolkit/docs/3.2/gridftp/ (2010). Accessed 5 Aug 2010
21. Banks, T.: Web Services Resource Framework (WSRF) – Primer v1.2, OASIS, 23 May 2006. http://docs.oasis-open.org/wsrf/wsrf-primer-1.2-primer-cd-02.pdf (2006). Accessed 5 Aug 2010
22. Graham, S., Karmarkar, A., Mischkinsky, J., Robinson, I., Sedukhin, I.: Web Services Resource 1.2 (WS-Resource), OASIS, 1 April 2006. http://docs.oasis-open.org/wsrf/wsrf-ws_resource-1.2-spec-os.pdf (2006). Accessed 5 Aug 2010
23. Globus Toolkit Security information. http://www.globus.org/toolkit/docs/4.0/security (2010). Accessed 12 Oct 2010
24. Alfieri, R., Cecchini, R., Ciaschini, V., dell'Agnello, L., Frohner, A., Gianoli, A., Lőrentey, K., Spataro, F.: VOMS, an authorization system for virtual organizations. Lecture Notes in Computer Science, vol. 2970/2004, pp. 33–40 (2004)
25. Aranda, C., Corcho, O.: Federating queries to RDF repositories. Monografia (Technical Report). Computer Faculty (UPM), Madrid, Spain. http://oa.upm.es/3302/ (2010). Accessed 7 Oct 2010
26. Antonioletti, M., Collins, B., Krause, A., Malaika, S., Magowan, J., Laws, S., Paton, N.W.: Web Services Data Access and Integration – The Relational Realization (WS-DAIR) Specification Version 1.0, Global Grid Forum (2006)
27. Antonioletti, M., Atkinson, M., Krause, A., Laws, S., Malaika, S., Paton, N.W., Pearson, D., Riccardi, G.: Web Services Data Access and Integration – The Core (WS-DAI) Specification, Version 1.0. http://www.ogf.org/documents/GFD.74.pdf (2006). Accessed 5 Aug 2010
28. Antonioletti, M., Hastings, S., Krause, A., Langella, S., Lynden, S., Laws, S., Malaika, S., Paton, N.W.: Web Services Data Access and Integration – The XML Realisation (WS-DAIX) Specification, Version 1.0. http://www.ogf.org/documents/GFD.75.pdf (2006). Accessed 5 Aug 2010
29. Taverna. http://www.taverna.org.uk/ (2010). Accessed 5 Aug 2010
30. ActiveBPEL. https://sourceforge.net/projects/activebpel/ (2010). Accessed 5 Aug 2010
31. Brito, M., Sato, L.: Extending OGSA–DAI Possibilities with a JDBC Driver, Computational Science and Engineering. In: 11th IEEE International Conference on Computational Science and Engineering, pp. 155–162 (2008). ISBN: 978-0-7695-3193-9. doi:http://doi.ieeecomputersociety.org/10.1109/CSE.2008.55
32. OGC Standards and Specifications. The Open Geospatial Consortium. http://www.opengeospatial.org/standards (2010). Accessed 5 Aug 2010
33. Jackson, M.J., Lloyd, A.D., Sloan, T.M.: Enabling access to federated grid databases: An OGSA–DAI ODBC driver. In: Proceedings of the UK e-Science All Hands Meeting, 2005
34. Google Maps API. http://code.google.com/apis/maps/index.html (2010). Accessed 5 Aug 2010
35. Jackson, M., Antonioletti, M., Chue Hong, N.P., Hume, A.C., Krause, A., Sugden, T., Westhead, M.: Performance analysis of the OGSA–DAI software. In: Proceedings of the UK e-Science All Hands Meeting, September 2004
36. Kottha, S., Abhinav, K., Muller-Pfefferkorn, R., Mix, H.: Accessing bio-databases with OGSA–DAI – A performance analysis, distributed, high-performance and grid computing in computational biology, Lecture Notes in Computer Science, vol. 4360, pp. 141–156 (2007)
37. Adamski, M., Kulczewski, M., Kurowski, K., Nabrzyski, J., Hume, A.: Security and performance enhancements to OGSA–DAI for Grid data virtualization: Research Articles, Selection of Best Papers of the VLDB Data Management in Grids Workshop (VLDB DMG 2006). Concurrency Comp. Pract. Ex. **19**(16), 2171–2182 (2007). doi: 10.1002/cpe.1165
38. Alpdemir, M.N., Gounaris, A., Mukherjee, A., Fitzgerald, D., Paton, N.W., Watson, P., Sakellariou, R., Fernandes, A.A.A., Smith, J.: Experience on performance evaluation with OGSA-DQP. In: Proceedings of the UK e-Science All Hands Meeting, September 2005
39. Dobrzelecki, B., Antonioletti, M., Schopf, J.M., Hume, A.C., Atkinson, M., Chue Hong, N.P., Jackson, M., Karasavvas, K., Krause, A., Parsons, M., Sugden, T., Theocharopoulos, E.: Profiling OGSA–DAI performance for common use patterns. In: Proceedings of the UK e-Science All Hands Meeting, September 2006

40. Wang, K., Xie, Y., Li, S., Wang, X.: Performance analysis of the OGSA–DAI 3.0 Software. In: 5th International Conference on Information Technology: New Generations, pp. 15–20, April 2008. ISBN: 0-7695-3099-0. doi: 10.1109/ITNG.2008.91

41. Santos, N., Koblitz, B., Distributed metadata with the AMGA metadata catalog. In: Proceedings of the Workshop on Next-Generation Distributed Data Management HPDC-15, Paris, France, June 2006

42. Fiore, S., Negro, A., Aloisio, G.: Data virtualization in grid environments through the GRelC data access and integration service. In: Proceedings 4th ICITST, IEEE, London, UK, pp. 817–822 (2009)

43. Fiore, S., Negro, A., Aloisio, G., The data access layer in the GRelC system architecture, Future Generation Computer Systems (2010). doi: 10.1016/j.future.2010.07.006

44. Aloisio, G., Cafaro, M., Fiore, S., Mirto, M., Vadacca, S., Grelc data gather service: a step towards P2P production grids. In: SAC, pp. 561–565 (2007)

45. gLite. http://glite.web.cern.ch/glite/ (2010). Accessed 7 Oct 2010

46. Moore, R., Rajasekar, A.: White Paper: IRODS: Integrated Rule-Oriented Data System. https://www.irods.org/pubs/DICE_iRODS_White_Paper-08.pdf (2008). Accessed Sept 2008

47. Storage Resource Broker (SRB). http://www.sdsc.edu/srb/index.php/Main_Page (2010). Accessed 7 Oct 2010

48. Malik, T., Szalay, A.S., Budavari, T., Thakar, A.: Skyquery: a web service approach to federate databases. In: Proceedings of the First Biennial Conference on Innovative Data Systems Research (CIDR 2003) Asilomar, USA, VLDB Endowment, pp. 188–196, January 2003. http://www.cidrdb.org/cidr2003/program/p17.pdf

49. Hastings, S., Langella, S., Oster, S., Saltz, J.: Distributed data management and integration framework: The Mobius project. In: Proceedings of Global Grid Forum 11 (GGF11) semantic grid applications workshop, pp. 20–38 (2004)

50. XAware. http://www.xaware.org (2010). Accessed 7 Oct 2010

51. IBM Information Integration. http://www-01.ibm.com/software/data/integration (2010). Accessed 7 Oct 2010

52. Virtuoso Universal Server, OpenLink Software. http://virtuoso.openlinksw.com (2010). Accessed 7 Oct 2010

Chapter 5
The DASCOSA-DB Grid Database System

Jon Olav Hauglid, Norvald H. Ryeng, and Kjetil Nørvåg

Abstract Computational science applications performing distributed computations using grid networks are now emerging. These applications have new and demanding requirements for efficient query processing. To meet these requirements, we have developed the DASCOSA-DB distributed database system. In this chapter, a detailed overview of the architecture and implementation of DASCOSA-DB is given, as well as a description of novel features developed to better support typical data-intensive applications running on a grid system: fault-tolerant query processing, dynamic refragmentation, allocation and replication of data fragments, and distributed semantic caching.

5.1 Introduction

During the recent years, there has been a trend toward applications deployed on increasingly larger distributed systems with need for advanced data management. A prime example of such applications is computational science applications that uses advanced computing capabilities to understand and solve complex problems. Such applications frequently requires powerful computing resources, for example, delivered through *grid computing services*.

While grid computing has gained maturity through the recent years, management of data in grid systems is less mature. Data storage and access is still mostly file oriented, and it is mostly left to users to manage files and their locations as needed. Although some support has emerged for metadata management, more advanced database features are not widely supported.

J.O. Hauglid (✉) · N.H. Ryeng · K. Nørvåg
Department of Computer and Information Science, Norwegian University of Science and Technology (NTNU), 7491 Trondheim, Norway
e-mail: joh@idi.ntnu.no; ryeng@idi.ntnu.no; noervaag@idi.ntnu.no
http://research.idi.ntnu.no/dascosa/

S. Fiore and G. Aloisio (eds.), *Grid and Cloud Database Management*,
DOI 10.1007/978-3-642-20045-8_5, © Springer-Verlag Berlin Heidelberg 2011

The goal of our research is a reliable *database grid* with location-transparent storage, that is, users/applications do not have to care about where data is stored and where queries are processed. The aim is sites cooperating on data storage and processing while retaining autonomy, that is, a grid-wide database system. It is important to note how our context differs from more traditional approaches. The focus is on applications where large amounts of data is created and used on the same site, and where parts of the data, in particular summary data, are accessed by other grid participants.

An example of such applications is weather forecasting, where the national weather forecasting institutions have large amounts of locally collected data, do forecast, and make the resulting data available. They also store historical data. Both the summary data and historical data will be of interest to, and used by, other weather forecasting institutions and environmental researchers.

In this chapter, we describe *DASCOSA-DB*, a distributed database system, which, in addition to providing location-transparent storage and querying, also includes novel features such as efficient partial restart of queries and redistribution of query operators in the context of failure, dynamic refragmentation, allocation and replication of data fragments, and distributed semantic caching. A detailed overview of the architecture and the implementation of DASCOSA-DB is given, as well as a description of some of the features developed to better support typical data-intensive applications running on a grid.

The rest of this chapter is organized as follows: In Sect. 7.6, we give a short overview of other similar systems. In Sect. 5.3, we present the system architecture of DASCOSA-DB. Section 7.3.2.2 describes how data and metadata management is handled, and Sect. 7.4 explains query processing, including semantic caching and partial restart of failed queries. Our distributed monitoring and management tool is described in Sect. 5.6. An experimental evaluation of the system is provided in Sect. 5.7. Finally, we summarize our work and describe future research directions in Sect. 5.8.

5.2 Overview of Related Systems

Distributed databases and query processing is not a new field. For an introduction to distributed databases, we refer to [15]. A survey of distributed query processing is given in [12]. In this section, we will give an overview of systems that are similar to DASCOSA-DB. This includes both storage systems without query capabilities and query systems without storage capabilities, as well as complete database systems.

Much of the more recent work is based on peer-to-peer (P2P) networks, both unstructured and structured. Especially, distributed hash tables (DHTs) have received much attention. A number of papers deal with focused issues such as query processing in DHT networks, including [2, 7].

OceanStore [13] is one of the storage systems without query capabilities. It provides an infrastructure for permanent storage and replication of objects, but no query system. Objects are accessed based only on their globally unique ID, and this ID has to be known to retrieve or update the object.

BigTable [5] is a large-scale distributed storage system with a model closer to relational databases. The storage model is similar to the relational model, but tuples are not stored or accessed as one unit. Instead, a row key and column key is used for both read and write operations. It does not provide more advanced query languages.

DASCOSA-DB does not provide its own storage infrastructure, but relies on an existing relational DBMS to store data. In that way, it is somewhat similar to the pure query engines that only provide a query processing service and no persistent storage.

Astrolabe [16] is one such system. Astrolabe is a distributed, hierarchical aggregation system designed for system monitoring. Astrolabe provides an interface that is similar to a database system, that is, it provides SQL queries and standard database programming interfaces such as ODBC and JDBC. To achieve scalability, updates are spread using a gossip protocol that guarantees eventual consistency. There is no guarantee that a client reads the most recent data, but if updates stop, all clients will eventually agree on the most recent value.

PIER [11] is a middleware query engine built on top of a DHT. PIER does not permanently store its data. Data sources publish their data in the DHT and update them regularly, and data that are not refreshed are removed. Typically, a PIER network will contain only object metadata (e.g., filenames, sizes, and tags) and a reference to the original data object. Clients will query the network to get the references to the objects of interest and retrieve the objects separately.

The difference between these query engines and DASCOSA-DB is that, although DASCOSA-DB has a middleware architecture like PIER, it provides persistent storage by using a local database on each site. It is not necessary to constantly republish data, as is the case with PIER.

Among the systems that provide a full DBMS, with both query processing and storage, are Hyperion [17], Orchestra [22], and Piazza [6]. All these systems allow each site to have its own schema, and use schema mediation techniques to allow cross-site querying. PeerDB [14] also falls into this category of systems with heterogeneous schemas, but the approach to schema mediation is different. Instead of relying on schema mediators, information retrieval techniques are used to find matching relations.

DASCOSA-DB does not use schema mediation. The systems mentioned above are meant to connect existing databases and provide a common query interface. Although DASCOSA-DB is a distributed database system with a high degree of site autonomy, it still behaves as one system, not many different systems with a common interface.

Other systems based on a common schema include APPA [1], Mariposa [21], and ObjectGlobe [4]. APPA provides a multilayered solution on top of a structured or super-peer P2P network, where the bottom layer is a simple key/value-store and the top level provides advanced services such as schema management, replication, and query processing.

Mariposa is a distributed database system that uses economic models to solve optimization problems. Mariposa sites buy and sell fragments and bid for the execution of queries. The trading and bidding makes sure queries are answered efficiently and that data are moved closer to where they are needed.

ObjectGlobe is a distributed query processing infrastructure that allows users to combine data sources and query operators from different providers at different sites to perform queries. Sites can sell data, query operators, computing power or a combination of these. The client combines these resources to a full query pipeline.

AmbientDB [3] is probably the system that bears the closest resemblance to DASCOSA-DB. AmbientDB is a system designed to provide full relational database functionality for standalone operation in autonomous devices that may be mobile and disconnected for long periods of time, while enabling them to cooperate in an ad hoc way with (many) other AmbientDB devices. A DHT is used both as a means for connection peers in a resilient way as well as supporting indexing of data.

Like AmbientDB, DASCOSA-DB is also constructed as a combination of middleware and federated databases, connecting the local databases of each site. The key difference is that AmbientDB is a system for mobile devices, which have low computational power and may frequently be disconnected from the network, while DASCOSA-DB is designed for sites that have the computational power necessary to do query processing and more stable network connections. DASCOSA-DB is also based on a DHT, like AmbientDB and PIER. However, the DHT is only used as a metadata catalog. Query processing uses point-to-point links following the query tree, more like Mariposa and ObjectGlobe. This is different from PIER, where the DHT is used extensively in query processing.

In terms of query capabilities, all sites of DASCOSA-DB are equal. There is no buying or selling of query operators or data. Data is fragmented, allocated, and replicated according to the needs of the combined load of all sites, trying to keep the costs of network communication low. Query operators are shipped out to sites to minimize network costs by trying to perform most query operations on local data.

Many of the systems mentioned above support SQL-like querying and presents data similar to a normal relational database system. DASCOSA-DB is fully a relational database system that supports standard SQL.

A brief description of a DASCOSA-DB demonstration is given in [9].

5.3   System Architecture

In this section, the architecture of DASCOSA-DB is described. DASCOSA-DB consists of a number of autonomous sites connected to form a distributed database system. First is described how sites are connected, how data is distributed, and how sites cooperate to execute queries and updates, and then the internal architecture of a single site is presented in more detail.

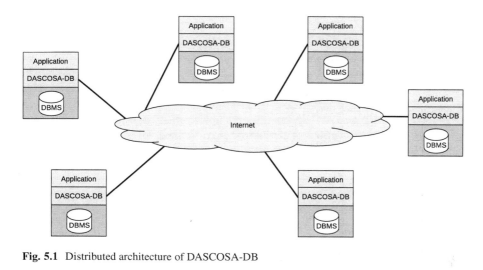

Fig. 5.1 Distributed architecture of DASCOSA-DB

5.3.1   Distributed Architecture

DASCOSA-DB is designed as a middleware layer that binds together local DBMSs running on different sites to make a distributed DBMS providing location transparency. Figure 5.1 shows the distributed architecture of DASCOSA-DB, as a middleware connecting local databases and applications to provide access to a large, distributed database. All sites are autonomous. There is no single site that controls the distributed DBMS. In this way, the sites act together as a P2P network.

All sites connect to form a DHT. This DHT is used to store the distributed catalog, which contains information on all tables, table fragments, replicas, and cache entries in the system. Currently, FreePastry[1] is used, but any other DHT may be used.

A new site wishing to join a running DASCOSA-DB system only needs to know the address of one connected site to join the DHT and thus be a part of the distributed database. When it has joined, it publishes information about its local metadata in the distributed catalog to make its local tables available to the rest of the system.

Sites communicate using messages. These messages can either be sent directly to a site if the address is known, or be routed to the target site using the DHT routing mechanism. The latter method is used for catalog lookups and updates.

DASCOSA-DB supports the relational model and bases its storage on a local relational database management system. The current implementation uses JavaDB,[2] but any relational database management system may be used. The back-end database system can be chosen freely at each site.

---

[1] http://freepastry.org/.

[2] http://www.oracle.com/technetwork/java/javadb/overview/.

The relational tables can be horizontally fragmented over a subset of the sites in the system. Each fragment can also be replicated. The distributed catalog maintains information about tables, fragments, and their replicas. Creation and removal of fragments and replicas can be done using DASCOSA-DB's automated refragmentation method. Based on logging of read and write accesses, fragments can be split or joined, or replicas can be created and removed. This is done to reduce overall communication costs by making more data available locally where it is used and scale the number of replicas by the amount of writes. For example, a site doing heavy reads on a table fragment will get a local replica once this pattern is detected.

When executing queries, DASCOSA-DB utilizes query shipping. After query optimization, different query operators are allocated and distributed to sites in the system. This allows different operators to be executed by different sites in parallel. DASCOSA-DB also includes support for distributed semantic caching to speed up query execution. During updates, replicas are kept up to date using synchronous replication and transactions are handled using the two-phase commit protocol.

5.3.2   Site Architecture

The overall architecture of a DASCOSA-DB site is illustrated in Fig. 5.2. As described above, sites communicate using direct messages or using the DHT. Together with modules handling broadcasting of messages to the network and request-response pairs of messages, these constitute the communication subsystem in DASCOSA-DB.

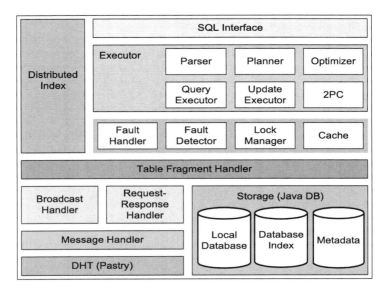

Fig. 5.2 High-level overview of the architecture of a DASCOSA-DB site

Local storage on a site consists of three parts. First, there is the relational data. Relational tables can have one or more fragments and each fragment has one or more replicas. Therefore, the unit of local storage is a table fragment replica. The second part of local storage is the indices for these replicas. Finally, each site stores a part of the distributed metadata catalog. Which part of the catalog a given site stores, is determined by the distributed hashing algorithm and the site's position in the DHT.

Which replicas a site stores locally can change at runtime. Based on an analysis of logged reads and writes, the local Table Fragment Handler can dynamically decide to change the fragmentation and allocation of replicas in one of four ways:

- Coalesce two fragments into one fragment. This means that all replicas of both fragments will have to be altered.
- Split a fragment into two fragments. As with coalesce, this will have global effect for all replicas of the fragment.
- Send a copy of a local replica to another site so that this site can get its own local replica to speed up local accesses.
- Delete a local replica. This will reduce the effort needed to keep all replicas of a fragment up to date and will therefore make sense in periods with many updates.

The Fault Detector and Fault Handler are used to implement partial restart of failed queries. If a site detects that another site designated to execute a subquery has failed, it can handle this fault transparently from the rest of the query execution. This is done by relocating the failed subquery to other sites. In many cases, this can be done efficiently by not having the new sites restart the subquery completely, but rather continue where the failed site stopped.

Each site in the system can receive SQL queries and updates, for example, using the provided user interface or using API calls. A received SQL statement is first parsed and transformed into relational algebra. If it is a query, a lookup in the distributed catalog is done to find location information about all involved tables. This information is then used by the Planner and Optimizer modules to generate a distributed query plan, including allocating the individual query operators to individual sites in the system. The operators are distributed to the involved site where the Query Execution module is responsible for the actual execution.

To facilitate easy interactive access to the system, as well as study configuration, distribution of data and query execution, DASCOSA-DB includes a monitoring tool that gives a live view of table fragments, replicas, catalog entries, and cache entries. It also provides a live view of query execution, including network traffic and currently running query operators.

5.4   Distributed Data and Metadata Management

Tables in DASCOSA-DB may be horizontally fragmented based on the primary key, and DASCOSA-DB provides an adaptive fragmentation and replication system [10] that automatically moves data between sites as needed. In this section, the fragmentation process and the replication of the fragments are described. Then it is described how metadata about fragments and replicas are retrieved from the local database when a site connects to the system and how it is published and subsequently retrieved from the global distributed catalog.

5.4.1 Fragmentation

A table may be stored in its entirety on one site, or it can be fragmented over a number of sites. An unfragmented table is treated as a table having a single fragment. Tables are fragmented horizontally based on the primary key. Each fragment of a table is given a fragment value domain (FVD) that defines which range of the primary key domain has been allocated to the fragment. The fragments are nonoverlapping, and the FVDs of all fragments of a table cover the whole primary key domain.

The FVD of a fragment may cover a much larger range than the range of actual tuples in the fragment. For example, a newly created table consists of one fragment with the whole primary key domain as its FVD, even though it does not store any tuples yet. As tuples are inserted, updated, read, and deleted, a larger part of the FVD is actually used, and the table may split into more fragments.

The traditional way of fragmenting and replicating tables in distributed database systems has been to use fixed value ranges or rules defined by database administrators. In DASCOSA-DB, fragments and replicas are created and migrated automatically by the system to accommodate the current query load. Based on access pattern monitoring, DASCOSA-DB will try to keep the number of accesses to remote sites as low as possible. The FVDs and fragment placements are not fixed, so fragments can be split, coalesced, and migrated automatically to adapt to changing workloads. Figure 5.3a shows a simple example of how two sites with different access patterns access the same table. Site S_2 has a few hotspots, while site S_1 accesses the whole table uniformly and infrequently. In this case, DASCOSA-DB will split (or merge if the table is already split) the table into six fragments, $F_1, F_2, \ldots, F_6$. F_1, F_3 and F_5 will be allocated to site S_2, while F_2, F_4 and F_6 will be allocated to site S_1.

To make informed decisions about useful fragmentation and replica changes, future accesses have to be predicted. As with most online algorithms, predicting the future is based on knowledge of the past. In our approach, this means detecting access patterns, that is, which sites are accessing which parts of which fragment. This is done by recording accesses to discover access patterns. Recording of accesses is a continuous process. Old data is periodically discarded so that statistics only include recent accesses. In this way, the system can adapt to changes in access patterns.

Given the available statistics, our algorithm examines accesses for each replica and evaluates possible refragmentations and reallocations based on recent history. The algorithm runs at given intervals, individually for each replica. Each site bases

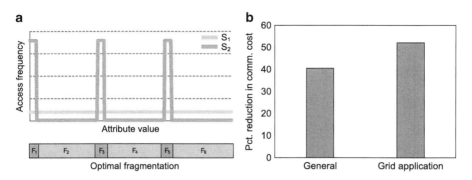

Fig. 5.3 (**a**) Access pattern and desired fragmentation. (**b**) Reduction in communication costs relative to static fragmentation

its decisions only on information available at that site, requiring no synchronization with other sites. With master-copy-based replication, all writes are made to the master replica before read replicas are updated. Therefore, write statistics are available at all sites with a replica of a given fragment. On the other hand, reads are only logged at the site where the accessed replica is located. This means that read statistics are spread throughout the system. To detect if a specific site has a read pattern that indicates that it should be given a replica, it is required that each site reads from a specific replica so that each site's read pattern is not distributed among several replicas.

There is a great potential for cost savings by improving fragmentation. Figure 5.3b shows the reduction in number of tuples transferred over the network in DASCOSA-DB for two different workloads. In the general workload, all sites access tuples uniformly across a selected range of the whole table. Approximately 80% of the accesses are read accesses and 20% are write accesses. The reduction in tuple transfers is more than 40%. In the grid application workload, each site alternates between read phases and write phases, changing hotspots for each phase. The grid application workload has more clearly separated phases, and the savings are more than 50%. The results clearly show that the cost of splitting, migrating, and replicating fragments pays off.

5.4.2   Replica Management

A table fragment is considered to be a logical entity. The physical entities stored in the local DBMSs are table fragment replicas. All fragments must therefore have at least one replica.

Replicas are kept up to date using synchronous replication. Every statement that changes the state of a fragment is sent to all sites with replicas. All replicas must be updated in order for a transaction to commit, and a two-phase commit protocol is used to ensure that all replicas agree on the decision to abort or commit the transaction.

Similar to the way fragments can be split or coalesced, replicas can be automatically created and deleted. Each site logs reads and updates to the locally stored replicas. A new replica is created at a given site if this site does a lot of reads. The idea is that the cost of transferring the replica to the site is less than having a constant stream of remote read requests. A local replica is deleted if there are few local accesses compared to the number of updates received. For both these mechanisms, the idea is to reduce the overall network traffic.

Not all replicas are treated equally. One replica is designated as the master replica. To ensure that automatic replica deletion does not delete all replicas, this replica is not eligible for deletion. The site containing the master replica has two special functions. First, it is the site where refragmentation decisions are made. This prevents two sites from independently and simultaneously deciding to, for example, split the same fragment. Only the site with the master replica is able to do this. Second, the site with the master replica acts as a lock manager for the table fragment. This allows us not to have a centralized lock manager, which could become a bottleneck in a large system. When the system first boots, the catalog site storing the catalog entry for a table decides for each fragment of the table which replica becomes the master replica, and thus also which site becomes the master replica site. A new master replica site can be selected if the current master replica site crashes. It is also possible for the current master replica site to transfer this status in case of refragmentation.

5.4.3   Metadata Management

DASCOSA-DB uses a DHT to store and access the metadata catalog. The DHT provides a reliable and robust routing and lookup mechanism. Due to the DHT routing, catalog lookups are fault tolerant. The DHTs hashing function also distributes responsibility for metadata storage. All sites in the system participate in the DHT, and when a metadata object is published in the DHT, the DHT places it on one of the sites according to a hash of the object. Using a uniform hashing function, metadata objects are uniformly distributed among the catalog sites.

When a site joins the DHT, it scans its local database and inserts information on local objects into the DHT. Catalog objects will time out if they are not renewed, and sites periodically republish their information before the objects time out and are removed. This is done to ensure that erroneous information that may appear due to sites crashing after publishing their metadata is cleaned up regularly.

The catalog keeps track of tables and their schemas. For each table, it stores information about the primary key and the name and data type of all attributes. The catalog also keeps track of how tables are fragmented and replicated, that is, how many fragments there are, the FVD of each fragment, and the number of replicas and their locations. Also, one replica of each fragment is designated the master replica, and the catalog stores this information.

The existence of caches of intermediate query results is also regularly published to the catalog in the same manner as table, fragment, and replica information. For each cached query result, the catalog stores a semantic descriptor, location, and timestamp. Information about cache entries is not looked up directly, but rather discovered as a side effect of table lookups. The cache lookup is included in table lookup requests and replies. This mechanism is described in more detail in Sect. 5.5.1.

5.5   Distributed Query Processing in DASCOSA-DB

DASCOSA-DB is a query shipping system where all sites store data and process queries. Queries may arrive from any site of the system, and the site that introduces a query to the system becomes the initiator site for that query. It is assumed that queries are written in some language that can be transformed into relational algebra operators, for example, SQL.

5.5.1   Query Pipeline

A query enters the system at one site. This site, called the initiator site, becomes the coordinating site for this query. The initiator site transforms the query into an algebra tree. During query planning, the different algebra nodes are assigned to sites. This requires catalog lookups to transform logical table accesses into physical localization programs, for example, a set of accesses to table fragment replicas. Sites can be assigned more than one algebra node so that one site can be assigned a whole subquery. As all sites have the capability to execute operators, sites storing table fragments used in the query are typically also assigned query operations on these fragments during planning. This tends to reduce network traffic as tuples can be processed locally. An example of an algebra tree with site assignment is shown in Fig. 5.4a. The initiator site plays the role of coordinator for this query and executes an initiator algebra node that is the endpoint of the query result.

DASCOSA-DB can cache the intermediate and final results of queries. Each site autonomously caches results of locally executed queries and subqueries and registers these in the distributed catalog so that the caches can be found by other sites. These catalog entries contain a semantic description of the cached query result, the address of the site that stores the cache entry, and a timestamp used to check cache entry validity.

As Fig. 5.4a indicates, the complexity of a query increases with the height of the query tree. The query $T * U * V$ is more complex than $T * U$. If some of the intermediate results, like $T * U$, are cached, the more complex queries may be

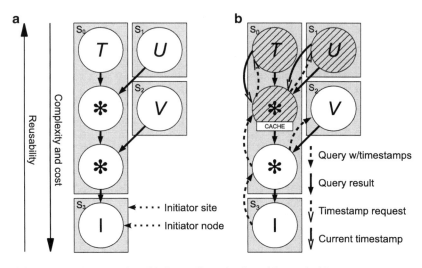

Fig. 5.4 (a) Example query plan. (b) Query dissemination with a cache hit

answered partly from these caches, saving both execution time and computational cost. More complex results in cache means larger savings when these caches are used. However, as the other arrow in Fig. 5.4a shows, the reusability of a result is higher for the less complex queries.

When a table is looked up in the catalog, the initiator site piggybacks a representation of the query to the lookup message. The catalog site that handles the lookup request sees this query representation and responds by piggybacking onto the response a list of suitable cache entries that might speed up query processing. Information about a cache entry is stored at the same site as one of the tables involved in the query that produced it. This means that after looking up all tables, the initiator site has been told about all caches involving the combination of these tables. During localization, the initiator site looks at these cache entries. If a relevant cache entry is found, the initiator site can rewrite the query to use the cache entry. This is done by including the query that produced the cached result as a subquery of current query and assigning the subquery to the site where it is cached.

After planning and possibly rewriting the query to use cached intermediate results, query dissemination begins by transmitting the algebra tree stepwise from the initiator site to the different sites involved. The root algebra node always stays at the initiator site. For each child of the root node, the initiator site sends out the subtree rooted at that child node to the child's assigned site. These sites, upon receiving query subtrees where the roots are assigned to them, loop through the children of the roots and ship them off to the sites to which they are allocated. This continues until all nodes have reached their destination. The result of this stepwise transmission is that each site knows the complete subquery for which it is the root.

However, if a site receives a subtree for a query it has in its cache, and if that cache entry is still valid, further dissemination of that subtree stops. Instead, the site prepares a special algebra node to produce the result from cache. To the sites higher up in the hierarchy, there is no way to tell if the result is served from cache or produced from scratch. This transparency allows the sites to make cache decisions without relying on central coordination. Figure 5.4b shows query $T * U * V$ with a cache hit on subquery $T * U$. Site S_0 checks the timestamp of the cache entry against the timestamps of T and U to see whether the cache is up to date. If it is, $T * U$ is delivered from cache, and the only query operator actually executed is the join of $T * U$ and V at site S_0. Site S_1 is never involved in the query processing, except when replying to the request for the timestamp of U.

Results of query operators are transferred between sites in tuple packets. The system supports stream-based processing of tuples, for example, joins performed by pipelined hash-join [23]. This means that an algebra node usually can start producing tuples before all the tuples are available from its operand nodes. This makes it possible for nodes downstream to start processing as soon as possible and therefore lets more nodes execute in parallel. This requires each site to be able to accept and buffer yet unprocessed packets, but it allows data transfers to be made without explicit requests, thereby improving response time. In case of limited buffer availability, flow control is used to temporarily halt packet transmissions.

The result of any algebra operator is a candidate for caching at the site where it is produced. Sites are allowed to use any cache replacement algorithm they want. A cache entry is usable as soon as it is created, but to enable the query planners to plan on using cached results, the cache entries must be registered in the distributed catalog. A site that has cached a result reports its existence to the same site that handles lookup request for one of the tables used to produce the result. For example, if the cache entry is the result of $T * U$, the catalog stores the information about this entry at either the site that stores the catalog entry for T or the catalog entry for U. Any site that later looks up both T and U to perform a join is guaranteed to find this entry.

5.5.2 Standard Query Operators

DASCOSA-DB supports the typical query operators. At the lowest level, the scan operator accesses the local DBMS and delivers tuples of a table fragment. To speed up execution, special scan nodes exist that push selection and projection down into the local DBMS.

Selection and projection operators also exist to be inserted into the query tree when the operations cannot be pushed down into the local DBMSs. The selection operators also support set operators, that is, IN and EXISTS, to compare against the result of subqueries.

The join operators include natural join, equijoin, and outer join. These are implemented as pipelined hash joins. An operator also exists to produce the Cartesian product. Other operators include sorting, limiting, aggregation (including grouping), duplicate removal (UNIQUE), and a skyline operator.

All operators, except the scan operators, have flow controlled input and output streams with a general interface. This makes it possible to connect them in any meaningful way to represent a query. This generalized interface also makes it easy to ship queries around, since the input and output streams are network transparent.

For most normal cases in-memory operators suffice, but for large operand sizes there are also variants of these operators that will use disk to avoid excessive memory consumption.

5.5.3   Fault-Tolerant Distributed Query Processing

The more sites that are involved in a query, the higher the probability of a site failing during query processing. Long queries and high churn rates in the system also increases the probability of site failures. The traditional way of handling failures focuses on update transactions, and the typical failure recovery is to do a complete restart of the failed transaction. Query failures have largely been overlooked. Complete query restart is an appropriate technique for small and medium-sized queries; however, it can be expensive for very large queries, and in some application areas there can also be deadlines on results so that complete restarts should be avoided. In some cases, various checkpoint-restart techniques have been employed to avoid complete restarts of operations, but these techniques have been geared toward update/load operations, and in many cases implies that a query will be delayed until the failed site is back online.

As an alternative to local checkpointing and complete restart, DASCOSA-DB supports partial restart of queries [8]. With partial restart, unfinished subqueries from failed sites can be resumed on new sites after failures. These restarted subqueries may also utilize partial results already produced before the failure – both results generated at nonfailing sites and results from failing sites that have already been communicated to nonfailing sites. The technique integrated in DASCOSA-DB can be compared to previous approaches like [20]. DASCOSA-DB's fault tolerant query processing (1) reduces query execution time compared to complete restart, (2) incurs minimal extra network traffic during recovery from query failure, (3) employs decentralized failure detection, (4) supports nonblocking operators, (5) handles recovery from multi-site failures, and (6) avoids duplicate tuples by deterministic delivery of tuples from base relations and operators. The query restart techniques can also be used to provide distributed suspend and restart of queries.

Figure 5.5a shows a system executing the query $T * U * V$. Originally, only sites S_1, S_2, S_4, S_5, and S_6 are involved, but sometimes during query processing S_4 fails. This is detected by site S_6, which is the recipient of the result of the failed algebra node. Site S_6 chooses S_3 to replace S_4, and reissues the query $T * U$ to this site. Site S_3 follows the normal query dissemination strategy and forwards the scan operators to sites S_1 and S_2. The particular challenges that have been solved in our approach relate to failure detection, selection of replacement site, and restart of the various relational algebra operators.

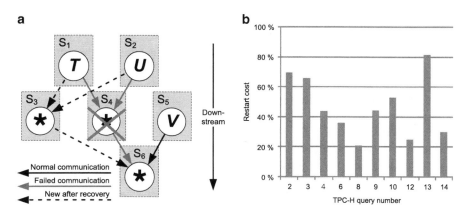

Fig. 5.5 (a) Example of query failure and restart. (b) Relative cost of restarted TPC-H queries

Failures during query processing are detected by using timeouts. There is no central failure detector. Instead, a site monitors all sites that produce the operands for query operators executing at that site. If a site failure is detected, a new site is selected for each of the failed operators. The impact of a failure is therefore localized – it only affects the sites receiving the results of the failed query operators. Other queries and subqueries executing at other sites continue as normal.

The replacement site selected to execute a failed query operator tries to pick off where the operator first failed. How this is done, depends on the operator. Two classes of operators can be identified: *stateless* and *stateful*. Stateless operators process tuples independently. Examples include projection and selection. For these operators, the number of operand tuples an operator has used to produce a given number of result tuples is stored. This number is transmitted with each packet of tuples sent in the network. Using this number, a replacement site knows where to start when resuming a failed operator. For example, assume that a failed site S_f was executing a selection. This selection was done on tuples received from another site S_o. The target site S_t for the selection, has received 500 result tuples when S_f fails. Assume that 800 tuples from S_o had been processed to produce those 500 result tuples. This fact will be known by S_t and transmitted to the replacement site S_r. S_r will then know that it should request S_o to resume sending tuples, skipping the first 800.

For stateful operators, on the other hand, each result tuple can be dependent on more than one operand tuple. Such operators include join and aggregation. When such operators are restarted, they must request operands to be replayed in full. However, they can still use the number of received operands before the failure to prevent sending duplicates. For example, a join must get its two operands completely, but it can skip sending the first result tuples up to and including the number of tuples received from the failed site.

For this partial restart technique to work correctly, tuples must be produced by an operator in a deterministic order. Note that this does not mean that this has to be a sorted order. For scan operators, it is required that tuples are retrieved from the local DBMSs in a deterministic order. Further, it is required that other operators are deterministic so that they produce tuples in a deterministic order given the same ordering of operand tuples. Thus, this requirement reduces to having operators consuming tuples in a deterministic order. This is achieved by having operators consume packets of operand tuples in a round-robin order sorted on the ID of the source site of an operand tuple packet.

The results in Fig. 5.5b show the cost of a restart for a representative collection of ten TPC-H queries. The average restart cost is 50%. The two queries with the least gain (query 2 and 13) were also the two shortest queries. There is a constant overhead in detecting site failure and restarting queries. For the longer queries, this constant overhead is relatively small, so these queries have a lower restart cost.

5.6   Distributed Monitoring and System Management

DASCOSA-DB includes an integrated distributed monitoring and management tool. Figure 5.6 shows the user interface which allows the user to issue SQL statements and monitor the state of the system in real time. It has proven very useful for the different research projects employing or extending DASCOSA-DB.

DASCOSA-DB supports running more than one site on the same physical computer. All these sites will still communicate as if distributed and have separate local DBMSs. Running more than one site locally allows the user to easily examine the execution of distributed queries as the monitoring tool can observe all these sites.

The available views show which table fragments are stored at each site and the schema for each of these. The catalog view for a site shows catalog entries stored at that site. Tables are listed with the number of fragments and replicas, and each fragment entry shows the FVD, the actual used ranged and the number of tuples in the fragment. The catalog view also shows cached query results.

Network traffic monitoring is made easy by using the network log, which will list all messages received and sent by a selected site. This allows the user to, for example, easily track the distributed execution of a query. Both query processing messages, catalog messages and other maintenance messages can be inspected.

The monitoring tool also allows the user to inspect running queries and follow the execution of algebra nodes as flow control changes the state of algebra nodes between processing and paused states. A complete view of all running queries and algebra nodes is provided.

Cache inspection is also provided. DASCOSA-DB has two caches: a restart cache that is used to provide fault tolerant query processing, and a semantic cache of intermediate query results. Each of these may be inspected through the management interface.

Finally, the management interface allows the user to simulate network failures and site crashes by toggling on or off message delivery to each site. When a site is disconnected, the rest of the system will notice its disappearance and adjust to the

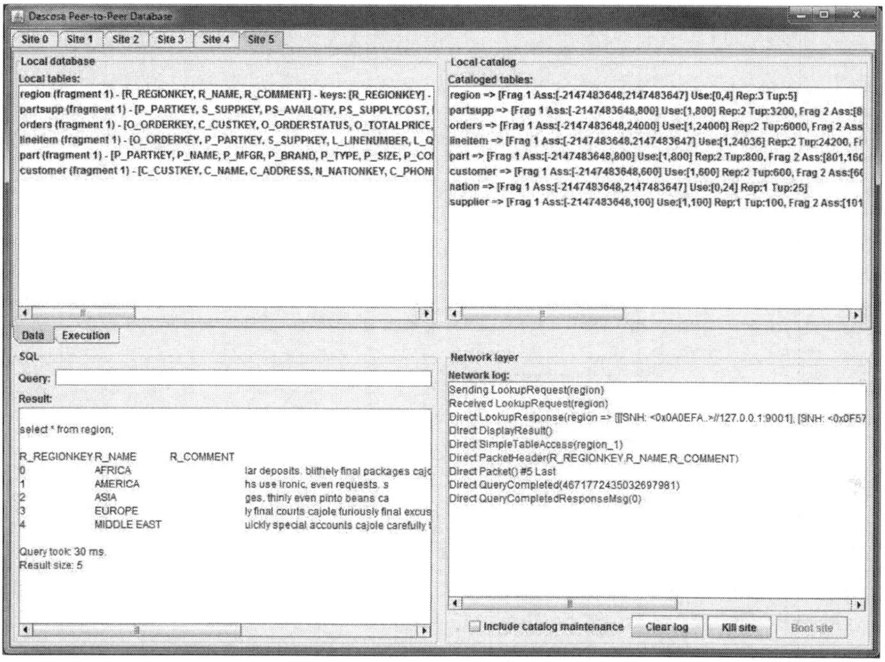

Fig. 5.6 Screenshot from the DASCOSA-DB system monitoring tool

new situation. Queries involving the failed site will restart, and new master replicas will be appointed.

5.7   Experimental Evaluation

The individual features of DASCOSA-DB have been evaluated experimentally in earlier papers [8, 10, 19]. In this section, it is showed how the system, as a whole, scales. Evaluation of additional DASCOSA-DB features not described in this chapter can be found in [18].

5.7.1   Experimental Setup

The system consists of ten interconnected sites running DASCOSA-DB. A TPC-H dataset is horizontally fragmented into five fragments. Each site stores one fragment, meaning that there are two replicas of each fragment. A set of 1,000 random TPC-H queries with random values for substitution parameters is used. An 80/20 distribution is used both for query and parameter selection.

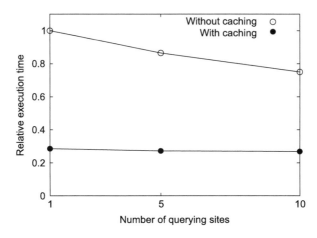

Fig. 5.7 Execution time relative to baseline

The number of sites that issue queries, and thereby the number of coordinator sites, is varied between 1, 5 and 10 to show how system performance increases with increased parallelism. Each querying site executes its queries in series, waiting for one to complete before issuing the next. The system is tested both with and without semantic caching enabled.

5.7.2 Results

The execution time of each experiment relative to a baseline is measured, where all queries were issued in sequence from a single site, without caching any query results. The results shown in Fig. 5.7 show that by increasing parallelism so that all sites issue queries, execution times are reduced by 25%. Since DASCOSA-DB allows queries to be issued from any site, the risk of the coordinator site becoming a bottleneck is reduced, and higher throughput can be achieved.

Further, semantic caching reduces the run time with up to 73%. This considerable improvement is possible because parts of the algebra tree for a query is similar to some parts of other queries. These parts are reused to provide a quicker response to the query, freeing up resources that otherwise would be used to process each query from scratch.

The execution time does not decrease as much with increasing number of querying sites as was the case without caching. The reason for this is that there is not much more time to save after the reduction in execution time caused by semantic caching. Also, caching is a means to improve execution time of a series of queries, not parallel queries. The result has to be cached before it is used. Still, our semantic caching method makes it possible to reduce execution time of multiple parallel querying sites since cache entries are shared with all other sites.

5.8 Summary and Future Challenges

The central point of the grid is to present the user with readily available computational power without the need to know where this power comes from. This should also be the central point for data storage used by the grid, and our DASCOSA-DB is designed with this in mind.

We have presented a middleware system that transparently provides access to data distributed throughout the grid. Based on the relational model, our query shipping database system efficiently queries data in situ, while constantly adapting to the shifting workload by moving table fragment replicas closer to where they are used and by replicating data that has to be read by many sites. Semantic caching reduces the need to compute everything from scratch and allows new queries to take advantage of the intermediate results of queries that have already finished, even if they came from different sites. In case of failures during query processing, DASCOSA-DB will restart only the failed subquery. DASCOSA-DB also provides a distributed monitoring and management system.

Although we now have a working distributed database system, there is no lack of remaining challenges. More advanced optimization in the presence of cached data is needed. We will also study rank-aware operators which are important for many of the intended application areas.

Acknowledgements The development of the DASCOSA-DB has been supported by grant #176894/V30 from the Norwegian Research Council.

References

1. Akbarinia, R., Martins, V., Pacitti, E., Valduriez, P.: Design and Implementation of Atlas P2P Architecture. In: Global Data Management, 1st edn. IOS Press, VA (2006)
2. Bauer, D., Hurley, P., Pletka, R., Waldvogel, M.: Bringing efficient advanced queries to distributed hash tables. In: Proceedings of LCN (2004)
3. Boncz, P., Treijtel, C.: AmbientDB: relational query processing in a P2P network. In: Proceedings of DBISP2P (2003)
4. Braumandl, R., Keidl, M., Kemper, A., Kossmann, D., Kreutz, A., Seltzsam, S., Stocker, K.: ObjectGlobe: ubiquitous query processing on the Internet. VLDB J. **10**(1), 48–71 (2001)
5. Chang, F., et al.: Bigtable: A distributed storage system for structured data. In: Proceedings of OSDI (2006)
6. Halevy, A.Y., Ives, Z.G., Madhavan, J., Mork, P., Suciu, D., Tatarinov, I.: The Piazza peer data management system. IEEE Tran. Knowl. Data Eng. **16**(7), 787–798 (2004)
7. Harren, M., Hellerstein, J.M., Huebsch, R., Loo, B.T., Shenker, S., Stoica, I.: Complex queries in DHT-based peer-to-peer networks. In: Proceedings of IPTPS (2002)
8. Hauglid, J.O., Nørvåg, K.: PROQID: Partial restarts of queries in distributed databases. In: Proceedings of CIKM (2008)
9. Hauglid, J.O., Nørvåg, K., Ryeng, N.H.: Efficient and robust database support for data-intensive applications in dynamic environments. In: Proceedings of ICDE (2009)
10. Hauglid, J.O., Ryeng, N.H., Nørvåg, K.: DYFRAM: Dynamic fragmentation and replica management in distributed databasesystems. Distributed and Parallel Databases **28**(2–3), 157–185 (2010)

11. Huebsch, R., Hellerstein, J.M., Lanham, N., Loo, B.T., Shenker, S., Stoica, I.: Querying the internet with PIER. In: Proceedings of VLDB (2003)
12. Kossmann, D.: The state of the art in distributed query processing. ACM Comput. Surv. **32**(4), 422–469 (2000)
13. Kubiatowicz, J., Bindel, D., Chen, Y., Czerwinski, S., Eaton, P., Geels, D., Gummadi, R., Rhea, S., Weatherspoon, H., Wells, C., Zhao, B.: OceanStore: An architecture for global-scale persistent storage. In: Proceedings of ASPLOS (2000)
14. Ng, W.S., Ooi, B.C., Tan, K.L., Zhou, A.: PeerDB: A P2P-based system for distributed data sharing. In: Proceedings of ICDE (2003)
15. Özsu, M.T., Valduriez, P.: Principles of Distributed Database Systems. Prentice-Hall, NJ (1991)
16. van Renesse, R., Birman, K.P., Vogels, W.: Astrolabe: A robust and scalable technology for distributed system monitoring, management, and data mining. ACM Trans. Comput. Syst. **21**(2), 164–206 (2003)
17. Rodríguez-Gianolli, P., et al.: Data sharing in the Hyperion peer database system. In: Proceedings of VLDB'2005 (2005)
18. Ryeng, N.H., Vlachou, A., Doulkeridis, C., Nørvåg, K.: Efficient distributed top-k query processing with caching. Proceedings of DASFAA. pp. 280–295 (2011)
19. Ryeng, N.H., Hauglid, J.O., Nørvåg, K.: Site-autonomous distributed semantic caching. In: Proceedings of SAC (2011)
20. Smith, J., Watson, P.: Fault-tolerance in distributed query processing. In: Proceedings of IDEAS (2005)
21. Stonebraker, M., et al.: Mariposa: A wide-area distributed database system. VLDB J. **5**(1), 48–63 (1996)
22. Taylor, N.E., Ives, Z.G.: Reliable storage and querying for collaborative data sharing systems. In: Proceedings of ICDE (2010)
23. Wilschut, A.N., Apers, P.M.G.: Dataflow query execution in a parallel main-memory environment. Distributed and Parallel Databases **1**(1), 103–128 (1993)

Part III
Cloud Data Management

Chapter 6
Access Control and Trustiness for Resource Management in Cloud Databases

Jong P. Yoon

Abstract Cloud computing is emerging as a virtual model in support of "everything-as-a-service" (XaaS). Service providers post XaaS of resources in a cloud database. There are numerous service providers such as feeders, owners, and creators, who are less likely the same agent. Consequently, resources in a cloud database cannot be securely managed by traditional access control models, and therefore cloud database services may be trustless. This chapter proposes a new security technique to measure the trustiness of the cloud resources. Using the metadata of resources and access policies, the technique builds the privilege chains and binds authorization policies to compute the trustiness of cloud database management. The contribution of this chapter includes a mechanism of the privilege chains that can be used to verify the legitimacy of cloud resources and to measure the trustiness of cloud database management.

6.1 Introduction

Cloud computing models consist of subjects and objects, the objects that can be created by or provided for the subjects. Subjects, as agent, can be a service provider (SP) or a service user (SU), where SPs provide objects to a cloud and SUs request objects from a cloud. The services provided by SPs can be everything, from the infrastructure, platform, or software resources. Each such service is called Infrastructure as a Service (IaaS), Platform as a Service (PaaS), or Software as a Service (SaaS). For example, Google Apps Engine (http://www.google.com/apps) or Microsoft Azure platform (http://www.microsoft.com/windowsazure/) is a PaaS, while Google Docs (http://docs.google.com) is a SaaS, and DropBox (http://

J.P. Yoon (✉)
Department of Mathematics and CIS, Mercy College, 555 Broadway, Dobbs Ferry,
NY 10522, USA
e-mail: jyoon@mercy.edu

S. Fiore and G. Aloisio (eds.), *Grid and Cloud Database Management*,
DOI 10.1007/978-3-642-20045-8_6, © Springer-Verlag Berlin Heidelberg 2011

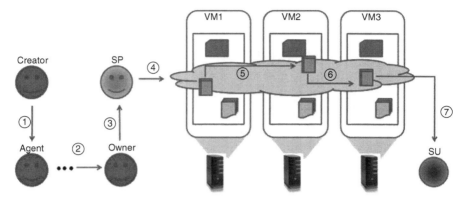

Fig. 6.1 Resource life cycle in cloud databases

www.dropbox.com) is an IaaS. A database deployed and virtualized in such an environment is called cloud database.

Cloud databases considered in this chapter assumes that SPs provide SUs with resources such as JPEG image files and crypto programs as {I|P|S}aaS. There are numerous SPs such as feeders, owners, granters, and creators. As illustrated in Fig. 6.1, a resource is created by a creator, who may then grant (① and ② in Fig. 6.1) the ownership to a new owner (③ and ④ in Fig. 6.1). A resource owner may grant the feedership to a cloud database and further to a resource (SP) feeder (⑤ and ⑥ in Fig. 6.1). While a resource is available in a cloud database server, a (SU) user may request for a usership of a resource from a server (⑦ in Fig. 6.1). Of course, it is also possible that a resource in one virtual machine (VM) is deployed to another VM (③ and ④ in Fig. 6.1) within the same cloud computing environment. Numerous agents with different roles [1, 2], such as creatorship, ownership, feedership, and usership, are involved to handle the same resource, which is available to be accessed.

Since in support of everything-as-a-service (XaaS), there are various operating systems such as Unix, Linux, or Windows, software packages such as DBMS, SAP, ERP, and CMS, and cloud resources available in one or more of those platforms. Each such platform has different mechanisms of authentication and authorization, from typical password-based or LDAP-based authentication [3] to RBAC [4, 5]. In the variety of cloud infrastructures, software packages, and platforms, a cloud resource previously accessed in one platform cannot be accessed by the same user in another platform, and vice versa.

Cloud databases facilitated with the features stated above authenticate who the SP or the feeder of a resource is in the cloud servers. Although a cloud authenticates a feeder, it does not ensure that a resource posted by the feeder is free from authentication spoofing, plagiarism, or virus attacks. In addition to that, it is also common that an information gap exists in between the creator and the feeder of a cloud resource [6, 7]. In the case of JPEG files, most likely in these days and absolutely in the future, the cloud resources contain the information about a creator or the owner. Currently, JPEG files from most digital cameras contain the creator information. It is likely that the creator of a resource is unnecessarily the same as the feeder of that resource. Since they are not the same, cloud databases exemplified below do not have some reasonable degree of trustiness:

- Databases that do not contain enough resources for a complete chain from the creator to the feeder (privilege chain will be defined later), or
- Databases that do contain resources fed by an insufficiently authorized feeder (insufficiently authorized feeder will be defined later).

One of the reasons being such information gaps are caused by the following. The possible cause and consequence are as follows:

- Although a feeder is properly signed up and in, the resources that the feeder posts may be not legally credited to it.
- It is likely that a digital asset as a cloud resource may not have the complete information of all agents, from the creator to the feeder.
- Although exists, authorization policies may not specify nor imply the complete information of all privilege permissions needed from creators to feeders.
- If a feeder of a resource does not have a legal feedership or relevant privilege granted nor implied all the way from the creator, there exists the legitimacy issue. Services with such resources are trustless.
- If anyone who is not certified is involved in a privilege chain (which will be discussed Sect. 6.4), the privilege chain is less likely trustworthy for the legitimacy of a cloud resource and the trustiness of services.

This chapter has contributions to security management of the cloud resources in general and to access control models for cloud database management in more specific. The contributions include:

- Chains of privileges, a graph, are constructed for authorization policies in a virtual machine memory (VMM) and for JPEG files of cloud database resources. The chain of privileges of JPEG files can be used for trustiness and security (access control) management of cloud resources.
- Reachability analysis is to reach any agent in the privilege chains (or together with a cloud social network), who is certified and may be the one the trustiness can be placed on. This reachability analysis can be used to trace for some reasonable degree of the legitimacy of cloud resources.

This chapter is organized as follows: Sect. 6.2 describes the formats of metadata of file resources available in cloud databases. Section 6.3 introduces the basic elements of the cloud memory, which can hold the user session information and the metadata of resources. Section 6.4 proposes the privilege chain for authorization policies, which will then be extended to the privilege graph. Another information chain to be used is for resource agents that are available in the metadata of cloud resources. The chain of metadata will be further extended to the resource metadata graph. Section 6.5 describes the notion of trustiness for cloud resources. If a cloud resource is trustless, external credentials and relationships among themselves, which is widely available in social networking domains, are used. Section 6.6 concludes this work.

6.2 Metadata of Files

Photographic images can be compressed by the Joint Photographic Expert Group
(JPEG, http://www.jpeg.org) compression algorithm. JPEG compression is used in a
number of image file formats, such as JPEG/Exif (Exchangeable Image File Format)
and JPEG/JFIF (JPEG File Interchange Format), and widely used for storing and
transmitting images on the Internet.

JPEG files contain metadata [8, 9], which consists of the data contained in
marker segments in a JPEG file. The image metadata object in the marker segments
between the start of image (SOI, #FFD8 in hex) marker and the end of image (EOI)
marker for the image, contains information about make and model of digital camera,
time and date the picture was taken, distance the camera was focused at, location
information (GPS) where the picture was taken, small preview image (thumbnail)
of the picture, firmware version, serial numbers, name and version of the image
manipulation program, name of the owner, etc.

Such metadata can be added by a user, software, or a digital camera. Some
of the software that can edit in the metadata segment includes EXIFcare, EXIF
Writer, EXIF Tool, MetaDataMiner, etc. For example, Fig. 6.2 illustrates metadata
information displayed in EXIF Viewer in (a) and EXIF Reader in (b), where there is
the column for "Owner Name," which is of our interest. It is also possible that any
attributes can be edited using software. For example, using EXIF Tool command,

```
exiftool -P -overwrite_oritinal -creator='Chris'

wedding1.jpg                                                        (6.1)
```

the original file creation date, *ownership*, *permissions*, type, *creator*, icon, and
resource fork can be overwritten. For security reason, this overwriting function can
be done one time only and read many. However, addition of new tags is always
possible. Therefore, if the file, wedding1.jpg, is granted to a feeder, say Steve,
then the following command is used:

```
exiftool -P -overwrite_oritinal -feeder='Steve'

wedding1.jpg                                                        (6.2)
```

Consequently, the file, wedding1.jpg, has the metadata that contains the following:

```
Creator: Chris
DateTimeCreated: 2009:06:28 10:03:42
Owner: Owen
DateTimeOwned: 2009:06:30 13:12:32
Feeder: Steve
DateTimeDeployed1: 2010:01:28 09:10:11
```

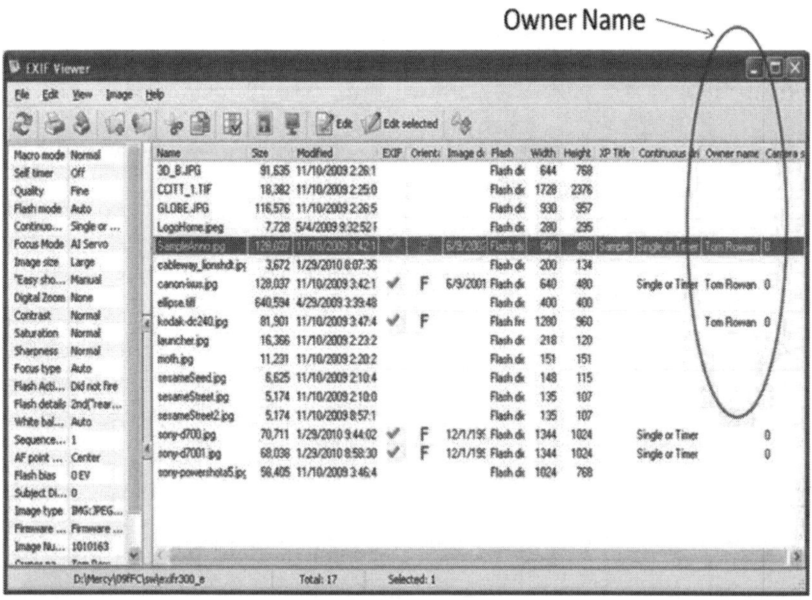

(a) EXIF Viewer

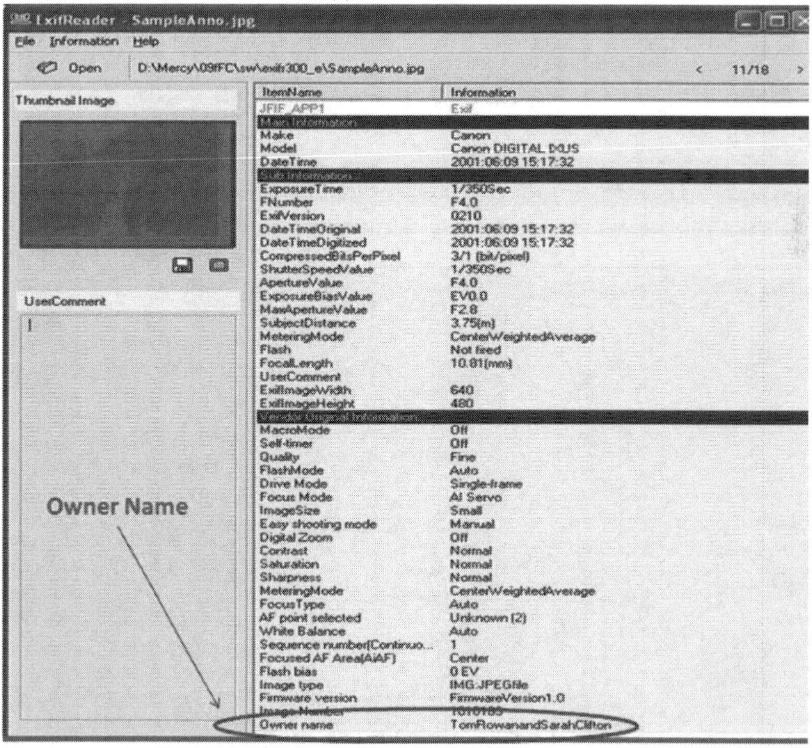

(b) EXIF Reader

Fig. 6.2 Metadata of JPEG

As illustrated above, the metadata contains a set of (attribute, value) pairs. The picture in the file, `wedding1.jpg`, is taken by Chris on 28 June 2009, owned by Owen on 30 June, then granted to the feeder, `Steve`, on 28 January 2010, and thereafter, the picture is up and available for user accesses. Each such agent is an important factor that this chapter can use for access control in cloud databases. Note that the metadata can be obtained and held in a VMM.

6.3   System-Context Information in Virtual Machines

In cloud computing architecture, the service provider (SP) implements the service logic and presents it to clients over the Internet (cloud). The service logic itself is typically composed of multiple components. The SP uses some virtualization abstraction, for example, a virtual machine, for service deployment. Several of these VMs, belonging to various independent SPs, can then be deployed on the infrastructure.

For simplicity, assume that a SP will deploy all resources on a single cloud infrastructure. An example is shown in Fig. 6.1, where the SP provides a VM for customers or service users (SUs). An SU may access the VM from another PC or possibly from a dumb terminal. The SP adds value by allowing reaming access to the infrastructure and possibly providing centralized management. There may be several architectures with various infrastructures possible [10, 11], more specifically, a cloud database with the memory components of VMs [1, 12].

As illustrated in Fig. 6.3, consider the memory for cloud computing in two-tiered architecture. Each SP provides its VMM which may serve one or more available cloud resources to SUs. There is a centralized management, where each VMM is addressed and monitored for namespaces bookkeeping by the cloud global memory (CGM). The CGM has the register set, stacks, and private storage area, which are

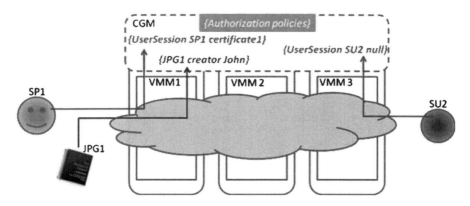

Fig. 6.3 Virtual machine memories and cloud global memory

known as the *context* of the memory. This chapter describes the VMM and CGM with their relationships with respect to trustiness management for cloud resources and its services.

6.3.1 Context for Cloud Resources

As an SP places a JPEG file in a VM for service, the metadata of the JPEG file is loaded in the CGM. Therefore, whichever VM holds a resource, the metadata of the resource can be posted in the CGM. For any cloud resources available in a local virtual machine, their metadata should be available in the CGM. To make the CGM simple, any metadata that are related to the trustiness are loaded in the CGM, while all other information is loaded in a VMM. (Note that JPEG metadata can be extracted from files as discussed in Sect. 6.2.)

For example, a JPEG file has the creator information, possibly together with the information about additional agents such as owners. The CGM context has the following memory context:

$$cgm\_context('fileEnv', 'creator')$$
$$cgm\_context('fileEnv', 'owner') \qquad (6.3)$$

What (6.3) means is that the CGM holds the value creator's name "Chris" bound to parameter "creator" in the context namespace "fileEnv." Therefore, when it comes to an invocation of the creator of a JPEG file, the function "cgm_context(<context>,<parameter>)" returns the value. For example, cgm_context('JPG1_Env', 'creator') will return "Chris."

6.3.2 Context for Service Providers and Service Users

The information about not only cloud resources but also service providers and service users will be available in the CGM. As an SP signs in, the session information about the SP's login and SP's certificates will be posted and loaded in the CGM.

$$cgm\_context('userEnv', 'session\_user')$$
$$cgm\_context('userEnv', 'network\_user') \qquad (6.4)$$

The CGM context namespace "userEnv" holds the parameter session_user of a logging in user in the parameter "session_user". For example, when John is logged in to post a resource in a VMM, then cgm_context('userEnv', 'session_user') will contain session_user "John" in the namespace userEnv. When cgm_context('userEnv', 'session_user') is invoked,

"John" will be returned. In addition to session_user and network_user, more parameters will be loaded into the CGM, such as authenticated_identity, client_identifier, and host, ip_address.

In the same manner, as an SU is signed in a VM, the credential-related information is loaded in the CGM, while all other user and session information loaded in a VMM. The context namespaces and parameters for SU's log-ins are very similar to (6.4).

6.3.3 Context for Authorization Policies

The security manager, if not the policy manager, creates and manages the policies that can be used to make access decisions. Such policies are called *authorization policy*. Typical authorization policies are defined over three elements (*subject*, *object*, and *signed action*), which means that subject is allowed to do action on object. Depending on the sign of actions, subject is permitted to do the action if plus sign, or denied otherwise. The format of such policies is $(s, o, \pm a)$, where s, o, and $\pm a$, respectively, are denoted as subject, object, and signed action [4, 5].

The authorization policies are loaded in the CGM as shown in Fig. 6.3. This format of the typical authorization policies will be extended in the following section.

Why the CGM is needed in a cloud computing environment although there are multiple VMMs are available? [13–15]. The reason for that is not because of provisioning for insufficient memory space, but aims at avoiding the propagation of subjects and privileges. In Fig. 6.3, VMMs without CGM, the metadata about JPEG files and user information should be propagated from one VMM, where the user is logged in or a JPEG file is provided in, to another VMM, where the user has access to JPEG files. Such propagation may cause to modify or temper the metadata information.

6.3.4 Interoperability of VMM and CGM

Since a cloud computing environment consists of multiple VMMs and a CGM, their interoperability is important and the control for interoperation may lie on each VMM or the CGM. As one of XaaS posted in a cloud database, its metadata (e.g., the feeder and creator information) should be extracted and properly located. Also, as a requester (or SU) is logged in, the session information of user's login will be recognized and the access policy for users and XaaS will be correspondingly located. To avoid inconsistency of using access policies, as mentioned in Sect. 6.1, assume that there will be one repository of access policies in the CGM. Depending on the control, in this subsection, three possible approaches are discussed.

- *Local control for interoperability*. Each VMM has its own control on sharing and managing system-context information about file metadata, and the CGM has the access control policies. As a file is posted, its metadata such as the feeder

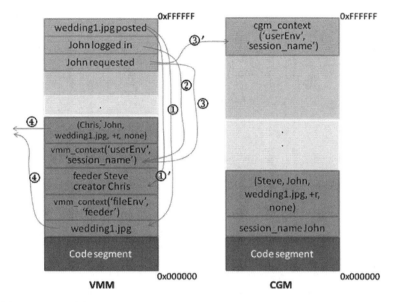

Fig. 6.4 Local control for interoperability

and creator information is extracted as long as available in the file, and stored in a local VMM. As a requester is logged in, context data for the user session information is extracted and stored in the CGM. For users and the requested file, the access policy will be matched and finally stored in the VMM.

For example, in Fig. 6.4, as the file `wedding1.jpg` is posted, it is stored in a VMM (see ①). For the file, the feeder information is stored in the VMM. From the file, the creator information is extracted and stored in the VMM (see ①′). When user `john` is logged in a (may be in a different) VM, the context data for his login session will be stored in both VMM and CGM (see ②). Based on the session availability, John can pose a request (see $f3$ and $3f'$). In the CGM, for the session context data, a corresponding access policy is matched and shipped back to the VMM. According to the policy shipped in, the VMM returns `wedding1.jpg` to `john` (see ④).

- *Global control for interoperability.* As a file is posted, its metadata such as the feeder and creator information is extracted as long as available in the file, and stored in a local VMM and the CGM. As a requester is logged in, context data for the user session information is extracted and stored in the CGM. For users and the requested file, the access policy will be matched and stored in the CGM. Therefore, the CGM holds the user session information, metadata of the resource, and the matched policy.

For example, in Fig. 6.5, the CGM has the feeder `Steve` and creator `Chris` for the file `wedding1.jpg` (see ① and ②), the session information for `john` (see ③ and ④), and the matched access policy (`Steve, John, wedding1.jpg,+r,`

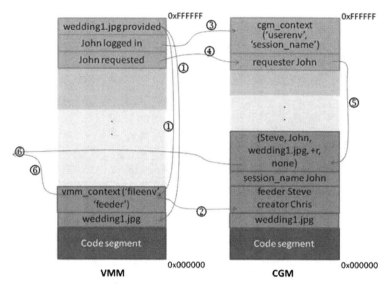

Fig. 6.5 Global control for interoperability

none) (see ⑤). With the information being available, the CGM will return the file to the requester (see ⑥), which is physically stored in a local VMM.

- *Federal control for interoperability.* As a file is posted, its metadata such as the feeder and creator information is extracted as long as available in the file, and stored in a local VMM only. As a requester is logged in, context data for the user session information is extracted and stored in the CGM. For users and the requested file, the access policy will be matched and stored in the CGM as well.

For example, in Fig. 6.6, a VMM has the feeder Steve and creator Chris information for the file wedding1.jpg (see ①), while the CGM has the session information for john (see ② and ③), and the matched access policy (Steve, John, wedding1.jpg,+r, none) (see ④ and ⑤). With the information available, the CGM will determine the authorization to request, and in response to the CGM's determination (see ⑥ and ⑦), the VMM returns the stored file to john.

6.4 Access Control Models

Why the CGM is needed in a cloud computing environment although multiple VMMs are available? The reason for that is not because of provisioning for insufficient memory space, but aims at avoiding the propagation of subjects and privileges. In Fig. 6.3, VMMs without CGM, the metadata about JPEG files and user information should be propagated from one VMM, where the user is logged in or a JPEG file is provided in, to another VMM, where the user has access to JPEG files. Such propagation may cause to modify or temper the metadata information.

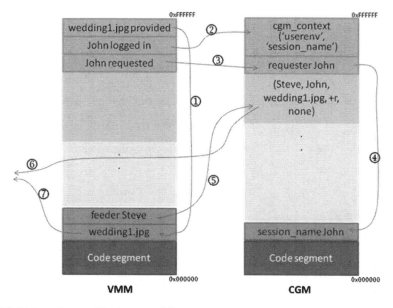

Fig. 6.6 Federated control for interoperability

Having the CGM with the VMMs of virtual machines available in a cloud database, this chapter proposes the chain of privileges and the chain of metadata. The former will be constructed based on the authorization policies, while the latter constructed based on the metadata of cloud resources.

Recall SP and SU introduced in Sect. 6.1. A resource is created by a creator, owned by an owner which may be the same as the creator, granted privileges to another by the owner, and finally placed for service in a cloud database. They are all SPs. A cloud resource then will be used by an SU, who accesses the cloud computing. Both SPs and SUs enter into a VMM. An SP posts a cloud resource in a VMM. As a cloud resource is posted, the metadata of the resource is extracted in a VMM and copied to the CGM. An SU requests a cloud resource from a VMM, the credential of an SU is extracted and held by the system context of the CGM as illustrated in Fig. 6.3. Here, assume that access policies are available in the CGM. With the metadata of a resource and the system-context information about users, an appropriate access policy will be enforced.

6.4.1 Access Policy Specification

The policy manager creates and manages the policies that can be used to make access decisions. Typical authorization policies are defined over three elements (subject, object, and signed action), which means that subject is allowed to do action on object. Depending on the sign of actions, subject is permitted to do the action if plus sign, or denied otherwise. The format of such policies is $(s, o, \pm a)$, where s, o, and $\pm a$, respectively, are denoted as subject, object, and signed action [4, 5].

The signed action specified for typical authorization policies is a privilege that can be applied to an object. This type of privileges is called *object privilege*. In addition to this, this chapter proposes to use another type of privileges, which can be applied to a system. This type is called *system privilege*. Some examples of the system privileges are "grant" or "admin."

Having these all together, the policies are specified over four elements:

$$(s, 0, \pm a, m), \tag{6.5}$$

where m denotes a system privilege, such as "`grant`" that the subject s is permitted to grant the privileges (the same object and system privileges) further to other subjects, or "`none`" that implies no further privileges but only the given object privilege $\pm a$.

For example, the policy (``john'', ``wedding1.jpg'',+r, ``grant'') implies that a request from jyoon is permitted to read the wedding1.jpg file and also permitted to grant the privileges to other subjects as well. On the other hand, the policy (``@cysecure.org'', ``wedding1.jpg'',+r, ``none'') means that requests from the host (``@cysecure.org) has no privilege of further granting to others but is permitted to read the file only.

Now, consider the delegation mechanism in access control. The delegation mechanism has been used to support decentralized administration of access policies [5, 15, 16]. It allows an authority (*delegator*) to delegate all or parts of its own authority or someone else's authority to another user (*delegatee*) without any need to involve modification of the root policy. In this context, there are two types of the subject (or agent) of actions: subject as an actor and as a target. An actor subject (s_a) permits a target subject (s_t) to do an action (a) on object (o), which is in the same context of delegation. The policy format in (6.5) is improved to the following:

$$(s_a, s_t, 0, \pm a, m), \tag{6.6}$$

where

- s_a denotes delegator (or actor subject). The creator (or the first agent) of a file will be of s_a. This actor subject may appear in the metadata of JPEG files.
- s_t denotes delegatee (or target subject). This target subject may appear in the metadata of JPEG files. The end user, the user who requests to access, of files will be of s_t.
- o denotes the target subject is called "direct object." One type of examples is JPEG files.
- $\pm a$ denotes a signed action. It can be "read," "write," "download," etc. that s_a and s_t may request.
- m, respectively, object, signed action, and privilege mode.

For example, the policy (``Steve'', ``john'', ``wedding1.jpg'', +r, ``grant'') implies that Steve grants john to read the wedding1.jpg with the grant system privilege.

6.4.2 Chains of Privileges and Metadata

Recall the metadata discussed in Sect. 6.2, such as creator, owner, feeder, requester, and user. From the creation of (e.g., JPEG file) resources to the service in a cloud database, there is one or more actor subjects (or SPs) involved. As an example of the case of resource wedding1.jpg, creator Chris delegates (sells) to owner Owen (in Fig. 6.7).

Along the sequence of SPs in the metadata of a cloud resource, there will be a chain from the creator to the feeder. Such a chain is called *chain for metadata* (CM). CM_i denotes the chain of the metadata for a resource i. As such, a chain can be constructed in a (linked) list. A node of the linked list represents an SP, any agents from creator to feeder. The head of a linked list is usually the creator of a resource, and the tail node is the feeder to a cloud. In the example of the JPEG file wedding1.jpg above, the head node is Chris and the tail is Steve, and the privilege chain is:

$$CM_{\text{wedding1.jpg}} : \text{Chris} \rightarrow \text{Owen} \rightarrow \text{Charlie} \rightarrow \text{Denny} \rightarrow \text{Steve} \quad (6.7)$$

The chain of this metadata is depicted in (a chain of the dotted boxes and lines) Fig. 6.7:

$$CM_{\text{w2.jpg}} : \text{Paul} \rightarrow \text{Peter} \quad (6.8)$$

$$CM_{\text{w3.jpg}} : \text{Chris} \rightarrow \text{Stew} \quad (6.9)$$

$$CM_{\text{w4.jpg}} : \text{Charlie} \rightarrow \text{Denny} \quad (6.10)$$

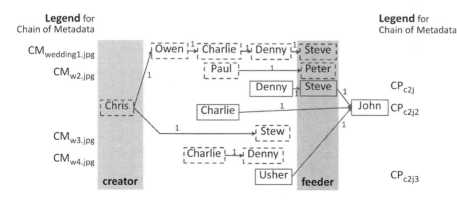

Fig. 6.7 Metadata graph (MG) and privilege graph (PG)

According to CM in (6.9) and Fig. 6.7, the resource w3.jpg has the feeder Stew who Chris agents to. In (6.8), w2.jpg has the feeder information but nothing for the creator's. Equation 6.10 shows that there is no information about the creator and the feeder of w4.jpg.

Similarly, consider access policies. An actor subject (s_a) in an access policy $(s_a, s_t, o, \pm a, m)$ is represented in a head node, while target subject (s_t) in a tail node. That is, $s_a \rightarrow s_t$. Furthermore, consider two access policies, p_i and p_j, $(s_{ai}, s_{ti}, o_i, \pm a_i, m_i)$ and $(s_{aj}, s_{tj}, o_j, \pm a_j, m_j)$, respectively. If $s_{aj} = s_{ti}$, then the node of p_i is linked to the node of p_j. That is, the privilege chain is $s_{ai} \rightarrow s_{ti}$ (or $s_{aj}) \rightarrow s_{tj}$. It means that p_i is the head node, while p_j is the tail node. Such a chain is called *privilege chain for access policy*, denoted by CP_j.

For example, consider the following access policies:

```
(''Denny'', ''Steve'', ''wedding1.jpg'',+{r,w,x},
''grant'')                                                       (6.11)
(''Steve'', ''John'', ''wedding1.jpg'',+r, ''none'')
                                                                 (6.12)
```

The privilege chain for the above policies is:

$$CP_{c2j} : \text{Denny} \rightarrow \text{Steve} \rightarrow \text{John} \qquad (6.13)$$

This is depicted in (a chain of the solid boxes and lines) Fig. 6.7. In Fig. 6.4, there are more CPs:

$$CP_{c2j2} : \text{Charlie} \rightarrow \text{John} \qquad (6.14)$$

$$CP_{c2j3} : \text{Usher} \rightarrow \text{John} \qquad (6.15)$$

Note that PG for John is constructed only while John is logged in the cloud.

In what follows the chains of privileges and metadata will be extended and such extension is called the privilege graph and the metadata graph.

6.4.3 Graphs of Privileges and Metadata

As an extension, this section models a set of privileges and a set of resource metadata as a directed graph. The privilege graph PG $= (V, E, L)$, where V is a set of vertices, each such vertex is from PC_i, $E \subseteq V \times V$ is a set of edges, and L is a set of weights. A value, known as weight, is associated with an edge.

Similarly, a directed graph for metadata of cloud resources, MG $= (V, E, L)$. It turns out that MG $= \{CM_j\}$ and PG $= \{CP_i\}$ for all i's and all j's. In Fig. 6.4, the chains of dotted boxes and lines are in MG, while those solid chains are in PG.

For PG and MG, some specific nodes are defined:

Definition 6.1. (*Creator Head and Feeder Tail*) In any chain of the graph, PG and MG, the head node of a chain is called *creator head node*, if the node is the creator of that resource. The tail node is called *feeder tail node*, if the node is the feeder of that resource.

Corollary 6.1. (*Complete or Broken* CM_i) CM_i is *complete* if it contains both creator head node and feeder tail node. Otherwise, it is of the following:

- *Rootless resource*: Chain with feeder tail node but no creator not owner
- *Unauthorized posting*: Chain with no feeder tail node

For example, in Fig. 6.7, $CM_{wedding1.jpg}$ is complete, while $CM_{w2.jpg}$ is a rootless resource and $CM_{w3.jpg}$ is unauthorized posting. $CM_{w4.jpg}$ is both unauthorized and rootless.

Corollary 6.2. (*Complete or Broken* CP_j) CP_j is *complete* if it contains both feeder head node and user tail node. Otherwise, it is *broken*. The following is of broken chain:

- *Less-authorized-granter:* where there is no creator granter
- *Old-granter:* the granter which is anyone but a feeder
- *Unknown-granter:* where there is no known granter

For example, in Fig. 6.7, all CPs are not complete. CP_{c2j} is a privilege chain with less-authorized granter. P_{c2j2} is an old granter chain because it has the user tail node, that is, other than the feeder head node. By old granter, what is implied is that the head node is not a feeder but any agent that appears earlier than the feeder in CM_i. CP_{c2j3} is an unknown granter chain. By unknown granter, it is likely that it is not found in any privilege chains for metadata.

The difference between old granter chain and unknown granter chain is that both have the user tail node but the unknown granter chain has the head node which is not found in MG.

6.4.4 Authorization Decision

Having both CM_i and CP_j available, as (1) *ensured the trustiness of resources* in Sect. 6.4.3, and this section (2) *makes the authorization decision*. Some cases are illustrated in Fig. 6.8, where each case has both CM and CP in dotted boxes and in solid, respectively. The authorization process can be made differently:

- Case (a) in Fig. 6.8 has both chains of subject. Since both chains are available, the subjects listed in one chain can be cross-checked with those in another. For example, the subjects, Chris, Owen, Charlie, Denny, and Steve in both CM_i and CP_j are the same. This is called "cross-verification." The two linked lists are back-tracking as far as they are identical. If no more nodes are to be tested, the verification is done. If they are not identical, further legitimacy verification is performed from the nodes that are not identical. This will be discussed in the following section.

- Case (b) illustrates a complete CM and a less-authorized CP. It means that there exists a full linked list for the metadata of the JFEP file, but a liked list for an access policy which has the granter Steve with no grant permitted from the creator. Although no perfect agent set is involved, the authorization decision is made as provided from the CGM.
- Case (c) illustrates a complete CM (note that it is still complete) and an old-granter CP. It means the granter Owen is not the feeder of the file but yet a possible agent who appears in CM. This is the case such that in CP permission is grated to s_t by s_a, where s_a is not the same as the tail node of CM. If there exists any node, in the privileged subject chain of metadata that is identical to s_a of access policies, then the authorization decision should be further negotiated [17] to see whether the feeder agrees on the permission. A scheme of negotiation is discussed in the next chapter.
- Case (d) illustrates a rootless resource that may be determined based on an access policy of less-authorized granter. Since there is no creator as the first granter in the chain of the file metadata, the resources (e.g., files in this chapter) are not fully trustful from the creator to the feeder. Also, the grantors of the access policies are not fully addressed from the creator to the feeder. This case sees whether the resource creator still can grant the permission of this resource to any agent in the chain, who will then may have the privilege to grant further all the way to the requester. How to do this from the resource creator will be discussed on the next section.
- Case (e) is similar to (d), except that the access policy states the grantor who is not the feeder but one of the old agents in the chain. The resolution is also very similar to the one proposed in (d).
- Case (f) is similar to (d), except that the access policy states the grantor who is not known nor available in the corresponding CM. In this case, the unknown grantor should be found externally or the permission is simply denied.
- Case (g) illustrates an unauthorized posting of resources and an access policy of a less-authorized granter. The difference between the cases (f) and (g) in Fig. 6.8 is that the granter agent User is unknown while Steve is known but now appearing in CM. In this case, the resolution is to negotiate between the tail node of CM and the header node of CP to authorize the permission. The negotiation process will be discussed in the next chapter.
- Case (h) illustrates the untrusted resource and illegal service. The resource posted in a cloud has lack of agent information found in their metadata. In addition to that, the granter of an access policy is unknown.

As discussed above, using both chains of privileged subjects from JPEG files and VMM can not only make authorization decisions but also control the legitimacy and quality issues of resources. If a resource is posted in one or more VMs, and multiple access policies for the same resource may be available in zero or more VMMs, then conflicts among the policies may exist. Such conflicts can be resolved in the CGM, but the details will not be discussed in this chapter.

6.5   Trustiness of Cloud Computing

Using both PG and MG, this section discusses how to verify the trustiness of cloud resources available in and their services from cloud databases.

6.5.1   Legitimacy of Cloud Resources

As described in Sect. 6.4.3, a chain of metadata can be complete, broken, or missing, while a chain of privileges can be complete, old granter, or unknown granter. With this in mind, the trustiness of cloud resources is defined as follows:

Definition 6.2. (*Traceability of Cloud Resources*) A resource i is traceable if any one below holds:

1. CM_i for the resource i is not unauthorized-posting, or
2. There exists a CP_j that has a node identical to any node of CM_i.

For example, in Fig. 6.7, there are four resources as shown in (6.7)–(6.10): wedding1.jpg, w2.jpg, w3.jpg and w4.jpg. The resources $CM_{wedding.jpg}$ and $CM_{w2.jpg}$ are traceable due to authorized-posting. $CM_{w4.jpg}$ is also traceable because although it is unauthorized-posting, one of its nodes (say Denny) is identical to a node in CP_{c2j}. Finally, $CM_{w3.jpg}$ is not traceable because it is unauthorized-posting and there is no node identical to the node in any CM.

Definition 6.3. (*Legitimacy of Cloud Resources*) If a resource is traceable, then it is legitimate with respect to PG and MG.

For example, in Fig. 6.7, wedding1.jpg, w2.jpg and w4.jpg are legitimate.

6.5.2   Trustiness of Services

Having both PG and MG available, as (1) *verified the legitimacy of resources* in Sect. 6.5.1, and this section (2) *measures the trustiness of cloud services*. For that, define the following:

Definition 6.4. (*Trustiness of Cloud Services*) If a legitimate resource is served, its service is trusted.

As discussed above, the legitimacy of resources and trustiness of their services are dependent on session logins, meaning that they are dynamically changing by those who are logged in. It is reasonable because if there are more authorization policies available due to more log-ins, more chains of privileges are available and so are more privileges granted. Depending on authorization policies, resources are traceable, so their legitimacy issues are technically verifiable. In the dynamic environment of cloud computing, over the course of time, a different set of VMs is involved, and so is different set of authorization privileges.

With this in mind, formulate the trustiness of service T to request j with resource i.

$$T(j,i) = \frac{\beta \cdot \left[\sum_j E(\mathrm{CP}_j) + \sum_i E(\mathrm{CM}_i) \right]}{2 \cdot \mathrm{MAX}(|V(\mathrm{CP}_i)|, |V(\mathrm{CM}_i)|)}, \tag{6.16}$$

where $E(\mathrm{C})$ and $V(\mathrm{C})$ denote the edges and the vertices of a chain C, and $|V|$ denotes the size (or the number) of vertices V. β is an integration factor that can be determined based on the number of VMs, the coverage of authorization policies, the number of common nodes of the chains, the number of certificates signed by a Certificate Authority (CA), etc. As an example, $\beta = 1$ if $(V(\mathrm{CP}_j) \cap V(\mathrm{CM}_i)) \neq \emptyset$; $\beta = 0.5$, if there exists any $\varepsilon \in V(\mathrm{CP}_j)$ or $V(\mathrm{CM}_i)$ is certified by the CA; otherwise, $\beta = 0$.

For example, in Fig. 6.7, $T(\text{c2j}, \text{wedding1.jpg}) = (2+4)/2 * \mathrm{MAX}(3,5) = 6/10 = 0.6$. Similarly, $T(\text{c2j}, \text{w2.jpg}) = 0 * (2+1)/2 * \mathrm{MAX}(3,2) = 0$ due to no common node, and $T(\text{c2j}, \text{w3.jpg}) = 0.5 * (2+2)/2 * \mathrm{MAX}(3,3) = 2/6 = 0.33$ under the assumption that Stew is CA-certified. $T(\text{c2j}, \text{w4.jpg}) = (2+2)/2 * \mathrm{MAX}(3,3) = 0.67$ due to the common node ''Denny,'' while $T(\text{c2j3}, \text{w4.jpg}) = 0$. It turns out that the cloud service with the resource wedding1.jpg to John is reasonably (60%) trusted.

6.5.3 Feeding Social Network Information for Trustiness Management

Recall that a cloud consists of one or more VMs. As XaaS, files are posted in a VM, and access policies are defined in a cloud database which contains the VM. The proposed technique is to assure that the missing gap between MG and PG can be bridged from external sources. Having both MG and PG built as discussed in Sect. 6.4.4, there may exist CM_i and CP_j such that a resource i is not traceable. For example, in Fig. 6.8, cases (d)–(h) are not traceable, meaning that there is no common node in CM and CP. This section uses a social network [18] to acquire possible information that can bridge the missing common node(s) in the linked list of CM and CP. Note that a social network is a social structure made up of individuals (or organizations) called "nodes," which are connected by one or more specific types

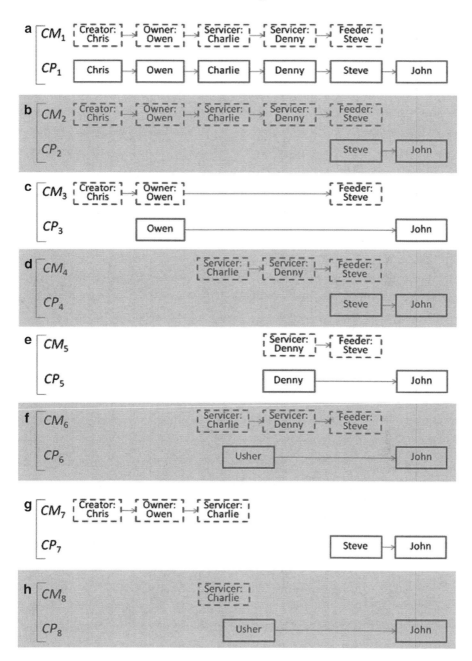

Fig. 6.8 Example of privilege and metadata chains

Fig. 6.9 Negotiation with social network

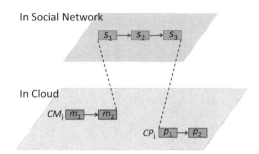

of interdependency. In Fig. 6.9, the upper plane represents a social network and the lower a cloud database.

Consider the linked list in Fig. 6.9. CM_i and CP_j are not complete and the resource is not traceable according to Definition 6.2. Assume that a social network provides the nodes that are matched with any nodes in CM_i and CP_j. Also, the social network information is "securely" fed to assure the trustiness of services.

For example, in Fig. 6.9, by feeding the linked list, $s_i \rightarrow s_{i+1} \rightarrow \ldots \rightarrow s_{j-1} \rightarrow s_j$, where s_i is identical to any node in CM_i and s_j is identical to any node in CP_j as well, according to Definition 6.2, it is possible to assure the traceability of the resource i in CM. It is likely that in a social networking, $s_i \rightarrow s_{i+1}$ exists if s_i follows s_{i+1}, in the Twitter scheme (www.twitter.com), and in reputation-based trust analysis [19] or linked data [20], s_i knows s_{i+1} and s_{i+1} is the author of s_{i+2}, and so on.

Using the PKI [21, 22], social network information can be fed into the resource trustness assurance process securely. Since information in social network is privacy-sensitive, secure transmission of social network information is needed. Consider the scenario such that a resource is in one VM, a request in another VM. A java program is to receive a social network linked list to verify the trustiness of a resource. In Fig. 6.10, M.jpg is in VMM1. A request A.class of B.jar package in VMM3 is a java program to take a linked list of social network information and bridges it between the CM and CP. One simple segment of A.java is the following:

```
import java.io.*;
public class A
{
public void bridge(LinkedList lL) throws
IOException {
// binding with CM and CP - code omitted
}
public static void main(String[] args)
throws IOException {
A myA= new A();
myA.bridge(args[0]);
}
}
```

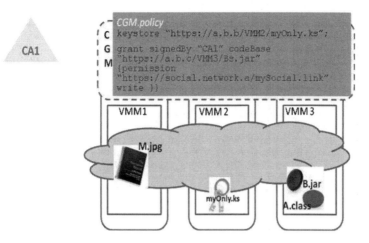

Fig. 6.10 Certificate-based negotiation

To receive a linked list of social network information, in this scenario, the following can generate private and public keys and store them in VMM2.

```
keytool -genkey -alias CA1 -keystore myOnly.ks
```
-storepass MercyCollege–keypass cysecure.org

–dname "CN = John Yoon, OU = CS, O = Mercy,

L = Dobbs Ferry, ST = NY,C = US";   (6.17)

Now, request the certificate authority CA1 to sign on the request jar file B.jar and generate the signed jar file Bs.jar.

```
jarsigner -keystore myOnly.ks
```
–storepass MercyCollege –signedjar Bs.jar B.jar CA1;   (6.18)

In the CGM, the access policy is defined to grant the write permission to the jar code Bs.jar to a linked list, mySocial.link as shown in Fig. 6.10. In the environment set as above, the following java is securely executed:

```
java -Djava.security.manager -Djava.security.policy
= CGM.policy -cp Bs.jar A mySocialNet
```
                                                          (6.19)

Hence, a linked list, mySocialNet, matched in a social network is securely transmitted into a cloud database.

6.6   Conclusion

To address the growing concern of security issues in cloud computing, this chapter has investigated a new approach to measure the legitimacy of cloud resources and the trustiness in cloud database management using the metadata- and privilege-based access control. Using metadata of files or XaaS and system-context user information, gained are numerous benefits including assurance of cloud resources for trust services. This chapter, although as an example JPEG files are only discussed, can extend the concept of using JPEG metadata to any types of resources and to any degree of trustiness in cloud databases as long as such resources have a header file which can contain metadata. This chapter contributed with the concept of using resource metadata and authorization policies to measure the trustiness of cloud services.

References

1. Christodorescu, M., Sailer, R., Schales, D.L., Sgandurra, D., Zamboni, D.: Cloud security is not (just) virtualization security. ACM Workshop on Cloud Computing Security (2009)
2. Vaquero, L., Rodero-Merino, L., Caceres, J., Lindner, M.: A break in the clouds: towards a cloud definition. ACM SIGCOMM Comp Commun Rev **39**(1) (2008)
3. Blezard, D., Marceau, J.: One user, one password: Integrating Unix accounts and active directory. In: ACM Conference on SIGUCCS (2002)
4. Ferraiolo, D., Kuhn, D., Sandhu, R.: RBAC Standard rationale: comments on "A Critique of the ANSI Standard on Role-Based Access Control". IEEE Secur. Priv. **5** (2007)
5. Joshi, J., Bertino, E.: Fine-grained role-based delegation in presence of the hybrid role hierarchy. In: ACM Symposium on Access Control Models and Technologies, 2006
6. Chow, R., Golle, P., Jakobsson, M., Shi, E., Staddon, J., Masuoka, R., Molina, J.: Controlling data in the cloud: Outsourcing computation without outsourcing control. In: ACM Workshop on Cloud Computing Security (2009)
7. Raj, H., Nathuji, R., Singh, A., England, P.: Resource management for isolation enhanced Cloud services. In: ACM CCSW (2009)
8. Haslhofer, B., Klas, W.: A survey of techniques for achieving metadata interoperability. ACM Comp. Surv. **42** (2010)
9. Pereira, F.: MPEG multimedia standards: evolution and future developments. In: ACM Conference on Multimedia, 2007
10. Security Guidance for Critical Areas of Focus in Cloud Computing, v.2.1, Cloud Security Alliance, 2009. http://www.cloudsecurityalliance.org/guidance/csaguide.v2.1.pdf
11. Lenk, A., Klems, M., Nimis, J., Tai, S., Sandholm, T.: What's inside the cloud? An architectural map of the cloud landscape. In: IEEE Conference on Software Engineering Challenges of Cloud Computing, 2009
12. Hao, F., Lakshman, T., Mukherjee, S., Song, H.: Enhancing dynamic cloud-based services using network virtualization. ACM SIGCOMM Comp. Commun. Rev. **40** (2010)
13. Cudre-Mauroux, P., Budura, A., Hauswirth, M., Aberer, K.: PicShark: mitigating metadata scarcity through large-scale P2P collaboration. Int. J. Very Large Data Base **17** (2008)
14. Ferraiolo, D., Atluri, V.: A meta model for access control: why is it needed and is it even possible to achieve? In: ACM Symposium on Access Control Models and Technologies, 2008
15. Kulkarni, D., Tripathi, A.: Context-aware role-based access control in pervasive computing systems. In: ACM Symposium on Access Control Models and Technologies, 2008

16. Ben Ghorbel-Talbi, M., Cuppens, F., Cuppens-Boulahia, N., Bouhoula, A.: Managing delegation in access control models. In: IEEE Conference on Advanced Computing and Communications, 2007
17. Lee, A., Winslett, M., Basney, J., Welch, V.: Traust: A trust negotiation-based authorization service for open systems. ACM SACMAT (2006)
18. Song, H., Cho, T., Dave, V., Zhang, Y., Qiu, L.: Scalable proximity estimation and link prediction in online social networks. ACM SIGCOMM Conference on Internet measurement conference (2009)
19. Srivaramangai, P., Srinivasan, R.: Reputation based two way trust model for reliable transactions in grid computing. Int. J. Comp. Sci. Issues **7**(5), 33–39 (2010)
20. Bizer, C., Heath, T., Berners-Lee, T.: Linked data – the story so far. Int. J. Semantic Web Inf. Syst. **5**(3), 1–22 (2009)
21. Huang, J., Nicol, D.: A calculus of trust and its application to PKI and identity management. In: ACM Symposium on Identity and Trust on the Internet, 2009
22. Zeng, W., Zhao, Y., Ou, K., Song, W.: Research on cloud storage architecture and key technologies. In: Conference on Interaction Sciences: Information Technology, Culture, 2009

Chapter 7
Dirty Data Management in Cloud Database

Hongzhi Wang, Jianzhong Li, Jinbao Wang, and Hong Gao

Abstract Data quality problem is caused by dirty data. Massive data sets contain dirty data in higher probability. As an important platform for massive data management, it is necessary to manage dirty data in cloud databases. Since traditional data-cleaning-based methods cannot clean dirty data entirely and are costly for massive datasets, a massive dirty data management method is presented in this chapter to obtain query result with quality assurance. To achieve this goal, a dirty database storage structure for cloud databases as well as a multi-level index structure for query processing is presented. Exploiting this index for a query on dirty data, candidates nodes in the cloud are selected to run and process the query efficiently. This chapter discusses the index structure and index-based query processing techniques. Experimental results show the efficiency and effectiveness of the presented techniques.

7.1 Introduction

Data quality plays an important role in modern information systems. From the reports sponsored by SAS and Merrill Lynch [1], enterprises in USA lose over 600 billion dollars due to data quality problems. For most enterprises, the collection and cleaning occupy the 50–80% budget of information integration.

Data quality problem is caused by dirty data, which refers to inconsistent, inaccurate, erroneous, redundant, and outdate data. Dirty data widely exists in data-centric systems. All of the steps of data processing may result in dirty data. For example, in the data collection step, errors will be resulted in by the noise of collection devices or typos of users; in the information integration step, inconsistency is led by the heterogeneity in the schema and by the integrity constraints of data sources;

H. Wang (✉) · J. Li · J. Wang · H. Gao
Harbin Institute of Technology, Harbin, China
e-mail: wangzh@hit.edu.cn; lijzh@hit.edu.cn; wangjinbao@hit.edu.cn; honggao@hit.edu.cn

S. Fiore and G. Aloisio (eds.), *Grid and Cloud Database Management*,
DOI 10.1007/978-3-642-20045-8_7, © Springer-Verlag Berlin Heidelberg 2011

in the information transmission step, incorrect data may be caused by network unreliability.

In massive data sets, dirty data exist with larger probability. There are two reasons. The first reason is that the difficulty of maintaining massive data and *errors in large storage devices* will lead to more errors in data. The second reason is that with massive data, the probability of inconsistent data becomes large.

Dirty data do harm to data utility. For example, as highlighted in [2], inaccurate data exist in 65% of the database of stock and results in around 10% loss in profit; the inconsistency existing in financial software will lead to the confusion in finance; the data duplication in a survey will lead to inaccurate results. Therefore, it is necessary to manage dirty data to reduce its harm and make effective use of data.

One traditional method for dirty data management is to clean data [3]. For massive data sets, data cleaning is not suitable in many cases. On one hand, data cleaning operation is often costly. When the massive data is updated frequently, the efficiency of the system will be affected by data cleaning. On the other hand, data cleaning techniques will not clean the dirty data exhaustively, and the cleaning of dirty data will result in the loss of information. Therefore, the techniques for dirty data management without cleaning data are in demand. Such techniques perform queries on dirty data directly and obtain query results with quality assurance. The query processing techniques on dirty data include [4–6]. However, they do not consider dirty data in massive data sets.

Building cloud databases is a feasible way to manage massive data sets. In a cloud database, there are three types of compute nodes: master, router, and slave. The master node is responsible for managing slave nodes. Router nodes are responsible to store index in cloud database. Slave nodes store data and process query locally. When a query is injected, it is sent to a router, and the router searches the index to find the set of slaves containing the query results. After index searching, the query is sent to a set of slave nodes. Slave nodes return local results, and these results are merged as final query results. Figure 7.1 shows a cloud database with one master node, two router nodes, and six slave nodes. Slave nodes store representatives and data index. The router stores node index to locate slave nodes which possibly contain query results.

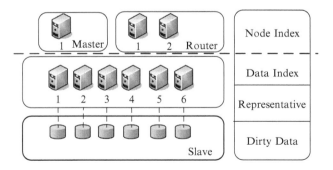

Fig. 7.1 Cloud database with one master, two routers, and six slaves

Table 7.1 Example relation T

id	Name	Address	Age	Gender
P1	Celine Dion	My heart will go on	42	F
P2	Celine Dian	My heart will go on	40	M
P3	Celine Don	My heart will go on	41	F
P4	Mariah Carey	Hero	40	F
P5	Enrique Iglesias	Ring my bells	35	M
P6	Enrique Iglesias	Ring my bells	35	M
P7	Enrique Iglesias	Ring my bells	35	M
P8	John Lennon	Imagine	70	M

An assumption for current data management in cloud systems is that the data are clean. Dirty data in such systems is never considered. This chapter presents efficient and effective dirty cloud database management techniques. Dirty data are distributed in cloud for efficient query processing. A 3-level index is constructed to efficiently locate nodes in cloud for a query and process queries on such nodes. For the efficiency of index-based query processing, a data partition strategy is designed. With indices, a query processing strategy is proposed to obtain query results with quality assurance on dirty data.

The remaining part of this chapter is organized as follows. Section 7.2 defines the query on dirty data and the measurements of the query result quality on dirty data. Data storage strategies and the index structure for dirty data are presented in Sect. 7.3. Query processing algorithms based on the indices are proposed in Sect. 7.4. Experimental results are shown in Sect. 7.5. Section 7.6 summarizes related work and Sect. 7.7 concludes the chapter.

7.1.1 Motivating Example

In this chapter, an example of some popular singers is used to illustrate the proposed techniques. The Table 7.1 contains some errors and inaccurate tuples.

7.2 Preliminaries

7.2.1 Metrics of the Quality of Query Results

The factors of data quality problems include errors, incompleteness, inconsistency, and inaccuracy. These factors have two aspects on query results. On one hand, since a tuple may have error or inaccuracy, a query may retrieve results which do not actually satisfy query constraints. Here, the term *satisfy* is used to indicate that a tuple's clean version matches the query. For example, the query "select name from T where age $= 40$" with semantics "retrieve names of singers with age 40," retrieves P2 and P6, but they are not the results. On the other hand, even though a tuple satisfies a query, it still possibly does not strictly match the constraint. For example, for the query "select name from T where gender $=$ M" with semantics "retrieve names of male singers," the tuple P2 is retrieved. However, it is obviously not the result of the query.

Therefore, the quality of query results is measured by two metrics: accuracy and recall. These metrics are defined as follows:

Suppose there is a dirty database D, whose corresponding clean database is D_C. Given a query q, the query result on D is denoted by D_q and the corresponding result set on D_C is denoted by $D_{C,q}$. $\forall t \in D_C$, the set of tuples in D corresponding to t is denoted by S_t.

Based on the above notations, the precision and recall of query q's result on database D are defined as:

$$\text{Precision}_{q,D} = \frac{|\bigcup_{t \in D_{C,q}} S_t \cap D_q|}{|D_q|}$$

$$\text{Recall}_{q,D} = \frac{|\bigcup_{t \in D_{C,q}} S_t \cap D_q|}{|\bigcup_{t \in D_{C,q}} S_t|}$$

7.2.2 Queries on Dirty Data

Using the metrics defined above, a query on dirty data is defined as (q, ϵ) with q as the query and ϵ as the threshold, which represents the difference between the tuple and the constraint.

Many methods have been designed based on the definition of the similarity or the difference between the tuples or values. These methods can be used to identify data objects referring to the same entity over inconsistent, inaccurate, duplicated data, or data with errors, so they can be applied in this system for matching tuples and query constraints. In this chapter, the linear function is chosen to measure the difference between two tuples as well as the difference between a tuple and a constraint.

For two tuples t_1 and t_2 with attribute set A, the difference between them is defined as:

$$\text{diff}(t_1, t_2) = \sum_{a \in A} \text{diff}_a (t_1[a], t_2[a]),$$

where diff_a is a difference function based on the type of attribute a. For example, if the type of a is string, diff_a is the edit distance; if the type of a is numeric type, diff_a is the absolute difference, otherwise if a is in category type, diff_a is a boolean function returning whether the two values are the same. For example, the difference between P2 and P3 in T is 7 with the difference in attribute name, title, age, and gender equals to 2, 1, 3, 1, respectively.

For a tuple t and a constraint C, with the consideration that C may be a complex constraint, the difference is defined as it follows:

$$\text{diff}(t, C) = \begin{cases} \text{diff}(t[a], c) & C = {}`t[a] = c' \\ \min_{t[a] \neq c} \text{diff}(t[a], c) & C = {}`t[a] \neq c' \\ \min_{t[a] > c} \{\text{diff}(t[a], c)\} & C = {}`t[a] > c' \\ \min_{t[a] < c} \{\text{diff}(t[a], c)\} & C = {}`t[a] < c' \\ \text{diff}(t, c_1) + \text{diff}(t, c_2) & c = c_1 \bigcap c_2 \\ \min\{\text{diff}(t, c_1), \text{diff}(t, c_2)\} & c = c_1 \bigcup c_2 \end{cases}$$

where a is the corresponding attribute in the atom constraint C. Note that for the constraint $C = \neg C'$, the negative operator is pushed to the subclauses in C'. For example, the difference between tuple P1 in T and C ="(age > 45 and age = M) or (age < 40 and age = F)" is 1 where diff(P1, "age > 45 and age = M") = 4 and diff(P1, "age < 40 and age = M") = 1.

ϵ can be obtained from the learning methods or defined by users. By tuning ϵ, the values of precision and recall are changed. Two extreme cases are that with $\epsilon \geq \max\{\epsilon_i\}$, the recall equals to 1, and that with $\epsilon \leq \min\{\epsilon_i\}$, the precision equals to 1.

In this chapter, only conjunction queries with equal constraints are considered.

An example of query on dirty data is $q = (c_q, 3)$, where c_q =("Celine Dien",_, 40,M). The query q on T will return P2.

7.3   Data Accessing Structure for Dirty Data in a Cloud Database

7.3.1   Storage Model for Dirty Data

Based on the inconsistency in dirty data, a cluster-based strategy is used to store the data. That is, entity identification is performed on the dirty data and the data objects possibly referring to the same real-world entity are clustered. For example, the tuples in table T is partitioned into {C1:{P1, P2, P3}, C2:P4, C3:{P5, P6, P7}, C4: P8}.

In a cloud database, dirty data are distributed on multiple slave nodes and partitioned based on the differences among tuples. During data partition, load balance is taken into consideration, since dirty data are divided into clusters with different size. The router node records data partition information and keeps a set of index $\langle p, id \rangle$, where p is the partition and id is the node id which keeps data in p. With the consideration of the effectiveness of index, the data should be partitioned reasonably. The data partition strategy will be discussed in Sect. 7.3.2.3.

7.3.2   Indexing Structures for Dirty Data in a Cloud Database

To perform query processing on dirty data in the cloud, indexing structures are designed in this subsection. The indexing structures have three tiers as it follows.

1. *Representatives*: The tuples in the same cluster are identified by a representative tuple. The query processing is performed on the representative directly greatly reducing scanning cost on real data.
2. *Data index*: For the columns in representatives, data indices are constructed to find proper representative for a query.
3. *Node index*: It is designed for efficiently locating all relative compute nodes which possibly contain query results.

In this section, the structures of the three indices are discussed, respectively. Query processing based on the indices will be discussed in Sect. 7.4.

7.3.2.1 Representative Construction

The operations on dirty data are mainly similarity operations and such operations are costly. To accelerate the processing of these operations, with the proposition that entity identification is performed on the similarity between tuples, in each cluster, a representative of all the tuples is constructed. For a query, the operations are performed on the representatives instead of all the tuples. For each cluster, the representative is the "center" such that its average similarity to all the tuples in the cluster is maximal. Since different types have different similarity definitions, columns representatives are constructed respectively depending on their tuples. The construction methods of columns with different types are as follows:

- *Numerous types*: For a set S of numbers in numerous types such as integer, float and double, the goal is to find a number n that satisfies $\min\{\sum_{n' \in S} |n - n'|\}$. Obviously, $n = \frac{1}{|S|} \sum_{n' \in S} n'$.
- *Category types*: Since the result of the comparison between two values in category types can only be same or different but not the quantitative similarity, the determination of representative value of a category type column is performed by voting and the majority is selected as the value.
- *String types*: Edit distance is often used to measure the similarity between two strings. The representative of a string set S is the string $s \in S$, which satisfies that $\sum_{s' \in S} \text{diff}(s, s')$ is minimal.

In the example, the center of C1 in table T is {"Celine Dion," "My Heart Will Go On," 41.6, F}, where 41.6 is the average of 42, 43, and 40. The center of C2 is {"Enrique Iglesias," "Ring My Bells," 36.67, M}.

7.3.2.2 Data Index

Data Index is used to efficiently retrieve the entities satisfying the constraint. Different from traditional index for clean data, the index for the dirty data on each computer node should be suitable for the query processing with quality assurance.

Therefore, for dirty data processing, the indices supporting queries with error bound are applied. In this section, two indices are introduced.

For numerous types, B-tree [7] is modified for approximate selections. Given a numerous value v and a threshold ϵ, the approximate selection is performed on it by converting the selection constraint $[a_1,a_2]$ to the constraint $[a_1 - \epsilon, a_2 + \epsilon]$ and performing it on the index.

For the string type, n-gram [8] is applied for similarity search on strings to estimate the upper bound of the difference between queried value. For example, with $n = 3$, the n-gram data index corresponding to the attribute "Representative works" in the representatives of the tuples in T is { _Ca: C2, _Di: C1, _Le: C4, ah_: C2, are: C2, ari: C2, Car: C2, Cel: C1, Dio: C1, e_d: C1, e_I: C3, eli: C1, enn: C4, Enr: C3, esi: C3, gle: C3, h_C: C2, hn_: C4, iah: C2, ias: C3,Igl: C3, ine:C1, ion: C1, iqu: C3, Joh: C4, Len: C4, les: C3, lin: C1, Mar: C2n_L: C4, ne_: C1, nno: C4, non: C4, nri: C3, ohn: C4, que: C3, rey: C2, ria: C2, riq: C3, sia: C3, ue_: C3}.

For category types, bitmaps are applied. For the type with a few categories, one bit represents one category, while for the type with many categories, one bit may also represent multiple categories. For example, for the type gender, a bitmap with two bits are used with each bit representing a gender.

7.3.2.3 Node Index

The node index is designed to select suitable nodes for processing a query. The basic idea is to disseminate similar entities into the same compute node and index the representatives of these entities.

The merging of the representatives should be compatible with the data indices in compute nodes, and such merging also depends on the type of columns. When the type is numerous, the merged value is the interval containing all the values of the column. If the type is category, the merged value is the set containing all the values in the column. For the string type, the merged value is the set as the union of q-grams index corresponding to the strings in the columns.

For example, suppose that there are two nodes with N_1 containing $C1$ and $C2$, and N_2 containing $C3$ and $C4$. The merged representatives of $C1$ is {{ Cel, _Di, Dio, e_d, eli, ine, ion, lin, ne_, _Ca, ah_, are, ari, Car, h_C, iah, Mar, rey, ria},{_Go, _On, _wi, art, ear, ero, Go_, Hea, Her, ill, l_G, ll_, My_, o_O, rt_, t_w, wil, y_H}, [40,41.6], {M,F} }, and the merged representatives of $C2$ is {{_Le, e_I, enn, Enr, esi, gle, hn_, ias, Igl, iqu, Joh, Len, les, n_L, nno, non, nri, ohn, que, riq, sia, ue_}, {_Be, _My, agi, Bel, ell, g_M, gin, Ima, ine, ing, lls, mag, My_, ng_, Rin, y_B}, [36.67,70], {M} }.

To find proper nodes for a query, the structure of the node index is similar as inverted index [9]. For each column c with value set S_c, a partial index ind_c is built with each entry (v, L_v), where v is the value and L_v is the list of nodes with the representative containing v. A secondary index is built to efficiently retrieve the lists associated with a query. For numerous type, an interval tree [10] is used to manage corresponding intervals and support both equal and range queries. For category types, the hashing index is used as the secondary index and for string types, trie [11] is used as the index for grams.

For the effective use of index and locating the query results on as few nodes as possible, similar tuples should be placed in the same node. Therefore, by modeling representatives for the clusters as a weighted clique with each representative r_v as a vertex v and the weight on each edge (u, v) as the $\frac{1}{\text{diff}(r_u,r_v)}$, the data partition problem is converted to the clustering problem on graph. It can be solved by various algorithms [12].

The differences of above merged representatives are defined in the similar way as those of tuples. That is, a linear sum of the difference of each attribute. Note that if the type of an attribute is string, since the corresponding gram set is used as the index, the difference of corresponding gram sets is used to measure the difference of two strings. For two gram sets S_1 and S_2, $1 - \frac{|S_1 \cap S_2|}{|S_1 \cup S_2|}$ is used to measure the difference between them.

7.4 Query Processing Techniques on Dirty Data in a Cloud Database

In this section, query processing algorithms on dirty data in a cloud database are presented. This chapter focuses on selection based on index. Join operation can be performed using techniques in [13, 14].

7.4.1 Algorithms for Locating Relative Compute Nodes

To process a query $Q = (V, \epsilon)$ on dirty data, the first step is to find the nodes possible to answer it. For a nodeset N and a constraint C, the goal is to find the nodes possibly with tuples satisfying C. The method is to estimate the upper and lower bounds of the difference of the values in column c in n and the atom constraint T_c in C on both the index and computed difference. Then the computed bounds are used to prune search space among compute nodes.

The interval computation method for atom constraint is based on the type of corresponding column. Some computation methods for major data types are shown as it follows.

N-gram is used to estimate the upper bound of strings. A property of edit distance estimation between strings [8] is that for two strings s_1 and s_2, if edit_distance(s_1, s_2) $< r$, then their corresponding n-gram sets G_1 and G_2 satisfy $|G_1 \cap G_2| \geq (|s_1| - n + 1)C r \cdot n$. Based on this property, the upper and lower bounds of the edit distance between a constraint string s and any value in the column c in a node N is $|s|$ and $\frac{|s|-n+1-|G_s \cap G_N|}{n}$, respectively, where G_s is the n-gram set of s and G_N is the n-gram set in the node index of N.

For numerate types, the constraint is converted to $[a_1, a_2]$ or (a_1, a_2). For the constraint $t = a$, the interval is $[a, a]$. Without generality, the case $[a_1, a_2]$ is considered. For a node N with interval $[l,u]$ for the column c in the node index and the interval $[a_1, a_2]$ as the constraint on column c, the lower and upper bounds of the values in column c and the constraints is defined as:

$$[LB, UB] = \begin{cases} [l - a_2, u - a_2] & a_2 \leq l \\ [0, 0] & a_1 \leq l \leq u \leq a_2 \\ [0, a_1 - l] & l \leq a_1 \leq u \leq a_2 \\ [a_1 - u, a_1 - l] & u \leq a_1 \\ [0, u - a_2] & a_1 \leq l \leq a_2 \leq u \\ [0, \max\{a_1 - l, u - a_2\}] & l \leq a_1 \leq a_2 \leq u \end{cases}$$

For category types, if the constraint is "$c = a$" where c is the name of a column and a is the name of a category and in the node index, the set corresponding to the column c of a node N is S_N, then the lower and upper bound is as in the following:

$$[LB, UB] = \begin{cases} [0, 1] & a \in S_N \\ [0, 0] & a \notin S_n \end{cases}$$

For a negative atom constraint $C = c \neq a$, the upper and lower bounds for the constraint $c = a$ are estimated. For a node N with the difference to C' in the interval $[l, u]$, the difference between N and C is computed as:

$$[L_{C,N}, U_{C,N}] = \begin{cases} [V_a, V_a] & l = 0 \wedge u = 0 \\ [0, V_a] & l = 0 \wedge u \neq 0, \\ [0, 0] & l \neq 0 \wedge u \neq 0 \end{cases}$$

where

$$V_a = \begin{cases} 1 & a \text{ is a category} \\ a & a \text{ is a number} \\ |a| & a \text{ is a string} \end{cases}$$

According to the definition of difference in Sect. 7.4, the difference between a complex constraint C and a merged representative of a node N is computed recursively as it follows:

$$[L_{C,N}, U_{C,N}] = \begin{cases} [L_{C_1,N} + L_{C_2,N}, U_{C_1,N} + U_{C_2,N}] & C = C_1 \wedge C_2 \\ [\min\{L_{C_1,N}, L_{C_2,N}\}, \min\{U_{C_1,N}, U_{C_2,N}\}] & C = C_1 \vee C_2 \end{cases}$$

For example, for the constraint (age $< 40 \vee$ gender $=$ F)$\wedge$ name $=$ 'Celine Dien' with threshold $\epsilon = 2$, the upper and lower bounds of the difference between this constraint and the representative r_1 of N_1 is $[\frac{1}{3}, 12]$ with ϵ larger than the lower bound, where the difference interval of r_1 to the clauses of 'age < 40', 'gender $=$ F', and name $=$ 'Celine Dien' are $[0,3]$, $[0,1]$, and $[\frac{1}{3}, 11]$, respectively. Therefore, N_1 should be selected to process the query.

The node selection algorithm is shown in Algorithm 1. It is assumed that C_Q, the constraint of Q, is in conjunctive normal form.

Algorithm 1 Node selection
---
input: a query $Q=(V, \epsilon)$
output: the node set S
for each atom constraint $C \in V$ **do**
    $S_C = $ get_range(C)
end for
$S_I = \emptyset$
for each clause $C \in C_Q$ **do**
    $S' = \emptyset$
    for each clause $C' \in C$ **do**
        $S' = $ merge($S', S_{C'}$)
    end for
    $S_I = $ intersection(S_I, S')
    $S_I = $ filter(S_I, ϵ)
end for
$S=$extract_node(S_I,ϵ)
return S

In Algorithm 1, the function get_range computes the intervals with upper and lower bounds of the atom constraint with each node, the function filter filters the intervals with lower bounds larger than ϵ, the function merge computes intervals corresponding to the same node for the conjunctive constraint, the function intersection computes intervals corresponding to the same node for the disjunctive constraint, and the function extract_node(S_I,ϵ) extracts the nodes corresponding to some interval in S_I with the lower bound in the interval smaller than ϵ.

7.4.2 Query Processing on Dirty Data in a Cloud Database

The query on a node is processed by data indices based on columns and corresponding representatives. Then with consideration of ϵ, tuples are retrieved as results.

As the first step, each single constraint c is processed separately on the corresponding data index. The tuples with difference smaller than ϵ are retrieved. The estimation method of the upper and lower bound of the difference is the same as Sect. 7.4.1. When the representatives are obtained, the tuples in the cluster are filtered based on the constraint.

An example is used to illustrate the query processing on a node. To perform the selection query with constraint (age < 40 $\vee$ gender $=$ F)$\wedge$ name $=$ 'Celine Dien' and $\epsilon = 2$ on N_1 with C1 and C2, with the data index, the difference between the constraint and C1 is estimated to be $[\frac{1}{3},\frac{1}{3}]$, while that between the constraint and C2 is $[3,11]$. Therefore, only the tuples in C1 are considered. Based on the computation of the distance, P2 is selected as the final result.

7.5 Experimental Results

To test the efficiency and effectiveness of the techniques proposed in this chapter, experimental results are shown in this section. The test is written in Java 1.6.0.13 and all the experiments run on an IBM Server with 8 virtual machines set up. One virtual machine (node 0) has 3 cores and 1,300M memory while the other 7 (node 1–7) are equipped with 1 core and 380M memory each. In the test, node 0 is the router node, and node 1 is the master node. The other are slave nodes, so 6 slave nodes are used. 100,000 tuples are loaded for per slave node, and the node index is built in router node while data index and representatives are stored in slave nodes.

The algorithms are tested on both real data set and synthetic data set. Information of publications are extracted from different web sites of ACM, Citeseer and DBLP as the real data set. The schema of the data is (author, title, year, conference). The data set has total 1,531,299 tuples, and up to 600,000 tuples from the real data set are used.

For the synthetic data set, a data generator is designed. The generator generates a relation with string, integer, and category as the types of columns. The data in schema (s_1, s_2, i, c) are generated, where s_1 and s_2 are in string type, i is in integer type and c is in category type. The generator generates some tuples with random integer numbers, fixed-length strings and categories as seeds at first. For each seed, some tuples are generated by editing each value several times randomly with operators of character insertion, deletion or conversion or revision. To simulate the situation in real world, the times of changing is in power distribution with a upper bound. The tuples generated from the same seed are considered as referring to the same entity. The data generator has five parameters: the number of seeds (#seed), the number of tuples generated from each seed (#tuple), the upper bound of the times of edit for the generation of s_1 and s_2 in each tuple (#edit), the max difference of the values of integer type (#diff). The default setting of the generator is #seed = 500, #tuple = 4,096, #col = 10, #edit = 4 and #diff = 5.

All indices and algorithms presented in this chapter are implemented. Recall and precision are used to measure the effectiveness of the algorithms and throughput is the metric to test the efficiency of the algorithms.

7.5.1 Experimental Results on Real Data

Experiments are performed on real data to test the effectiveness and efficiency of the proposed algorithms. Queries of different types are performed on the data set, including selection on string, number, and category attributes. The experiments on efficiency for the queries with string, number, and category constraints are shown in Figs. 7.2–7.4, respectively.

In this test, 1 router, 1 master, and 2,3,4,5,6 slave nodes are used. Index files and data files are generated offline and loaded into corresponding nodes when the system

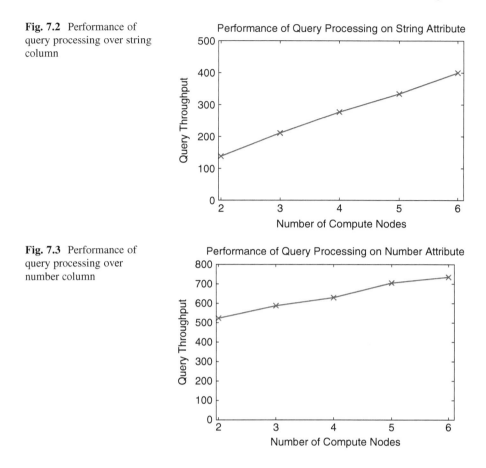

Fig. 7.2 Performance of query processing over string column

Fig. 7.3 Performance of query processing over number column

starts. After all nodes are ready, a query is injected into each slave node and query processing starts running. After a slave node finishes to process a query, it fetches another query to process from the master node, and the master node records the number of processed queries. Query processing is tested for 150s and the average value of throughput is used as the final result.

The results show that the throughput grows when the number of slave nodes increases. Among the tests, the string query achieves the best scalability. The throughput of the number query and the category query grow slower with the number of nodes. The reason is that the string query processing is cpu-intensive. It means that the computation on the q-gram index dominates the total running time. The number and the category query processing cost much less during searching index and local data; so network communication takes up more time. In the experiment environment, all virtual nodes share one message on the same network interface of the server. This makes network communication slower than

Fig. 7.4 Performance of query processing over category column

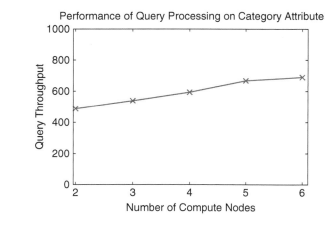

communications among individual machines. Hence, the number query and the category query processing do not scale very well. On the other hand, string query processing scales very well because cpu computation takes up more time, and communication is not the bottleneck.

The precision and recall are always 100%. The reason is that the tuples in the three data sources referring to the same entity are similar. Based on the index structure, all possible clusters are always found. Therefore, the recall is assured. By the filtering in the cluster, the precise results can be found.

7.5.2 Experimental Results on Synthetic Data

To test the algorithms with more flexible setting, experiments are performed on synthetic data. Since for the efficiency the trends are the same as those of real data, this section focuses on the effectiveness.

To test the impact of data variance on the effectiveness and efficiency, the maximal edit distances of strings are varied from 0 to 20, and edit distance threshold is set to be 2. The precision is kept 1.0 and the recall is shown in Fig. 7.6. From the experimental results, it is observed that with the increase in maximal edit distance, the recall decreases slightly. It is because that the larger the difference in values is, the more difficult it is to distinguish the tuples referring to the same clean tuple.

To test the efficiency, data with 60,000 clean tuples and ten dirty tuples for each clean tuple are generated. The same experiment setting as that of real data is used. Each slave node loads 100,000 dirty tuples. Figures 7.5–7.7, number query and category query processing have quite similar throughput with test on real data. The only difference is that string query processing has greater throughput than that of real data. The reason is that each clean data has ten dirty tuples in synthetic data, so the router node stores less node indices. This makes index searching in the router node faster, thus increasing the throughput.

Fig. 7.5 Performance of
query processing over string
column

Fig. 7.6 Recall vs. max edit
distance

Fig. 7.7 Performance of
query processing over
number column

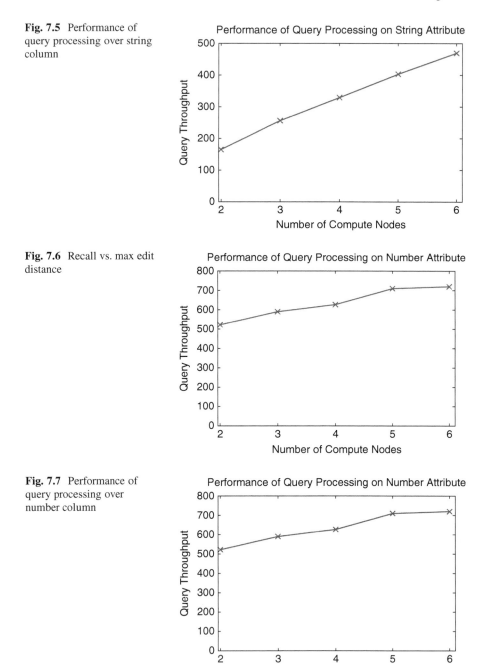

Fig. 7.8 Performance of
query processing over
category column

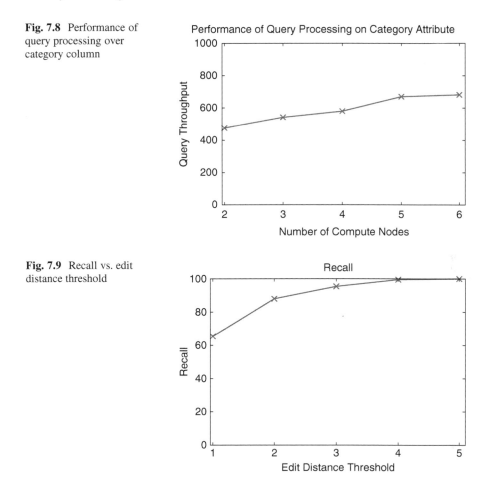

Fig. 7.9 Recall vs. edit
distance threshold

To test the effect of the difference threshold, the thresholds are varied from 1 to 5 and the maximal edit distance of string is set to be 5. The results of recall is shown in Fig. 7.9. It is observed from the results that the recalls increase with the threshold. It is because that when the threshold is smaller, the constraint becomes tighter and some tuples with large distance from the clean tuple will be neglected.

7.6   Related Work

Internet companies have developed distributed storage systems to manage large amounts of data, such as Google's GFS [15], which supports Google's applications. BigTable [16] is a distributed storage model for managing Google's structured data. While such systems are not available for researchers, there are some open source implementations of GFS and BigTable such as HDFS [17], HBase and HyperTable, which are good platforms for research and development. Yahoo proposed PNUTS [18], a hosted, centrally controlled parallel and distributed database system for Yahoo's applications. In these systems, data are organized into chunks, and then randomly disseminated among clusters to improve data access parallelism. Some central master servers are responsible for guiding queries to nodes which hold query results. Amazon's Dynamo [19] is a readily available key-value store based on geographical replication, and it can provide eventual consistency. Each cluster in Dynamo organizes nodes into a ring structure, which uses consistent hashing to retrieve data items. Consistent hashing is designed to support key-based data retrieval. Commercial distributed storage systems such as Amazon's S3 (Simple Storage System) and Microsoft's CloudDB tend to have little implementation details published. Some other systems such as Ceph [20], Sinfonia [21], etc. are designed to store objects, and they aim to provide high performance in object-based retrievals instead of set-based database retrieval. MapReduce [22] was proposed to process large datasets disseminated among clusters. MapReduce assigns mappers and reducers to process tasks, where mappers produce intermediate results in parallel and reducers pull the intermediate results from mappers to do aggregations. Recent work, such as [23] and [24], attempt to integrate MapReduce into database systems.

In another aspect, because of its importance, data quality becomes one of the hot spots in database research. However, most of the techniques focus on data cleaning [3], which is to remove the dirty part from the data. However, for massive and frequently updating data, data cleaning is not feasible because of its inefficiency and the loss of information. The authors of [4–6] process queries directly on inconsistent data and eliminate inconsistent part in the final results. These methods are only designed for single computer and not suitable for massive dirty data management in cloud environments.

Existing systems do not provide functionality to manage dirty data in cloud environments, thus building dirty data management system in cloud platform is a necessary complementary work. The work in this chapter stands on this point.

7.7   Conclusions

In many applications, data quality problem results in severe fails and loss. Data quality problems are caused by dirty data. Massive data sets have larger probability to contain dirty data. In cloud database as a promising system for massive data management, dirty data management brings technical challenges. Traditional methods to process dirty data by cleaning cannot clean dirty data entirely and often results in decreasing the efficiency. Therefore, in this chapter, dirty data techniques are designed for cloud databases to keep dirty data in the database and obtain the query results with quality assurance on the dirty data. To manage dirty data in cloud databases efficiently and effectively, a data storage structure for dirty data management in cloud database is presented. Based on the storage structure, a three-level index for the query processing is proposed. Based on the index, efficient query processing techniques of finding the proper nodes for query processing and query processing techniques on single nodes are presented. Experimental results show that the presented methods provide effectiveness and efficiency, and thus they are practical in cloud databases.

Acknowledgements This research is partially supported by National Science Foundation of China (No. 61003046), the NSFC-RGC of China (No. 60831160525), National Grant of High Technology 863 Program of China (No. 2009AA01Z149), Key Program of the National Natural Science Foundation of China (No. 60933001), National Postdoctoral Foundation of China (No. 20090450126, No. 201003447), Doctoral Fund of Ministry of Education of China (No. 20102302120054), Postdoctoral Foundation of Heilongjiang Province (No. LBH-Z09109), and Development Program for Outstanding Young Teachers in Harbin Institute of Technology (No. HITQNJS.2009.052).

References

1. Eckerson, W.W.: Xml for analysis specification. Technical Report, The Data Warehousing Institute. http://www.tdwi.org/research/display.aspx?ID=6064, 2002
2. Raman, A., DeHoratius, N., Ton, Z.: Execution: The missing link in retail operations. Calif. Manag. Rev. **43**(3), 136–152 (2001)
3. Rahm, E., Do, H.H.: Data cleaning: Problems and current approaches. IEEE Data Eng. Bull. **23**(4), 3–13 (2000)
4. Fuxman, A., Miller, R.J.: First-order query rewriting for inconsistent databases. In: ICDT, pp. 337–351 (2005)
5. Fuxman, A., Fazli, E., Miller, R.J.: Conquer: Efficient management of inconsistent databases. In: SIGMOD Conference, pp. 155–166 (2005)
6. Andritsos, P., Fuxman, A., Miller, R.J.: Clean answers over dirty databases: A probabilistic approach. In: ICDE, p. 30 (2006)
7. Garcia-Molina, H., Ullman, J.D., Widom, J.: Database system implementation. Prentice-Hall, NJ (2000)
8. Li, C., Wang, B., Yang, X.: Vgram: Improving performance of approximate queries on string collections using variable-length grams. In: VLDB, pp. 303–314 (2007)
9. Baeza-Yates, R.A., Ribeiro-Neto, B.A.: Modern information retrieval. ACM, NY (1999)
10. Cormen, T.H., Leiserson, C.E., Rivest, R.L., Stein, C.: Introduction to algorithms, 2nd edn. MIT, MA (2001)
11. Fredkin, E.: Trie memory. Commun. ACM **3**(9), 490–499 (1960)
12. Schaeffer, S.E.: Graph clustering. Comp. Sci. Rev. **1**(1), 27–64 (2007)
13. Sarawagi, S. , Kirpal, A.: Efficient set joins on similarity predicates. In: SIGMOD Conference, pp. 743–754 (2004)
14. Xiao, C., Wang, W., Lin, X., Yu, J.X.: Efficient similarity joins for near duplicate detection. In: WWW, pp. 131–140 (2008)
15. Ghemawat, S., Gobioff, H., Leung, S.-T.: The Google file system. In: SOSP 2003, pp. 29–43
16. Chang, F., Dean, J., Ghemawat, S., Hsieh, W.C., Wallach, D.A., Burrows, M., Chandra, T., Fikes, A., Gruber, R.E.: Bigtable: a distributed storage system for structured data. ACM Trans. Comput. Syst. **26**(2) (2008)
17. Apache Hadoop http://hadoop.apache.org/

18. Cooper, B.F., Ramakrishnan, R., Srivastava, U., Silberstein, A., Bohannon, P., Jacobsen, H.A.: PNUTS: Yahoo!'s hosted data serving platform. PVLDB 1(2), 1277–1288 (2008)
19. DeCandia, G., Hastorun, D., Jampani, M., Kakulapati, G., Lakshman, A., Pilchin, A., Sivasubramanian, S., Vosshall, P., Vogels,W.: Dynamo: Amazon's highly available key-value store. In: SIGOPS, pp. 205–220 (2007)
20. Weil, S.A., Brandt, S.A., Miller, E.L., Long, D.D.E.: Ceph: a scalable, high-performance distributed file system. In: SODI, pp. 307–320 (2006)
21. Aguilera, M.K., Merchant, A., Shah, M., Veitch, A., Karamanolis, C.: Sinfonia: A new paradigm for building scalable distributed systems. In: SOSP 2007
22. Dean, J., Ghemawat, S.: Mapreduce: Simplified data processing on large clusters. In: OSDI 2004
23. Yang, H.-C., Dasdan, A., Hsiao, R.-L., Parker, D.S.: Map-reduce-merge: Simplified relational data processing on large clusters. In: SIGMOD, pp. 1029–1040 (2007)
24. Abouzeid, A., Bajda-Pawlikowski, K., Abadi, D.J., Rasin, A., Silberschatz, A.: Hadoopdb: An architectural hybrid of mapreduce and dbms technologies for analytical workloads. PVLDB 2(1), 922–933 (2009)

Chapter 8
Virtualization and Column-Oriented Database Systems

Ilia Petrov, Vyacheslav Polonskyy, and Alejandro Buchmann

Abstract Cloud and grid computing requires new data management paradigms, new data models, systems, and capabilities. Data-intensive systems such as column-oriented database systems will play an important role in cloud data management besides traditional databases. This chapter examines column-oriented databases in virtual environments and provides evidence that they can benefit from virtualization in cloud and grid computing scenarios. The major contributions involve: (1) the experimental results show that column-oriented databases are a good fit for cloud and grid computing; (2) it is demonstrated that they offer acceptable performance and response times, as well as better usage of virtual resources. Especially for high selectivity, CPU- and memory-intensive join queries virtual performance is better than nonvirtualized performance; (3) the performance data shows that in virtual environments they make good use of parallelism and have better support for clustering (parallel execution on multiple clustered VMs is faster than on a single VM with equal resources) due to data model, read-mostly data optimizations and hypervisor-level optimizations; and (4) analysis of the architectural and system underpinning contributing to these results.

8.1 Introduction

Cloud applications need semantic metadata for the purposes of composition, analysis, discovery, and correlation. Such applications comprise a self-describing set of services or are themselves appearing as a service. To allow easy composition and reuse, cloud applications and their interfaces should be exposed as well-described entities. Cloud applications, on the contrary, operate on large heterogeneous data

I. Petrov (✉) · V. Polonskyy · A. Buchmann
Databases and Distributed Systems Group, TU-Darmstadt, Germany
e-mail: ilia.petrov@dvs.tu-darmstadt.de; polonskyy@dvs.tu-darmstadt.de;
buchman@dvs.tu-darmstadt.de

S. Fiore and G. Aloisio (eds.), *Grid and Cloud Database Management*,
DOI 10.1007/978-3-642-20045-8_8, © Springer-Verlag Berlin Heidelberg 2011

sets, comprising a wide range of data types (e.g., semistructured data, documents, or multimedia data) and loose typing. Analyzing these data properly (especially in the user context), discovering related data and correlating it as a major task that is very metadata intensive. The Semantic Web initiative from W3C [36] is therefore indispensable. In addition, progressively more Semantic Web data is being published on the Web. Some very prominent examples are: Princeton University's WordNet (a lexical database for the English language) [35]; and the online Semantic Web search tool Swoogle [32] that contains 3,316,943 Semantic Web documents and over 1×10^{10} triples at the time of the submission of this paper. In addition, there are several Semantic Web Query language proposals [27, 30].

The database community has investigated different approaches to handle and query semantic data. Some examples are: XML stores, relational databases, and column-oriented databases. Several promising approaches have been developed to store and process RDF data in relational systems [10, 12, 16, 34]. Recently, the influence of partitioning and different schemata on relational RDF stores is investigated [2, 29]. They also proposed using native column-oriented databases for RDF data and investigated their performance advantages. In this paper, the approach proposed by [2, 29] is assumed to investigate the influence of virtualization on the original assumptions in terms of benchmark, dataset, read-mostly behavior, and system selection. Conclusions are drawn and the applicability on such scenarios on the elastic cloud model is investigated.

There are also multiple research proposals to use RDF semantic data together with cloud technologies such as MapReduce [14] or Hadoop [18, 20]. However, these mainly focus on braking up the RDF set of triples into smaller chunks to minimize expensive processing of large tables. Column stores due to mechanisms such as compression and late materialization (Sect. 8.2) can provide very good performance for a significantly wider range of query types. The approach under investigation is semantic data management in cluster environments or in cloud scenarios based on the Amazon EC2 model (which offers virtual computing facilities placed on the Amazon Cloud infrastructure). Interestingly enough, it can also be realized in clustered virtualized environments, where the user installs its virtual machines and the cluster infrastructure takes care of distributing them. The users book virtual resources and pay for their use, and a certain quality of service level.

The goal of this chapter is to examine the behavior of column stores storing semantic data and running semantic queries in virtual environments and prove that they can benefit from virtualization in cloud and grid computing scenarios. The major contributions involve:

1. The performance data shows that column-oriented databases are good fit for cloud and grid computing due to their data model and support for clustering;
2. It is demonstrated that column-oriented databases offer acceptable performance and response times, as well as better usage of virtual resources. Especially for high selectivity, CPU-intensive join queries virtual performance is better than nonvirtualized performance;

3. The experimental results show that in virtual environments they make good use of parallelism and support better clustering (parallel execution on multiple clustered VMs is faster than on a single VM with equal resources) due to data model, read-mostly data optimizations, and hypervisor layer optimizations;
4. An analysis of the architectural and system underpinning contributing to these results.

8.1.1 Data-Model, Requirements

The Semantic Web data model, called "Resource Description Framework" [25], views the data as a set of statements containing related information about resources. The complete set of statements can be formally represented as a graph, in which the nodes are resources or simple literals/values, and the edges are the so-called properties or predicates. Every statement is a triple of <"subject","predicate/property", "object">. (Possible ways of storing them in a database system are described in Sect. 6.3.1). The information stored in the RDF graph can be structured in many different ways. To promote expressing strict schemas that will standardize descriptions within certain domains, W3C introduced a set of widespread standards such as RDFS [26] and OWL [24]. These abstract descriptions have to be processed by tools. Concretely, most of the semantic data is represented in XML using a standardized RDF/XML binding. The present data model, although being very general and allowing for flexibility, can potentially cause significant performance degradation.

8.1.2 Usecases and Scenarios

The authors believe that semantic metadata will be used for tasks such as: (1) exploration of a set (quite possibly graphically), (2) as an entry point into data exploration – discovery; (3) for inference and data set correlation. The list of scenarios is not exhaustive and quite possibly scenarios that may become relevant in the near future are missing. The listed ones, however, have many features in common. These execute multiple concurrent queries are mostly interactive. Many queries – especially the discovery ones – are property-centered, requiring heavy metadata usage as well as association navigation. Interestingly enough, most of those queries are reflected by the benchmark introduced in [2] and summarized in Sect. 6.3.1.

On the contrary, a direct consequence of the chosen cloud computing model is that the semantic data set will be spread across multiple virtual machines. A central question to be answered in this paper is how are these to be configured.

8.2   Column-Oriented Database Systems

Column-oriented database systems (or shortly "column stores") are data storage systems that store information column-wise in contrast to traditional row stores that store information row-wise in tuples. The column-wise storage of data leans upon the fully decomposed data model that has been first introduced in [13]. Column stores are said to perform an order of magnitude better on analytical loads.

At present, there are several column store products available from industry (MonetDB/X100 [9] Sybase IQ, SAP BI Accelerator, Vertica, etc.) and academia (C-Store [31], MonetDB [6, 23].

Figure 8.1 presents an example of how data is stored in column stores. The table Triples contains semantic triples in their relational representation. In column stores, every column is stored separately as separate binary table. These "column" tables are called Binary Association Tables (BAT) in MonetDB terminology. Therefore, separate BATs are present for each column (one for the Subject column, another one for the Predicate column of the Triples table and so on).

The structure of every BAT is simple and strictly defined. BATs contain two columns (called Head (H) and Tail (T) in MonetDB). The first column (the Head) contains artificial keys. The second column (the Tail) contains the respective value. To preserve the correspondence of the respective values in the different BATs, the artificial key in all entries belonging together are identical. For example, the attribute values of triple <s2, p1, o1> from the Triples table all have the same artificial key (002) in the respective BATs. This mechanism allows for seamless reconstruction of the original rows.

Several authors, among others Abadi et al. and Harizopoulos et al. [1, 15], have compared the advantages of column stores over traditional row-wise storage, for

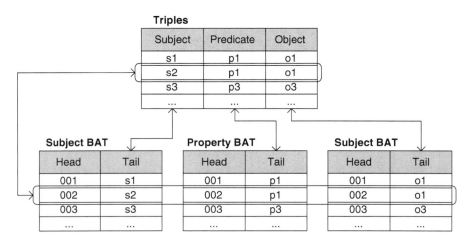

Fig. 8.1 Binary table decomposition in column stores

example, relational database systems. Abadi et al. [1] summarizes the advantages column-stores offer for data warehousing and analytical workloads:

1. Compression – compression on the column values (e.g., RLE) is a key strength of column stores. It is extensively used to minimize the IO and memory consumption increasing the performance by a factor of 10 [1]. In Fig. 8.1, RLE can be seen in the Object and Predicate BATs: the entries for artificial IDs 001 and 002 can be encoded by the respective value and the number of occurrences, for example, p1, 2x, instead of p1, p1.
2. Projection – is performed at no cost; in addition, only the relevant columns are read thus minimizing the IO costs;
3. Aggregation performance is better on column stores;
4. Higher IO through column partitioning – column data can be stored in different files decreasing contention and increasing IO parallelism;
5. Late materialization – column data is federated in row-form as late as possible in the query execution plan yielding performance increase of a factor of 3;
6. Block iteration – multiple column values are passed as a block yielding performance increase of a factor of 1.5.

Harizopoulos et al. and Holloway and DeWitt [15, 17] in addition investigate a series of optimizations that can be performed in read-mostly database systems such as column-stores. These range from the operator structure and efficient methods for processing intermediate results to the physical page format. In addition, [19] proposes adaptive segmentation and adaptive replication mechanisms to partition data or reuse query results. The bottom line is that using column store appropriately can provide up to $10\times$ better performance. The fact that they are appropriate for handling semantic data has already been proven by [2, 29].

Even though column stores offer best performance with read-only loads, they need to handle updates. The best way to update column store data is in bulk mode. The scenarios presented in Sect. 6.1.2 fit well since Semantic Web data are first crawled, indexed, and prepared for the column store update. Next, the bulk update can be performed simultaneously, which is a comparatively fast operation. It can be performed at predefined time slots.

8.3 Performance Influence of Virtualization on Column-Oriented Databases

This chapter introduces the proposed experimental approach, the benchmark, and the system used. The experimental search space and experimental design are described together with the attained results and their analysis.

8.3.1 Benchmark

The performance of column-oriented database systems depends strongly on the chosen schema and the type of partitioning. Column-stores support different ways of organizing the data into tables, which are then represented in the column format. As mentioned earlier, two alternative schema designs are comparatively examined: a single triples table and a vertically partitioned schema. Since the data has a read-mostly character sorting/partitioning it according to different criteria can significantly increase the performance. The benchmark and dataset used in [29] are reused; the experimental analysis is partially based on the findings provided in that paper. Three partitioning types (PSO, SPO, and Vertical (Vert) partitioning) are examined further below. As the name implies, PSO yields sorting the table first according to the Properties/Predicates, next according to the Subjects, and finally according to the Objects (Fig. 8.2). Alternatively, SPO implies: Subjects first, Properties second, and Objects third (Fig. 8.2).

Vertical partitioning (Fig. 8.3) creates a single table for each distinct property value. The table contains the respective values for the subjects and objects of all triples containing the respective property.

The benchmark as proposed by [2] and implemented by [29] comprises eight queries (all of which are listed in Table 8.1). The sequence of queries is designed to assist the graphical exploration of RDF datasets based on the Logwell tool [21]. The benchmark is fully described in [2], here a brief summary is provided.

The user entry point is a list of all values of the *type* property (such as *Text* or *Notated Music*) – Query 1 (Q1). Upon a selection, a query is executed to retrieve all subject values associated with that subject together with the popularity/frequency of

	Raw Data			PSO			SPO	
Subj.	Prop.	Obj.	Subj.	Prop.	Obj.	Subj.	Prop.	Obj.
ID5	type	CDType	ID2	artist	"Orr, Tim"	ID1	author	"Fox, Joe"
ID5	title	"GHI"	ID1	author	"Fox, Joe"	ID1	copyright	"2001"
ID5	copyright	"1995"	ID1	copyright	"2001"	ID1	title	"XYZ"
ID1	copyright	"2001"	ID2	copyright	"1985"	ID1	type	BookType
ID2	type	CDType	ID5	copyright	"1995"	ID2	artist	"Orr, Tim"
ID2	title	"ABC"	ID6	copyright	"2004"	ID2	copyright	"1985"
ID3	language	"English"	ID2	language	"French"	ID2	language	"French"
ID4	type	DVDType	ID3	language	"English"	ID2	title	"ABC"
ID4	title	"DEF"	ID1	title	"XYZ"	ID2	type	CDType
ID6	type	BookType	ID2	title	"ABC"	ID3	language	"English"
ID6	copyright	"2004"	ID3	title	"MNO"	ID3	title	"MNO"
ID1	type	BookType	ID4	title	"DEF"	ID3	type	BookType
ID1	title	"XYZ"	ID5	title	"GHI"	ID4	title	"DEF"
ID1	author	"Fox, Joe"	ID1	type	BookType	ID4	type	DVDType
ID2	artist	"Orr, Tim"	ID2	type	CDType	ID5	copyright	"1995"
ID2	copyright	"1985"	ID3	type	BookType	ID5	title	"GHI"
ID2	language	"French"	ID4	type	DVDType	ID5	type	CDType
ID3	type	BookType	ID5	type	CDType	ID6	copyright	"2004"
ID3	title	"MNO"	ID6	type	BookType	ID6	type	BookType

Fig. 8.2 PSO and SPO partitioning and nonpartitioned Triples table

Triples

Subj.	Prop.	Obj.
ID5	type	CDType
ID5	title	"GHI"
ID5	copyright	"1995"
ID1	copyright	"2001"
ID2	type	CDType
ID2	title	"ABC"
ID3	language	"English"
ID4	type	DVDType
ID4	title	"DEF"
ID6	type	BookType
ID6	copyright	"2004"
ID1	type	BookType
ID1	title	"XYZ"
ID1	author	"Fox, Joe"
ID2	artist	"Orr, Tim"
ID2	copyright	"1985"
ID2	language	"French"
ID3	type	BookType
ID3	title	"MNO"

Vertical Partitioning

ARTIST	
Subj.	Obj.
ID2	"Orr, Tim"

AUTHOR	
Subj.	Obj.
ID1	"Fox, Joe"

COPYRIGHT	
Subj.	Obj.
ID1	"2001"
ID2	"1985"
ID5	"1995"
ID6	"2004"

LANGUAGE	
Subj.	Obj.
ID2	"French"
ID3	"English"

TITLE	
Subj.	Obj.
ID1	"XYZ"
ID2	"ABC"
ID3	"MNO"
ID4	"DEF"
ID5	"GHI"

TYPE	
Subj.	Obj.
ID1	BookType
ID2	CDType
ID3	BookType
ID4	DVDType
ID5	CDType
ID6	BookType

Fig. 8.3 Vertical partitioning of the unpartitioned table Triples

that property– Query 2 (Q2). A list of Object filters for each property is also shown. The user can then select a filter producing list of popular frequent objects with high counts – Query 3 (Q3). The user can then select a specific object associated with that property, thus narrowing the selectivity – Query 4 (Q4) is thus similar to Q3. The system then retrieves all corresponding subjects by inference considering the selected property object pairs – Query 5 (Q5) – and updating all other graphical displays with the newly selected information – Query 6 (Q6).

Query 7 (Q7) finally displays more information about a property knowing a subject (termed "end") and a set of related properties. Last but not least [29] extends the benchmark by adding an eighth query allowing to "join" triples on a common subject by knowing a start object and having a nonintersecting subject list. The SQL text of the queries is provided below:

Clearly, queries 1 through 8 are formulated for the single table solutions (unpartitioned triples table or SPO, PSO partitioned triples table). The queries for the vertically partitioned schema will have a UNION form. What all queries have in common is the significant amount of joins. Mostly, self-joins and equi-joins on the Triples table but also joins on the Triples and Properties table. The majority of the queries use aggregation. It should also be noted that Q8 uses negation (B.subj ! = 'conferences') in the section condition, which makes it very IO intensive.

The authors of [29] have performed a complete analysis of the query space reaching the conclusion that the proposed queries offer good coverage: the query types (point, range, aggregation, and join) and alternatives (entry, inference, and discovery) can be considered representative.

Table 8.1 Description of the semantic benchmark queries

```
Q1    SELECT A.obj, count(*) FROM triples AS A WHERE A.prop ='<type>'
      GROUP BY A.obj;
Q2    SELECT B.prop, count(*) FROM triples AS A, triples AS B, properties P
      WHERE A.subj = B.subj AND A.prop = '<type>'
        AND A.obj = '<Text>' AND P.prop = B.prop GROUP BY B.prop;
Q3    SELECT B.prop, B.obj, count(*) FROM triples AS A, triples AS B,
        properties P WHERE A.subj = B.subj AND A.prop = '<type>' AND A.obj =
        '<Text>'
      AND P.prop = B.prop GROUP BY B.prop, B.obj HAVING count(*) > 1;
Q4    SELECT B.prop, B.obj, count(*) FROM triples AS A, triples AS B, triples
        AS C, properties P WHERE A.subj = B.subj AND A.prop = '<type>'
      AND A.obj = '<Text>' AND P.prop = B.prop AND C.subj = B.subj
      AND C.prop = '<language>' AND C.obj = '<language/iso639--2b/fre>'
      GROUP BY B.prop, B.obj HAVING count(*) > 1;
Q5    SELECT B.subj, C.obj FROM triples AS A, triples AS B, triples AS C
      WHERE A.subj = B.subj AND A.prop = '<origin>'
      AND A.obj = '<info:marcorg/DLC>' AND B.prop = '<records>'
      AND B.obj = C.subj AND C.prop ='<type>' AND C.obj !='<Text>';
Q6    SELECT A.prop, count(*) FROM triples AS A, properties P, (
                (SELECT B.subj FROM triples AS B
                 WHERE B.prop = '<type>' AND B.obj = '<Text>'
            )UNION (
                SELECT C.subj FROM triples AS C, triples AS D
                WHERE C.prop = '<records>' AND C.obj = D.subj AND
                D.prop = '<type>' AND D.obj = '<Text>'
            )) AS uniontable
      WHERE A.subj = uniontable.subj AND P.prop = A.prop
      GROUP BY A.prop;
Q7    SELECT A.subj, B.obj, C.obj FROM triples AS A, triples AS B, triples AS C
      WHERE A.prop = '<Point>' AND A.obj = '''end''' AND A.subj = B.subj
      AND B.prop = '<Encoding>' AND A.subj = C.subj AND C.prop ='<type>';
Q8    SELECT B.subj FROM triples AS A, triples AS B WHERE A.subj =
      'conferences' AND B.subj != 'conferences' AND A.obj = B.obj;
```

8.3.2 Dataset

The publicly available Barton Libraries dataset [7] serves as experimental basis in [2, 29] and is used as experimental dataset in this paper to ensure comparability and repeatability. Key characteristics of the data set are summarized in Table 8.2. It comprises approximately 50 million tuples and approximately 18 million classes and literals. An analysis performed by [29] shows that the most frequent property is type, occurring in 24% of all triples and that 13% of the properties account for 99% of all tuples. No indices are created.

8.3.3 Testbed and Experimental Design

The experimental testbed used is depicted in Fig. 8.4. All experiments have been performed on a Sun Fire × 4,440 server, with 64 GB RAM and four quad-core

Table 8.2 Barton dataset characteristics

Characteristic	Value
Total triples	50,255,599
Distinct properties	222
Distinct subjects	12,304,739
Distinct objects	15,817,921
Distinct subjects that are also objects (and vice versa)	9,654,007
Strings in dictionary	18,468,875
Data set size [MB]	1,253

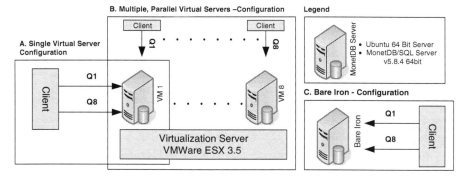

Fig. 8.4 Experimental testbed and experimental configurations

Opteron 8356 2.3 GHz processors. The server uses as host-based storage a RAID 10 array over eight 10,000 RPM SAS drives. As virtualization solution VMWare ESX server 3.5, update 4 is used. In addition, the server has an installation of Ubuntu Server 64-bit to measure the performance of the nonvirtualized system (this configuration is termed "bare iron").

To compare the performance of virtualized column stores the performance on the Bare Iron (BI) configuration (configuration C, BI) is measured first. The bare iron measurements serve as baseline for comparison to other configurations.

The next step is to configure a virtual machine to measure the performance of the virtualized MonetDB. The leading principle is that the virtual machine should have a configuration identical to bare iron, that is that the virtual resources (SUM(CPU), SUM(Memory)) should be equal to the physical resources. The resulting configuration is termed "single virtual server" (configuration A (1 VM), Fig. 8.4).

There is at least another possible virtualized configuration. Consider Configuration B (Multiple Virtual Servers (8VM), Fig. 8.4). Multiple identical virtual machines (one for each benchmark query) are created, under the resource equality principle.

The benchmark is executed in two modes on all three configurations: sequential and parallel. In the sequential mode, every benchmark query is executed alone against the MonetDB database. In the parallel mode, all eight queries are executed together against the benchmark dataset. In Configuration B (multiple virtual servers), all queries execute alone in their own VM; however, they are issued in parallel to the respective VMs, and hence run in parallel. While the sequential execution times give us exact information about the execution time of a query, the parallel execution gives us indirect information about the resource contention and the degree of parallelism.

8.3.4 Experimental Results

The average query execution times are shown in Tables 8.3 and 8.4. Each query is run three times to ensure stable average values through acceptable standard deviation. For every experiment, these tests are performed under "cold" and "hot" conditions. In cold experiments, the virtual machine is restarted after every execution to avoid caching effects, whereas in hot experiments each of the three experimental runs is performed directly after the previous one, thus accounting for caching effects (there is an initial test run to warm up the system; its results are discarded).

Table 8.3 contains the results of the sequential experiments; it illustrates the execution times for each query for the different virtualization and partitioning alternatives. The goal of this experiment is to: (a) study the query behavior in

Table 8.3 Sequential execution times (s) for different queries, hot and cold runs

Hot runs		Q1	Q2	Q3	Q4	Q5	Q6	Q7	Q8	Sum
	BI	0.64	3.95	4.90	3.63	1.42	8.95	0.31	15.22	23.79
	1VM	0.69	4.02	5.01	3.74	1.81	11.30	0.45	25.17	27.02
PSO	8VM	0.70	4.10	5.07	4.19	2.24	17.86	0.41	36.09	34.58
	BI	1.80	2.76	3.86	1.78	3.22	8.95	1.49	13.80	23.86
	1VM	2.00	2.78	3.95	1.88	3.58	11.05	1.87	20.27	27.10
SPO	8VM	2.30	2.88	4.19	2.38	4.81	20.67	12.66	19.05	49.90
	BI	0.64	0.58	0.85	0.62	1.32	2.91	0.26	32.73	7.17
	1VM	0.68	0.58	0.87	0.66	1.65	3.34	0.39	41.73	8.18
Vert	8VM	1.90	2.32	2.64	2.12	2.37	6.87	0.36	39.78	18.58
Cold Runs										
	BI	1.08	9.04	4.93	3.63	1.45	9.43	0.32	15.79	29.87
	1VM	1.13	14.19	5.16	3.74	1.80	12.17	0.50	25.54	38.68
PSO	8VM	40.01	42.35	43.27	29.77	55.70	26.54	66.16	66.16	303.8
	BI	3.22	4.60	3.81	2.35	4.04	8.86	1.46	13.66	28.35
	1VM	4.76	4.67	4.14	3.36	5.34	11.12	1.91	21.07	35.30
SPO	8VM	4.42	19.45	20.48	28.53	48.85	50.21	58.41	35.82	230.4
	BI	0.94	2.60	1.54	0.65	1.37	2.94	0.32	35.51	10.36
	1VM	1.02	2.61	1.58	0.67	1.65	3.56	0.46	45.02	11.55
Vert	8VM	1.22	10.00	12.61	12.73	9.28	17.04	4.64	61.24	67.51

Table 8.4 Parallel execution times (s) for different queries, hot and cold runs

Hot runs		Q1	Q2	Q3	Q4	Q5	Q6	Q7	Q8
	BI	1.05	4.63	9.97	9.94	7.99	15.54	5.23	21.23
	1VM	0.90	7.98	8.32	10.57	11.93	26.35	6.04	46.40
PSO	8VM	0.69	4.05	5.21	4.28	3.15	26.40	0.37	41.60
	BI	2.41	3.36	5.17	7.86	8.02	14.07	3.88	16.53
	1VM	3.75	7.40	9.32	10.36	17.36	23.75	6.24	26.55
SPO	8VM	1.97	2.97	4.26	2.40	7.42	24.32	5.90	18.17
	BI	0.74	2.93	2.74	2.05	3.80	6.25	1.63	35.60
	1VM	1.11	3.86	5.55	5.67	4.60	8.75	0.79	48.57
Vert	8VM	0.69	2.06	2.68	2.19	1.87	6.93	0.30	39.12
Cold runs									
	BI	5.26	12.70	14.24	13.48	5.73	20.10	9.25	18.44
	1VM	13.52	22.37	28.30	27.11	26.34	36.25	22.93	33.94
PSO	8VM	6.49	172.21	168.4	175.1	138.7	161.2	259.0	185.5
	BI	9.79	11.68	10.96	9.87	10.38	19.90	8.78	15.37
	1VM	15.15	16.32	18.30	17.56	14.49	29.41	19.78	25.47
SPO	8VM	16.43	81.81	70.82	81.03	127.3	129.7	162.8	76.11
	BI	1.99	6.56	7.09	6.73	3.76	10.06	5.74	39.70
	1VM	2.11	7.33	8.18	7.35	5.67	12.54	5.83	51.83
Vert	8VM	2.23	23.52	33.21	27.31	13.65	31.54	30.96	79.03

isolation; (b) study the effect of partitioning; (c) estimate the overhead incurred through virtualization; and (d) be able to compare all those. The bare iron results for all data partitioning alternatives are already described and examined in detail in [2, 29]. Here, these are reported since they serve as baseline for comparison and because the numbers from [2, 29] cannot be used directly since those were obtained on different testbeds (hardware, OS). The performance differences especially to [29] are due to different hardware. The Bare Iron configuration is instrumented with significantly more memory and CPU resources than [29] to keep the resource equality conditions (and be able to compare to virtualized configurations).

By considering the hot runs, it becomes evident that the vertically partitioned schema yields best results except for the extremely IO-intensive Q8. Comparing the Bare Iron (BI) results for the different partitioning schemes, the influence of data partitioning on the query execution can be clearly seen. The authors of [29] provide detailed analysis, but the major reason is that the different join, and sort algorithms used are faster depending on the used columns and the way they are presorted.

Based on the experimental data in Table 8.3, the difference between the execution times of BI and the single virtual machine for queries Q1 through Q7 is not as significant. The column Sum indicates the sum of these execution times, that is SUM(Q1..Q7). Clearly, vertical partitioning not only provides the best performance, but also the smallest difference between bare iron and configuration A (single virtual machine). In conclusion, virtualization is a very viable option for column stores.

As already mentioned, Q8 represents a special, very IO-intensive case due to the negation, which requires special attention since IO is a major bottleneck in virtualization scenarios. The hot experiments offer such a good performance because they make intensive use of caching. Once the data are not cached along the IO chain, the significance of IO as bottleneck becomes evident. This is why all experiments are repeated in cold state: before a query is executed, the server or the whole virtual machine is restarted to eliminate the influence of caching completely. In all virtualized scenarios, the performance decrease is significant. Furthermore, it should also be mentioned that cold runs represent a fairly exotic case, since production systems are running with high levels of availability; hence, absence of cached data is considered infrequent.

The results of the parallel query execution experiment are shown in Table 8.4. In this experiment, all eight queries are executed together. In the BI and 1VM configurations, all queries are executed against a single MonetDB database. In the 8VM configuration, each of the eight queries is executed against its own column database, running in a separate virtual machine; all virtual machines are running simultaneously.

The query execution times from the hot runs show an interesting general trend: most of the queries execute faster in the 8VM configuration than on Bare Iron. *This statement forms one of the central conclusions of this paper.*

The main cause for this result is that hardware resources cannot compensate for the concurrency on database level. Since these are hot runs, it can be assumed that the majority of the required data is cached in memory (however, not all – a point proven by Q8). In the 8VM, therefore, with high probability, a repeated execution consumes data already in the MonetDB cache. Given the present type of read-only load and the dedicated execution (one query per VM), the buffer is reused for the next run. In the other two configurations BI and 1VM, all queries execute together against the database. Even though these configurations have the same hardware resources as the 8VM configuration, the results differ. Some of the factors contributing to this result are: processing of intermediate results, buffer management, multi-threading, and IO-management.

Most of the queries rely on aggregation and self-joins. These operations produce significant amount of intermediate results, which cannot be handled efficiently altogether by MonetDB.

The buffer manager implements a replacement strategy holding certain pages in the memory and replacing them for others when needed. Least-recently used (LRU) is widely spread and used by MonetDB Server. There is natural variance among the query execution times of the eight queries. Due to parallel execution, the LRU strategy replaces pages needed by already executed queries by pages required by the current queries. Therefore, effect of page locality cannot be utilized as efficiently as in the 8VM configuration. The reader should also consider the fact that the experiments are designed in such a way that in all configurations there is enough buffer space to cache the whole database. Therefore, there is essentially no need to replace cached pages.

The parallel execution of all queries in the 1VM and BI configurations relies on multi-threading. Most queries are CPU intensive. Even with the hypervisor CPU overhead, the 8VM execution is faster than the BI execution. This effect can only be attributed to latching (thread synchronization) causing threads to wait and use CPU inefficiently.

Last but not least, the IO-management on hypervisor level offers some interesting optimizations, which speed up the parallel VM execution, but do not affect the single VM scenarios. [28, 33] describe hypervisor level optimizations based on efficient IO command queuing on hypervisor level. These only engage when there are multiple concurrent VMs. These contribute to the better 8VM configuration performance.

Since the IO is a major bottleneck in virtualized scenarios, the cold runs (Table 8.4) of the parallel execution show execution time that are significantly worse than the bare iron. Clearly, IO is the dominant factor. In production systems, however, cold execution is a rare case.

8.3.5 Influence of Data Partitioning

As described in the previous section, the choice of data partitioning has a significant influence on the query execution times. Partitioning the dataset with respect to different criteria is plausible from a performance point of view due to the read-only nature of the data-set reported among others by [2, 29]. The evaluation of some queries is heavily property-based, of other queries subject- or object-based. Partitioning the data with respect to subjects, object or predicates favors different types of queries. Hence, this is a very important design-time decision.

Query 7 is a very good example in this respect. Its execution times vary highly depending on the chosen type of data partitioning and the type of virtualization configuration. A detailed investigation is provided by examining the query execution plan and the costs for the different operators in different configurations.

Query 7 (Q7) is a triple selection self join query without aggregation. It retrieves information about a particular property (e.g., Point) by selecting subjects and objects of triples associated with Point through related properties. Associated means: to retrieve subject, Encoding, and Type of all resources with a Point value of "end." The result set indicates that all such resources are of the type Date, which is why they can have "start" and "end" values: each of these resources represents a start or end date, depending on the value of Point.

A condensed version of Q7's execution plan is shown in Table 8.5. The column "operation" describes the operators and the operator tree. The number, for example "_1" indicate intermediate results defined as BAT [8]. MonetDB defines operator tree in terms of the MIL language [8]. While the semantics of operators such as *join* and *semijoin* is intuitive, an operator such as *uselect* requires a short explanation. The *uselect* operator selects values from the respective BAT. For example, the result of the operation *uselect(property* = *"Encoding")* will be the BAT _5 containing all triples with a property value *Encoding*.

Table 8.5 Query execution plan and operator run-times (ms) for different configurations

Operation	SPO	SPO VM	PSO	PSO VM
_1: uselect(object="End")	24.2	34.4	3.4	3.6
_2: semijoin(property, _1)	11.2	12.1	3.3	3.4
_3: uselect(_2="Point")	3.3	3.3	3.1	3.3
_4: join(_3, subject)	4.1	4.5	1	1
_5: uselect(property="Encoding")	488.1	492.8	0.02	0.03
_6: join(_5, subject)	65.9	61.4	13.1	14.5
_7: join(_4, _6)	112.1	115	17.1	17.5
_8: join(_4, _7)	1.1	1.1	0.02	0.02
_9: uselect(property="type")	931.1	1, 109.1	0.02	0.02
_10: join(_9, subject)	135.1	167.2	54.8	56.9
_11: join(_8, _10)	42.9	44.8	39.5	41.1
_12: join(_11, _8)	0.5	0.5	0.02	0.03
_13: join(_5, object)	57.9	61.3	0.01	0.02
_14: join(_7, _13)	1.9	2	1.9	2
_15: join(_11, _14)	0.02	0.03	0.02	0.02
_16: join(_9, object)	143.2	146.9	0.01	0.01
_17: join(_11, _16)	3.4	3.7	2.7	2.9

The influence of data partitioning is clearly visible comparing the execution times for the *uselect* operators, for example, the ones computing intermediate results _5 and _9 but also on some of the *join* and *semijoin* operators, for example, _2, _10, _13, _16, all those operators rely heavily on property data. SPO partitioning sorts the triples with respect to the subjects; hence, the evaluation requires significant IO resulting in higher times. PSO sorts the triples data with respect to (1) properties and (2) subjects, yielding significantly higher performance. The creation of intermediate BATs _5 and _9 results in a large sequential read on PSO and in multiple small reads on SPO. In addition, internal column store optimizations such as run-length encoding on the property values are inefficient in SPO, but have significant performance influence on PSO.

Here, the plan for vertical partitioning is left out because it is very different due to the different schema. It relies on many union operations. The execution is very fast since MonetDB can select from the tables for the respective properties (e.g., *End, Point, Encoding, Type* see Sect. 6.3.1). Even though it shows very high performance, some scalability issues may arise when the number of unique properties increases.

Since column-stores are commonly viewed as read-mostly data stores, the data partitioning decision is a viable one; it has no negative implications due to the absence of modification operations. Since most of the queries are also known at design-time, such an optimization can be performed statically and increase performance.

8.4 Conclusions: Column-Oriented Databases' Parallelism in Virtual Environments

Cloud computing and semantic data represent an important combination [22]. Some existing semantic applications rely on cloud approaches (Google MapReduce [14], BigTable [11] or open source analogs, for example, Hadoop [3] or extensions such as Pig [5] or HBase [4]). The authors believe that column stores in virtual environments are an important addition. Assuming that the complete dataset is replicated at each node, its scale out can be improved to meet that of cloud models such as Amazon EC2.

As already concluded in Sect. 6.3.3, parallelism is very advantageous with column-stores in virtual environments. Running on multiple virtual machines in parallel increases significantly the overall performance. This is due to several factors ranging from column store properties through hypervisor level optimizations. Those were described in Sect. 6.3.

Due to the flexibility of virtual resources, the hypervisor can even offer the possibility of dynamically migrating virtual machines for one host to another whenever there are not enough physical resources to ensure certain quality of service. Such scenarios are very relevant in the realm of grid, cluster, and cloud computing.

In the elastic cloud or cluster computing scenario, parallelism is naturally given and required. The present experiments prove that virtual computers with less virtual resources can achieve better performance running in parallel. Such machines are easy to handle and migrate. They are also easy to cluster depending on the type of load.

Acknowledgements The authors thank Martin Karsten and Romulo Gonzalez for providing us with the semantic benchmark and for their kind assistance with setting it up and configuring MonetDB.

References

1. Abadi, D.J., Madden, S.R., Hachem, N.: Column-stores vs. row-stores: How different are they really?. In: Proceedings of the 2008 ACM SIGMOD international Conference on Management of Data. SIGMOD. ACM, pp. 967–980 (2008)
2. Abadi, D.J., Marcus, A., Madden, S.R., Hollenbach, K.: Scalable semantic web data management using vertical partitioning. In: Proceedings of VLDB 2007, Vienna, Austria, 23–27 Sept 2007, pp. 411–422 (2007)
3. Apache Hadoop. http://hadoop.apache.org/core (2010)
4. Apache HBase. http://hadoop.apache.org/hbase (2010)
5. Apache Pig. http://incubator.apache.org/pig (2010)
6. Architecture of MonetDB. http://monetdb.cwi.nl/projects/monetdb/MonetDB/Version4/Documentation/monet/index.html (2010)
7. Barton Library Catalog Data. http://simile.mit.edu/rdf-test-data/barton/ (2010)

8. Boncz, P.: Monet, a next-generation DBMS kernel for query-intensive applications, Doctoral Dissertation. CWI, 2002
9. Boncz, P.A., Zukowski, M., Nes, N.: MonetDB/X100: Hyper-Pipelining query execution. In: Proceedings of the Biennial Conference on Innovative Data Systems Research (CIDR), pp. 225–237, Asilomar, CA, USA, January 2005
10. Broekstra, J., A. Kampman, F. van Harmelen.: Sesame: A generic architecture for storing and querying RDF and RDF Schema. In: ISWC, pp. 54–68 (2002)
11. Chang, F., Dean, J., Ghemawat, S., Hsieh, W.C., Wallach, D. A., Burrows, M., Chandra, T., Fikes, A., Gruber, R.E.: Bigtable: a distributed storage system for structured data. In: Proceedings of the 7th USENIX Symposium on Operating Systems Design and Implementation, vol. 7, Seattle, WA, 06–08 Nov 2006
12. Chong, E.I., Das, S., Eadon, G., Srinivasan, J.: An Efficient SQL-based RDF querying scheme. In: VLDB, pp. 1216–1227 (2005)
13. Copeland, G.P., Khoshafian, S.N.: A decomposition storage model. In: Proceedings of the 1985 ACM SIGMOD international Conference on Management of Data, Austin, Texas, USA. SIGMOD '85, pp. 268–279 (1985)
14. Dean, J., Ghemawat, S.: MapReduce: Simplified data processing on large clusters. In: Proceedings of the 6th Conference on Symposium on Operating Systems Design & Implementation, vol. 6, San Francisco, CA, 06–08 Dec 2004
15. Harizopoulos, S., Liang, V., Abadi, D.J., Madden, S.: Performance tradeoffs in read-optimized databases. In: Dayal, U., Whang, K., Lomet, D., Alonso, G., Lohman, G., Kersten, M., Cha, S.K., Kim, Y. (eds.) Proceedings of the 32nd international Conference on Very Large Data Bases, Seoul, Korea, 12–15 Sept 2006. Very Large Data Bases, VLDB Endowment, pp. 487–498 (2006)
16. Harris, S., Gibbins, N.: 3store: Efficient bulk RDF storage. In: Proceedings of PSSS'03, pp. 1–15 (2003)
17. Holloway, A.L., DeWitt, D.J.: Read-optimized databases, in depth. In: Proceedings of VLDB Endow, vol. 1, issue 1, pp. 502–513. Aug 2008
18. Husain, F.M., Doshi, P., Khan, L., Thuraisingham, B.: Storage and retrieval of large RDF graph using Hadoop and MapReduce. In: Proceedings of International Conference on Cloud Computing, Beijing, China, 01–04 Dec 2009. LNCS, vol. 5931, pp. 680–686. Springer, Berlin (2009)
19. Ivanova, M., Kersten, M., Nes, N.: Self-organizing strategies for a column-store database. In: Proceedings of the 11th International Conference on Extending Database Technology (EDBT 2008), Nantes, France, 25–30 Mar 2008
20. Liu, J.F.: Distributed storage and query of large RDF graphs. Technical Report. The University of Texas at Austin, Austin, TX, USA (2010). http://userweb.cs.utexas.edu/~jayliu/reports/Query_of_Large_RDF_Graphs.pdf
21. Longwell website. http://simile.mit.edu/longwell/
22. Mika, P., Tummarello, G.: Web semantics in the clouds. IEEE Intell. Syst. 23(5), 82–87 (2008)
23. MonetDB. http://monetdb.cwi.nl/
24. OWL Web Ontology Language. Overview. W3C Recommendation. http://www.w3.org/TR/owl-features/ (2004)
25. RDF Primer. W3C Recommendation. http://www.w3.org/TR/rdf-primer (2004)
26. RDF Schema. W3C Specification Candidate. http://www.w3.org/TR/2000/CR-rdf-schema-20000327/ (2000)
27. RDQL – A Query Language for RDF. W3C Member Submission 9 January 2004. http://www.w3.org/Submission/RDQL/ (2004)
28. Scalable Storage Performance. VMWare Corp. White Paper. http://www.vmware.com/files/pdf/scalable_storage_performance.pdf (2008)
29. Sidirourgos, L., Goncalves, R., Kersten, M., Nes, N., Manegold, S.: Column-store support for RDF data management: not all swans are white. In: Proceedings of VLDB Endowment, vol. 1, issue 2, Aug 2008, pp. 1553–1563 (2008)

30. SPARQL Query Language for RDF. W3C Working Draft 4 October 2006. http://www.w3.org/TR/rdf-sparql-query/ (2006)
31. Stonebraker, M., Abadi, D.J., Batkin, A., Chen, X., Cherniack, M., Ferreira, M., Lau, E., Lin, A., Madden, S., ONeil, E., ONeil, P., Rasin, A., Tran, N., Zdonik S.: C-store: A column-oriented DBMS. In: Proceedings of the 31st VLDB Conference (2005)
32. Swoogle. http://swoogle.umbc.edu/
33. VMware vSphere 4 Performance with Extreme I/O Workloads. VMWare Corp. White Paper. http://www.vmware.com/pdf/vsp_4_extreme_io.pdf (2010)
34. Wilkinson, K., Sayers, C., Kuno, H., Reynolds, D.: Efficient RDF storage and retrieval in Jena2. In: SWDB, pp. 131–150 (2003)
35. Wordnet rdf dataset. http://www.cogsci.princeton.edu/wn/
36. World Wide Web Consortium (W3C). http://www.w3.org/

Chapter 9
Scientific Computation and Data Management Using Microsoft Windows Azure

Steven Johnston, Simon Cox, and Kenji Takeda

Abstract Cloud computing is the next stage in the evolution of computational and data handling infrastructure, establishing scale out from clients, to clusters to clouds. With the use of a case study, Microsoft Windows Azure has been applied to Space Situational Awareness (SSA) creating a system that is robust and scalable, demonstrating how to harness the capabilities of cloud computing. The generic aspects of cloud computing are discussed throughout.

9.1   Introduction

Cloud computing offerings come in many variations and mean different things to different people. In general, cloud computing is defined as a dynamic, scalable, on-demand resource [6]; a utility computing resource which can be purchased and consumed much the way electricity is purchased (utility computing [40, 41]). Generally, cloud providers offer computation and storage facilities as part of their offering, in addition to other proprietary capabilities. This chapter focuses mainly on the data and computational resources since both are fundamental across different cloud architectures.

Two key benefits of a cloud-based architecture are high availability (24/7) and super-scalability. Scalability is an inherent design from the underlying cloud architecture, and it is this capability that provides a mechanism to scale, for example, a web base application, from hundreds of users to millions without changing the codebase [1].

Cloud or utility computing is not a new concept; it has evolved from work in areas such as Grid computing and other large distributed applications, in areas such as engineering [18, 23, 53] and biology [35, 49].

S. Johnston (✉) · S. Cox · K. Takeda
Faculty of Engineering and the Environment, University of Southampton, Southampton, UK
e-mail: sjj698@zepler.org; s.j.cox@soton.ac.uk; ktakeda@soton.ac.uk

S. Fiore and G. Aloisio (eds.), *Grid and Cloud Database Management*,   169
DOI 10.1007/978-3-642-20045-8_9, © Springer-Verlag Berlin Heidelberg 2011

In this chapter, cloud capabilities are discussed in general terms, listing provider specific examples where appropriate. Companies/organisations are referred to as cloud providers (Microsoft, Amazon) supplying a cloud offering (Microsoft Windows Azure [16,37,42], Amazon EC2 [3,46]), consisting of cloud resources (Azure tables, Amazon SimpleDB).

The advantages of a cloud-based architecture can be categorised into a number of key areas: burst capability (predictable and unpredictable); scalability; development life cycle reduction of, e.g., algorithms; disparate data aggregation or dissemination. The rest of this chapter introduces a Windows Azure case study demonstrating the advantages of cloud computing (Sect. 9.2), and the following section describes the generic resources that are offered by cloud providers and how they can be integrated to produce a large dynamic data store (Sect. 9.3). The economics of cloud offerings are discussed in Sect. 9.3.7, and Sect. 9.4 concludes with a critique of today's cloud computing ecosystem.

A key principle of cloud providers is that building large data centres benefits from economies of scale to reduce the overall cost; more so if the cloud data centre is highly used. Cloud computing is one business model which sells or rents parts of large data centres, resulting in cost savings. Often, this implies shared-resources (multi-tenanted) and machine virtualisation, but is not a requirement.

The level of interaction with cloud data centres defines the type of cloud offering, and can be divided into the following categories:

- Infrastructure as a Service (IaaS).
  Cloud IaaS sells/rents out infrastructure such as servers, virtual machines and networking and is an alternative to physically owning infrastructure. For example, renting a virtual machine on Amazon EC2 [46].
- Platform as a Service (PaaS).
  Often build upon IaaS, cloud PaaS offerings include an Operating System (OS) and perhaps a software stack (.Net, Java). For example, Microsoft Windows Azure Workers [37].
- Software as a Service (SaaS).
  SaaS offers an end-user application and can be built upon IaaS and PaaS. For example, SalesForce CRM [8,55].

As you progress from IaaS to SaaS, the cloud provider assumes more responsibility, e.g., OS software and virusguard updates. This helps reduce maintenance costs and also reduces flexibility, for example upgrading from one SaaS application version to the next has to be done on the cloud providers schedule.

9.2 Cloud Computing Case Study

This section demonstrates how a cloud-based computing architecture can be used for planetary defence and Space Situational Awareness (SSA) [34], by showing how utility compute can facilitate both a financially economical and highly scalable solution for space debris and near-earth object impact analysis. As our ability to track smaller space objects improves, and satellite collisions occur, the volume of objects being tracked vastly increases, increasing computational demands. Propagating trajectories and calculating conjunctions becomes increasingly time critical, thus requiring an architecture which can scale with demand. The extension of this to tackle the problem of a future near-earth object impact, and how cloud computing can play a key role, is also described.

9.2.1 Background

Space situational awareness includes scientific and operational aspects of space weather, near-earth objects and space debris [38, 39]. This work is part of an international effort to provide a global response strategy to the threat of a Near Earth Object (NEO) impacting the earth [7], led by the United Nations Committee for the Peaceful Use of Space (UN-COPUOS). The impact of a NEO (an asteroid or comet) is a severe natural hazard but is unique in that technology exists to predict and to prevent it, given sufficient warning. The International Spaceguard survey has identified nearly 1,000 potentially hazardous asteroids greater than 1km in size, although NEOs smaller than 1 km remain predominantly undetected and exist in far greater numbers; impacting the Earth more frequently. Impacts by objects larger than 100 m (twice the size of the asteroid that caused the Barringer crater in Arizona) could be devastating (see Fig. 9.1). The tracking and prediction of potential NEO impacts are of international importance, particularly with regard to disaster management.

Space debris poses a serious risk to satellites and space missions. Currently, Space Track [58] publishes the locations of about 15,000 objects, these include satellites, operational and defunct, space debris from missions and space junk. It is believed that there are about 19,000 objects with a diameter over 10 cm. Even the smallest space junk travelling at about 17,000 miles per hour can cause serious damage; the Space Shuttle has undergone 92 window changes due to debris impact. There are over 500,000 objects over 1 cm in diameter and there is a desire to track most, if not all of these. By improving ground sensors and introducing sensors on satellites, the Space Track database will increase in size. Tracking and predicting space debris behaviour in more detail can reduce collisions as the orbital environment becomes ever more crowded. Cloud computing provides the ability to trade computation time against costs. It also favours an architecture which inherently scales, providing burst capability. By treating compute as a utility, compute cycles are only paid for when they are used. This case study demonstrates the benefits of a cloud-based architecture, the key scenarios are outlined below and further discussed in Sect. 9.3.

- Predictable Burst.
  Data must be processed in a timely manner twice per day.

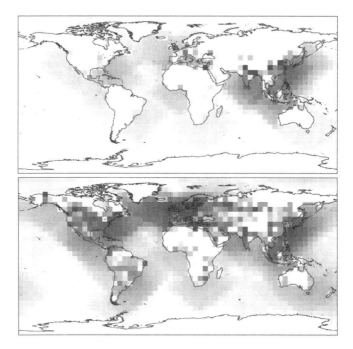

Fig. 9.1 Map showing the relative consequences of the impact of a 100 m diameter asteroid at 12 km s^{-1} into global grid cells with shading denoting (*top*) casualty generation and (*bottom*) infrastructure damage [7]

- Scalable demand.
  In the event of a collision event or future launch, additional load must be handled flexibly and scalable.
- Algorithm Development.
  New algorithms are required to understand propagation of space debris and collision events and to cope with scaling from tracking 20,000 objects (at 10 cm resolution) to 500,000+ (at 1 cm resolution).
- Data sets in the cloud.
  Whilst the original source data are obtained from published sources, the resulting output data could be made available as a cloud data set for aggregation from multiple sources and better dissemination to multiple consumers.

9.2.2 Satellite Propagation and Collision

The following cloud application framework is designed to tackle space debris tracking and analysis and is being extended for NEO impact analysis. In this application, propagation and conjunction analysis produces in peak compute loads

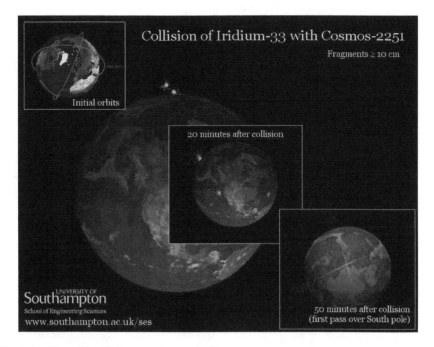

Fig. 9.2 Iridium-33 collision with Cosmos-2251

for only 20% of the day with burst capability required in the event of a collision when the number of objects increases dramatically; the Iridium-33 Cosmos-2251 collision in 2009 resulted in an additional 1,131 trackable objects (see Fig. 9.2). Utility computation can quickly adapt to these situations consuming more compute, incurring a monetary cost but keeping computation wall clock time to a constant. In the event of a conjunction event being predicted, satellite operators would have to be quickly alerted so they could decide what mitigating action to take. This work migrates a series of discrete manual computing processes to the Azure cloud platform to improve capability and scalability. The workflow involves the following steps: obtain satellite position data, validate data, run propagation simulation, store results, perform conjunction analysis, query and visualise satellite object.

Satellite locations are published twice a day by Space Track, resulting in bi-daily high workloads. Every time the locations are published, all previous propagation calculations are halted, and the propagator starts recalculating the expected future orbits. Every orbit can be different, albeit only slightly from a previous estimate, but this means that all conjunction analysis has to be recomputed. The quicker this workflow is completed the quicker possible conjunction alerts can be triggered, providing more time for mitigation.

The concept project uses Windows Azure as a cloud provider and is architected as a data-driven workflow consuming satellite locations and resulting in conjunction alerts, as shown in Fig. 9.3. Satellite locations are published in a standard format

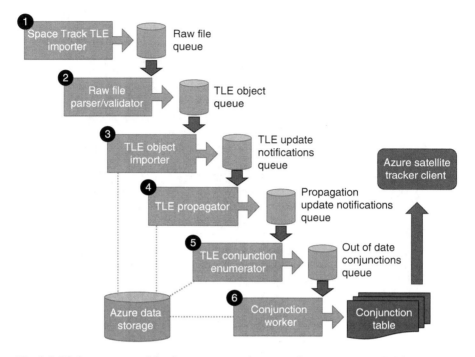

Fig. 9.3 Workers consume jobs from queues and post results onto queues, chaining workers together forms data-driven workflows. This provides a simple plug-in framework where workers can be substituted for example to test newer algorithms or to compare with different algorithms

known as a two-line element (TLE) that fully describes a spacecraft and its orbit. Any TLE publisher can be consumed, in this case the Space Track website, but also ground observation station data.

In Fig. 9.3, the workflow starts by importing a list of TLEs from Space Track (step 1); the list of TLEs are first separated into individual TLE Objects, validated and inserted into a queue (step 2). TLE queue objects are consumed by workers which check to see whether the TLE exists; new TLEs are added to an Azure Table and an update notification added to the Update Queue (step 3). TLEs in the update notification queue are new and each requires propagation (step 4); this is an embarrassingly parallel computation that scales well across the cloud. Any propagator can be used. Currently, only the NORAD SGP4 propagator and a custom Southampton simulation (C++) code [38] are supported. Each propagated object has to be compared with all other propagations by calculating all possible conjunction pairs (step 5); an all-on-all process which can be optimised with filtering out obviously independent orbits. A conjunction analysis is performed on each pair to see whether there is a conjunction or predicted close approach (step 6). Any conjunction source or code can be used, currently a basic implementation; plans are to incorporate more complicated filtering and conjunction analysis routines as they become available. Conjunctions result in alerts which are visible in the Azure Satellite tracker client. The client uses Microsoft Bing Maps to display the orbits. Each step in the workflow is an independent worker. The number of workers can be increased to speed up the process or to cope with burst demands.

Ongoing work includes expanding the Bing Maps client as well as adding support for custom clients by exposing the data through a REST interface. This pluggable architecture ensures that additional propagators and conjunction codes can be incorporated. The framework demonstrated here is being extended as a generic space situational service bus to include NEO impact predictions. This will exploit the pluggable simulation code architecture and the cloud's burst computing capability to allow refinement of predictions for disaster management simulations and potential emergency scenarios anywhere on the globe.

9.2.3 Summary

It has been shown how a new architecture can be applied to space situational awareness to provide a scalable robust data-driven architecture which can enhance the ability of existing disparate analysis codes by integrating them together in a common framework. By automating the ability to alert satellite owners to potential conjunction scenarios, it reduces the potential of conjunction oversight and decrease the response time, thus making space safer. This framework is being extended to NEO trajectory and impact analysis to help improve planetary defences capability.

9.3 Cloud Architecture

Cloud offerings provide resources to perform computation and store data, both relational and non-relational, on a pay as you use basis (see Fig. 9.4). The benefits of a cloud-based architecture may be divided into the following scenarios:

Burst capability. This includes tasks that have load peaks, either predictable (e.g., those related to Christmas or end of tax year) or unpredictable (e.g., the Slashdot effect [22]). When sizing a data centre for such a scenario, it has to be able to cope with the peak load; for the majority of the time this hardware remains unused, resulting in poor utilisation hence increased costs. Moving such an application to a cloud provider ensures that you only incur costs for the resources used. An alternative is to run a data centre at or near maximum load and offload any peaks to a cloud resource; a cloud-hybrid approach (see Sect. 9.3.6.1).

Scalability. It is difficult to judge how *popular* an application will become so there is an inherent need to make all applications scalable. In addition to the application scaling the underlying resources need to scale. As with the burst capability, cloud computing offers instant scalability (quicker than purchasing physical machines

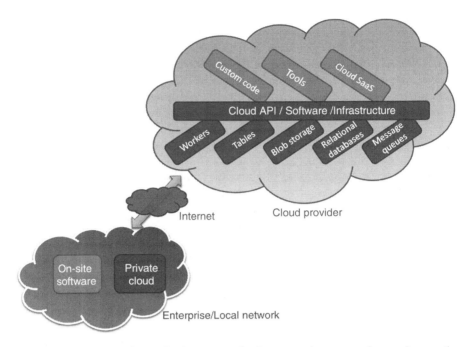

Fig. 9.4 Local or enterprise applications access cloud resources (compute and storage) across the internet. Applications can be hosted locally, partially (hybrid cloud) or completely by a cloud provider

[40]) allowing an application to operate throughout the hype cycle, failure to scale in such a scenario often results in the failure of useful applications. More importantly, as an application declines or plateaus in popularity, cloud computing permits the scaling back of resources; currently, this is very difficult to accomplish with physical hardware.

Reducing the development life cycle. For tasks where the development cycle is large and compute intensive such as algorithm development, cloud offerings can be used to reduce the development cycle, thus reducing the overall time to completion.

Disparate data aggregation or dissemination. Cloud offerings are inherently global, highly available and have large bandwidth capabilities, making them ideal for data aggregation and dissemination. Using multiple cloud data centres reduces the load, and moving data physically closer to where it is consumed reduces latency; for example, by copying data to multiple data centres or using a cloud caching technology [29, 54]. Once a dataset resides in a globally accessible cloud resource, it too becomes a resource. This infrastructure facilitates the publication and consumption of data by third parties providing *mashup* capabilities previously impossible or difficult to achieve using conventional architectures [30, 44, 51].

This section describes the capabilities of key cloud resources, detailing how they can combine to produce a logical robust architecture, including pricing strategies and their influence on architectural decisions.

9.3.1 Computational Workers

All cloud computing vendors offer users the ability to perform computation, either by running code [26] or by supporting applications; a unit of computational activity is referred to a worker.

One of the advantages of cloud computing is that it is super-scalable, which is achieved by having large numbers of workers; scaling-out. This is in contrast to traditional scaling-up where the number of workers is constant but where the workers are made more powerful.

It is not always easy to scale-out workloads, large volumes of simple jobs which can be run in parallel are easy to convert, but tightly interconnected or serial jobs may not be appropriate for cloud scale-out scenarios. To achieve super-scalability of cloud-based applications, a solution will inevitability involve some re-architecting. An alternative may include running an application, in parallel multiple times with different input parameters (parameter sweep) to achieve throughput. Hence, consider scaling-out workers rather than scaling-up workers.

Some cloud providers facilitate scale-up solutions by providing different sized workers [16, 46]; this helps with scalability issues but will not result in super-scalability. By offering to host workers on more powerful machines (more RAM, more cores), user applications will perform better. Cloud pricing is geared towards utility pricing and so a worker which is twice as powerful will cost twice as much. Consider the case where a *large* worker is eight times as powerful as a *small* worker; will the application have eight times the throughput? In utility computing, it is often beneficial to achieve throughput by having a larger number of slower workers where the hardware is 100% used, than to have a single powerful worker which may under-utilise the hardware; processes tend to be CPU, I/O or memory bound. For example, a CPU-bound application which is moved to a *large* worker will contain proportionally more RAM. If the application cannot make use of this RAM, the resource will be underutilised, and therefore more costly.

The more the components you add to a system, the higher the probability of a failure, Cloud computing guarantees failure; always handle failure/retry scenarios gracefully, particularly for 24/7 applications. It is insufficient to rely on SLAs since breaches often result in refunds or bill reductions that do not cover the full cost of an application failure [59].

One mechanism to overcome worker failures is to ensure that workers are idempotent [52]. For example, consider a worker which consumes tasks from message queues (see Sect. 9.3.5) and processes the message. The result should remain the same regardless of how many times the message is executed.

Idempotency can be achieved using storage tables (see Sect. 9.3.3) to log which messages are in progress or completed; hence, halting the repeated execution of a message. Alternatively, an eventual consistency method can be adopted whereby repeated tasks are eventually cancelled out; for example, using compensating transactions [50].

Idempotency can assist in scenarios where it is critical to process messages reliably within a fixed time period. If workers are idempotent and a single worker can process the message within the time period, executing the same message with multiple workers will mitigate cloud infrastructure failures. This will result in additional costs but provides a mechanism to directly trade reliability against cost.

9.3.2 Blob Storage

To facilitate the storage of large volumes of unstructured data, cloud providers offer a variety of blob storage capabilities. Blob storage effectively allows users access to unlimited storage capabilities. As conventional filesystems cannot scale to cope with the data volumes that cloud providers are capable of storing, most cloud blob storage is accessed using REST or custom, proprietary APIs. Since these differ between providers, it makes cloud mobility more difficult.

By abstracting blob storage away from the underlying hardware, cloud providers can offer data redundancy and high availability. For example, Microsoft Windows Azure and Amazon S3 guarantee that data are replicated a minimum of three times [14].

For regulatory reasons, some data cannot exist outside geographical regions. This is partially catered for by offering users different geographical locations for data to reside; for example, North America, Europe and Asia. These address some regulatory issues, but it is unclear how country-specific data can be handled. Location-specific data centres can improve performance by ensuring that data are *close* to consuming applications.

Transferring data into and out from the cloud storage is expensive, both in terms of cost and time. Once data are in the cloud, it is efficient to access using workers, but inefficient to access from remote or corporate sites. Consider carefully which data ase transferred to the cloud. It could be more efficient to generate the data in the cloud using a worker. In cases of hybrid cloud solutions (see Sect. 9.3.6.1), storing a large datasets both onsite and in the cloud can reduce transfer costs and improve performance.

9.3.2.1 Sparse Files, Virtual Disks and Versioning

Using blob storage, APIs can be limiting as it requires application code changes to use the storage API, which is not standard across cloud providers. There are various mechanisms which facilitate mounting blob storage as if it were a physical device, therefore permitting the use of existing I/O libraries. This greatly assists with cloud application portability and cloud mobility. Cloud workers can run existing applications in the cloud without having to make changes to the applications.

Virtual disks. In virtual infrastructures, it is common for HDD partitions to be represented using virtual harddisks [57]. The virtual disk is just a large file which usually lives in the virtual infrastructure's underlying filesystem. For example, a 100 GiB file can be mounted as a harddrive device, formatted with a file system (NTFS, Reiser, Ext3) and used as if it were a physical disk; the I/O performance may be reduced slightly [57]. Some cloud providers support the mounting of virtual disks inside workers [12]. By attaching virtual disks to a workers, non-cloud aware application can store data to blob storage via the virtual disk.

When creating a virtual disk, it can be difficult to calculate a reasonable starting size. Expanding the disk after it has been created is possible (up to the maximum single blob size), but the virtual disk filesystem must support expansion. To use virtual disks on blob storage efficiently, the cloud provider has to support sparse files. A sparse file only stores data for the parts of the file which contain data. For example, an empty 100Gib virtual disk will not incur any storage costs, but will still appear as a 100GiB partition. As the partition is filled, blob storage charges will apply. If the virtual disk is mounted from a destination where billing applies, consider the number of I/O transactions in addition to the data volumes, as transaction costs can be considerable, especially if a cache is not used [12].

Storing data in virtual disks opens opportunities to leverage some blob storage features, such as snapshotting to maintain different versions of the same virtual hard disk (VHD). A virtual disk can be mounted as read–write by one worker but read-only by many workers; handy for sharing a common codebase.

Mounting. Using virtual disks on blob storage adds another level of indirection, which has resizing, maximum size and performance implications. A different method is to mount cloud blob storage as if it was a physical disk which improves cloud application portability and migration. This does not limit the total volume size and eliminates resizing issues. This method is not identical to physical disk or VHD, for example in Amazon EC2, the user mode filesystem for S3 is not Posix [10] compliant, and may not perform as expected for all cloud applications.

Scratch disk. Where performance is important, there is no substitute for physical disk or locally attached virtual disk. Cloud workers have scratch disk space which is a fast local disk and can be used by applications. This storage is not permanent and needs to be synced to persistent storage. As workers fail or disappear, dealing with failed workers and non-persisted data becomes an additional responsibility of the cloud application.

Local tools. Alternatively, if an application supports the blob storage API, accessing the generated data by mounting it locally assists users with importing and exporting data. In addition to the tools methods listed above, there are tools that allow users to drag and drop files into cloud storage locations [13, 60].

9.3.3 Non-Relational Data Stores

Cloud-based non-relational databases offer super-scalability but are different from relational databases; for cloud-based RDBs, see Sect. 9.3.4. Relational databases are scalable, but they potentially become a bottleneck for super-scalable applications, in-part because of features such as transactions and row/table locking. Non-relational cloud data stores take the form of large tables, which have a fixed or limited number of indices [28]. The table is divided across multiple sets of hardware, and as the table grows more hardware can be added, either automatically (Azure table) or by configuration (Amazon tables). Be sure to select data index keys carefully as the non-relational table efficiency relies on index keys for load distribution.

The features supported vary from vendor to vendor, and there are no standard interfaces, thus reducing cloud mobility and portability [11]. As the market matures, cloud consumers can expect to see more standards emerge between vendors.

Azure table storage [15, 28] indexes a single row using two keys, a partition and a row key. Transactions are supported across rows which are in the same partition. A row is identified by the partition and row key, a partition contains many rows. As the data volume expand the Azure fabric increases the hardware, reducing the number of partitions that a single machine has to manage; Amazon SimpleDB is spread across domains which can be manually increased [14].

Super-scalability raises data consistency issues across very large tables. Often applications rely on data consistency, but this need not be a requirement. For example, an online shop displaying the number of items in stock only gets interesting when stock gets low. Providing the stock levels are high, displaying out of date stock levels is not important. As the stock level approaches zero, data consistency can be enforced but may take longer to calculate the stock levels. This approach speeds up page queries without affecting end users. Amazon SimpleDB supports eventual consistency, in which row data that is out of date will become consistent over time, usually under a second [61].

9.3.4 Cloud-Relational Database Storage

Relational databases will never scale to the data volumes that non-relational table storage can manage; however, they have a much richer feature set and are fundamental to the software ecosystem [19].

Most cloud providers offer a relational database which will work seamlessly with existing RDB code, applications and tools. Amazon RDS [46] and SQL Azure [42] are both cloud-hosted relational database offerings. Since they are compliant with existing SQL platforms, they do not require changes to existing applications. SQL applications often require frequent, low latency I/O. Moving a database to a cloud provider and keeping the applications onsite can have performance costs as well as monetary implications.

The cloud database offerings vary but in general are managed, upgraded and backed up automatically, drastically reducing TCO. For example, SQL Azure ensures that data are committed to three Microsoft SQL database instances to guarantee high availability.

Many relational databases support processing inside the database, ranging from simple calculations (standard mathematical operations) to complete code execution (Microsoft CLR integrations [2]). If your application requires these advanced features, ensure that your cloud provider fully supports code execution. Current pricing models do not acknowledge that the database can be used to perform processing, others actively restrict it, for example Microsoft SQL Azure reserves the right to terminate long running or CPU intensive database transactions [42].

Cloud Database Sharding. In keeping with the cloud philosophy of commodity hardware and super scalablability, a single database instance will quickly prove to be a bottleneck. Scaling up the database hardware will result in better performance but is limited by hardware capabilities, another approach is to scale out the database data this is known as database sharding [17].

Database sharding involves taking a large database and breaking it into a number of smaller database instances, ideally with no shared data. Sharding the data results in smaller databases which are easier to backup and manage. The load is distributed which results in better scalability, availability and larger bandwidth, although partitioning data involves extra work. This is different from more traditional scaling approaches, such as replication [19] (master–slave), in which each shard is loosely coupled with minimal dependencies and all instances support read–write.

Deciding how to shard a database schema is domain specific. The overall concept is to look at high transaction or high update areas and break them into multiple shards. This is currently a manual process and requires custom implementations, as this approach becomes more popular the tooling and support will increase [33]. Despite the sharding process adopted, it is important to remember that as a shard becomes overloaded (or underutilised) there is a requirement to rebalance the shards so that each has a similar workload. Although this may appear a complex task, consider the case where scalability is only required for short periods throughout the year. One approach may be to split a master database across many shards during a peak period and then consolidate the shards back into a master database as traffic subsides. This method leverages the cloud scalability features and does not require shard re-balancing; this only works for predictable workloads such as ticket sales [32, 48].

9.3.5 Message Queues

As described in Sect. 9.3.1, workers are units of processing which can occur in large numbers for super scalability but are also potentially unreliable; they can fail.

Passing data into and out of these workers reliably is accomplished using message queues. A message queue is a FIFO queue which is accessible by worker processes, messages are guaranteed to be delivered to workers. Eliminating the need to address workers by port and IP address, or managing file/data locks which are potential bottlenecks.

Messages do incur a cost (data charges and I/O charges) and time overhead, in some cases it is acceptable to bunch messages. For example, messages can be collected into groups of ten and stored as a single message. This method reduces I/O costs and speeds up workers where queue performance is an issue. It is particularly useful for large volumes of messages. Care must be taken to ensure that the message groups do not exceed the maximum message size. Grouping messages results in a worker having to processes all messages in a single group, which can lead to inefficient usage of worker processes, particularly when the queue is nearly empty.

In a production application, it is easy to foresee a situation where a queue is full of messages but there is a need to upgrade the processing worker which uses a newer message protocol version. Waiting for the queue to empty, then upgrading the workers is not always possible, especially in a 24/7 application. It is for this reason that all messages must include the protocol version. This allows workers of different versions to run side-by-side during an upgrade.

Removing messages from queues can be problematic and requires careful consideration, for which there are several patterns.

In general, when a worker pops a message from the queue it becomes invisible to other queue consumers. Once the worker has processed a message, it is then deleted from the queue, however, there is a need to deal with the case that the worker fails, often silently. Worker failure can be divided into two categories. Those that fail due to underlying cloud fabric issues and those that fail due to message content issues. Cloud fabric issues can be overcome by restarting the worker or sending the message to another worker. Message issues, often referred to as poison messages [9], can crash workers and need to be removed from the queue.

A poison message can be detected by counting the number of times the message has been popped from the queue, and setting the maximum number of times a message can be popped before it is declared poisonous. It is good practice to move poison messages to a separate queue where they can be dealt with either manually or by another worker. If a pop counter is not provided by the cloud infrastructure, then it has to be implemented separately taking care not to add extensive overhead or computation bottlenecks.

Managing messages timeouts is specific to the message contents and worker task. For example, if each message contains a task for the worker to perform, frequent short running tasks must be handled differently to long running in-frequent tasks.

Let us consider the following patterns for message timeouts:

Fixed message timeout. Windows Azure and Amazon SQS provide mechanisms to set a message queue timeout. Once the message timeout expires and the message has not been deleted, it automatically becomes visible to other workers. If a message timeouts whilst a worker is still processing, the message will become visible to another worker, resulting in a single message being processed more than once. It is for this reason that idempotent workers should be considered (see Sect. 9.3.1). Frequent short running tasks with idempotent workers can easily be managed by having a fixed message timeout. Long running tasks become problematic as workers have to wait for the message timeout to expire before failures are detected, this is often too long.

Variable message timeout. Some cloud offerings permit updates to message timeouts on a per message basis [46]. In this case, long running tasks can continuously update the message timeout, for example a task that normally takes 8 h to complete can consume a message which has a timeout of 5 min. Before the timeout expires, the worker can update the message timeout period extending it for a further 5 min. This method ensures that failed workers are discovered more rapidly; however, this requires that the worker can safely detect ongoing tasks. Often this requires the worker to have knowledge of the task being preformed, for example it can be difficult to distinguish an infinite loop from a large loop. Receiving messages multiple times should be infrequent, and idempotent workers will result in data consistency but for long running tasks repetition of work should be avoided as it can increase the costs.

Custom queue timeout. By extending cloud queue implementations, it is possible to monitor for both timeouts and poison messages. Every time a message is popped from the queue, it is then logged to permanent storage (table or blob storage) and then deleted from the queue. The permanent storage can record the number of times the message has been popped as well as a timestamp. There are two methods for detecting failed workers, (1) as with the variable message timeout method, the worker can update the message last active timestamp, (2) if all the tasks in a message queue are similar, then tasks can be restarted if the processing time exceeds, say four standard deviations.

This method requires that the message queue can detect failed tasks, which increases processing and storage requirements of queues. This may not be suitable for queues with high volumes of traffic.

Unlike most cloud offerings which add timed-out messages to the *head* of the queue, this method has the ability to add message timeouts to the *head* or *tail* of the queue; useful if the queue is experiencing starvation due to continually failing jobs (i.e., poison messages). Multiple poison messages that are continually added to the head of the queue will eventually block all messages, resulting in starvation.

9.3.6   Integrating Blobs, Tables and SQL Databases

Selecting an appropriate storage mechanism for cloud resident data is very important. When poor choices are made, it can result in higher data costs, e.g., storing blob data in a relational database, or it can result in difficult changes in the future, e.g., selecting table storage for highly relational data and then having to migrate to a relational database.

As the pricing and capabilities of cloud storage offerings are constantly changing, it is difficult to produce a definitive storage guide; however, there are some generic issues worth considering. As with a conventional relational database, it is advisable to store large blobs outside the database; databases are good with relational data, file systems are good with large blobs (database filestreams blur this divide [56]). With cloud offerings there are multiple storage offerings, each with benefits and limitations. Where possible look at the strengths of each storage resource and divide up the dataset into those pieces better suited to the storage resource (be aware of possible synchronisation and locking issues).

Blob storage behaves like a file system and is the most cost-effective mechanism for storing binary data. It is highly available and fault tolerant, suitable for extremely large data (see Sect. 9.3.2 for variants).

Cloud table storage (not relational database tables, see Sect. 9.3.3) is super scalable without the performance degradation that is seen in a relational database as it is scaled across multiple machines (in some cases automatically). It does not have the full relational database feature set, but will support limited transactions and row locking, blob storage and one/two keys. This makes it suitable for data that are not very relational and where large transactions are not important; table joins tend to be difficult. The benefits are usually scalability and pricing.

Relational database storage is for all intents and purposes a SQL database. As this tends to be the most expensive storage, consider carefully if all the features are required. By dividing up a dataset, it may be possible to store a small volume of relational data in a relational database and the bulk of the data in cloud tables or blob storage. When migrating an application, consider which storage capabilities are required for the dataset and carefully choose the most appropriate cloud storage resources; even if this involves splitting the dataset.

9.3.6.1 Hybrid Approach: Bridging the On-Premise, Cloud Divide

When cloud offerings are discussed it is often to replace, rather than to complement existing systems. However, there are very few cases where a single cloud provider will meet all your needs and ever fewer cases where your application will migrate in its entirety to the cloud.

It is for these reasons that many cloud applications will be a hybrid of both onsite and cloud resources. Moving an application in pieces also provides a good opportunity to explore the cloud offerings and decide whether they are beneficial to the application without having to migrate the entire application.

A good approach to building a hybrid application is to start by migrating simple standalone features such as backups and data storage, then progress to replicating capability onsite and in the cloud, e.g., offloading peak demand. This ensures that the application runs onsite but offloads some of the computation intensive work to a cloud offering, providing an insight into performance and expenditure. Where onsite and cloud workers operate on a single dataset consider replicating the dataset at both locations to reduce I/O charges and latency (be aware of synchronisation issues). If the dataset is large, it may be quicker and less costly to regenerate the dataset using a cloud worker rather than copying.

Finally, consider any standalone computation that can be fully migrated to a cloud offering; this may just mean turning off onsite capability that is replicated in the cloud.

Depending on which cloud provider is chosen, there are several technologies that assist with hybrid cloud architectures. Cloud offerings which provide a virtual machine (in the general term) often result in security difficulties, e.g., joining a cloud virtual machine with an onsite domain, and authenticating users. Some solutions include fully managing the virtual machines as if they were onsite using a virtual private network (VPN) [5] or using federated security solutions such as Active Directory Federated Services v2 (ADFS v2) [31].

For data security concerns, performance issues and data ownership legal reasons, some applications are not suited to the cloud offerings available today; public clouds. There is an emergence of private cloud offerings which run a cloud providers software stack on a local (onsite) data centre; private clouds [43]. Private clouds may not benefit from a completely managed stack (including hardware), and therefore have the potential to be more costly. They will offer a greater level of control over where data resides and reduced I/O latency. At the time of writing private clouds only look beneficial where data location, I/O performance [24] or specific hardware is paramount; for medium to large data centres.

9.3.7 Economics

Cloud resource pricing is still in its infancy and as the market matures pricing strategies will change [20, 27]. Cloud offerings contain a variety of pay-as-you-go options, bulk purchasing and spot-based pricing.

In general, users pay for specific cloud services (e.g., message queues), computational resources (e.g., virtual machines) and data transfer into and out of data centres; often internal data transfers are not billable. Purchasing resources on this basis removes initial start up capital expenditure and makes the total cost of ownership more transparent. Applications which are cyclic or have unpredictable peak workload can benefit from this pricing model as it supports rapid dynamic resource allocation and more importantly caters for resource reductions.

Cloud providers are bulk buying hardware, electricity and network bandwidth, and in some cases custom-specifying (in conjunction with third parties/OEMs) to produce data centres which have minimal energy and staffing requirements [36]. On this basis, cloud offerings can compete with in-house data centre costs incurred by most small to medium businesses.

Paying for resources from an unlimited pool must be managed closely to prevent Economic denial-of-service (EDoS) attacks. Denial-of-service (DoS) [47] attacks overwhelm resources making them unavailable to intended users, whereas an EDoS attack abnormally increases service usage burdening the cloud user with high cloud resource bills.

As pricing strategies mature, finer grained pricing models will emerge, such as spot pricing [4, 11], whereby users buy resources on a live auction or stock market. Users can prioritise processing demands to meet budget requirements. If all providers offer live pricing, it is easy to foresee the need to dynamically move computation from one provider to another. Currently, moving computation between providers requires multiple application implementations, although there are calls to standardise APIs [21]. Since the cloud computational market is still in its infancy, to assist with future cloud feature adoption and cloud mobility, it is advisable to keep cloud specific features separate from core applications or custom code.

The introduction (Sect. 9.1) discusses burst and scalability features of a cloud-based architecture and shows how these features relate to the economics of cloud computing.

In the case where workloads are cyclic or have predictable/unpredictable bursts, a conventional data centre has to be sized to cope with peak demand. In between high peak loads, the data centre is underutilised; hence, each CPU hour becomes more expensive (Fig. 9.5). Using a utility compute model hardware can be scaled up or down on demand, this ensures that the utilisation of hardware remains high, hence reducing the cost of a CPU hour.

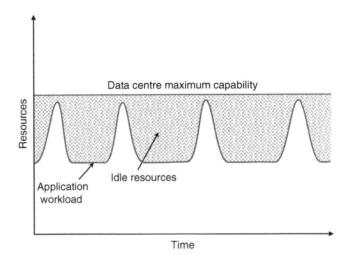

Fig. 9.5 A data centre designed to handle a predictable cyclic load. In between peak load, the data centre utilisation drops as hardware remains unused. The *shaded area* shows underutilised resources

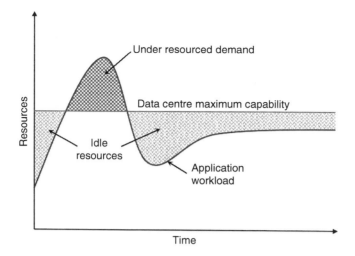

Fig. 9.6 The load characteristics of an application as it goes through the hype cycle. An optimally sized data centre cannot meet peak demand

Conventionally, a new Internet-based service would require hardware, and as it became heavily consumed, more hardware would have to be procured and added. Selecting the initial volume of hardware can prove difficult and projects which fail still incur the initial hardware capital expenditure. Adding hardware is time consuming and results in steps of data centre capacity. Since peak loads have to fall within the data centre's capacity, there is an incentive to over purchase hardware. In addition, the popularity of many applications follows the hype curve [25], which has a peak in demand before the usage plateaus. Designing a data centre to cope with the peak in hype is often impossible (and a financially bad decision), users then experience a poor service and become disgruntled (see Fig. 9.6). The ability of utility computing to expand and contract ensures that an application can ride the hype curve to success. If the application is a failure, it can fail fast, incurring minimal expenditure and providing an opportunity to try more application ideas.

Applications can be too successful, resulting in over demand (see Fig. 9.7). Scaling a data centre to cope with popular applications can become difficult especially where the application exceeds the hardware procurement cycle. Ultimately, a very popular application can outstrip the data centre's upgrade capability. Using a cloud-based architecture provides that ability too add resources on demand thus ensuring smooth, cost-effective scalability.

Whichever way cloud providers decide to price resources, architects should resist the temptation to base key architectural decisions on pricing models. Consider pricing, calculate costs and use good architecture practices to produce flexible systems; remember cloud providers can change prices overnight, re-architecting an application takes considerably longer.

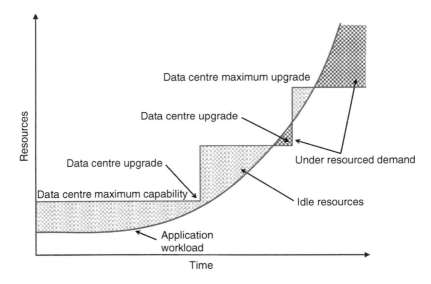

Fig. 9.7 Scaling an application to cope with widespread popularity. There are limits to the upgrade capacity of a given data centre, limiting the expansion of an application. Application growth can out perform hardware procurement resulting in under resourced demand

9.4 Discussion

A cloud-based solution provides a good opportunity to consider scale-out options, providing a better than Moore's law performance improvement [45], although not all applications are suitable cloud candidates. Local laws can prohibit the storage of data outside the country of origin, and there are issues of ownership and access to data stored in a location with a different jurisdiction. Critical applications can operate in the cloud but are at the mercy of the public network in between, for both reliability and security. In the event of a cloud provider failing to meet an SLA, the compensation may not reflect the true cost of the failure.

Sharing resources with other users (multi-tenanted) can result in security and performance issues. There is a need to ensure that data are secure and performance cannot be compromised by erroneous or malicious users using a combination of standards, QoS agreements and SLAs. Security is even more important for cloud providers as they are attractive targets for hackers since they are a single source of large volumes of valuable data.

Although cloud pricing models are yet to evolve, it is difficult to predict costs either on a monthly basis or per application/job. There is a requirement to offer a greater range of pricing options ranging from spot pricing to monthly invoicing with maximum limits or fixed price options.

Total cost of ownership calculations are too complex. As data centres are upgraded, some providers increase the size of a base unit of processing power.

This ensures that the cloud computation cost reflects the true cost of computation. Cloud providers do not publish pricing roadmaps, leaving users at the mercy of the cloud providers pricing strategy with no insight to future pricing. Self hosting cloud infrastructure (private clouds) may address some pricing issues as they offer a base capability for a fixed price and are perceived to overcome some of the security and privacy issues.

Licensing issues remain unaddressed for both applications and the operating system. For example, it can be difficult to use bulk operating system licenses on a cloud infrastructure. Some licenses of third party applications may be sold on a per core basis making burst applications prohibitively expensive.

Only a few providers support automatically scaling of cloud resources as demand increases, currently it is manual (web or interface driven), which has to be carefully managed to control costs.

Cloud computing provides a unique opportunity to co-locate compute and data addressing co-ownership issues whilst supporting global collaborations. Data centre consolidation and the drive to make them more efficient fits well with the need for organisation to become accountable for their energy demands.

9.5   Conclusions

This chapter looks at the generic capabilities of cloud computing, describing the advantages and disadvantages of a cloud-based architecture, using a case study to demonstrate how Microsoft Windows Azure can be used to create a scalable and extensible Space Situational Awareness solution. Capability across the key cloud scenarios has been demonstrated with a super-scalable plug-in framework, which supports public data access and reduces the algorithm development cycle.

Cloud computing will continue to evolve and establish a position in the market addressing some of today's concerns. It is because of the dynamic nature of cloud computing that the application architecture becomes evermore important. An application has to be fault tolerant, to cope with worker failure, modular to assist with parallelising a task and dynamic enough to take advantage of pricing promotions and eliminate vendor lock-in. As the volumes of data stored in a cloud application increase, vendor lock-in affects the data as well at the codebase, so consider the cost of retrieving data from a provider (the cost and time to copy data can be large).

Not withstanding concerns about vendor lock-in and security/privacy, there is great potential for cloud-based architectures to revolutionise our ability to deliver cost-effective solutions and tackle large-scale problems in science, engineering and business.

Acknowledgements Much of this work has been made possible because of contributions from Hugh Lewis and Graham Swinerd at the University of Southampton. We gratefully acknowledge support from Microsoft and Microsoft Research.

References

1. Abbott, M.L., Fisher, M.T.: The art of scalability: Scalable web architecture, processes, and organizations for the modern enterprise, 1st edn. Addison-Wesley, MA (2009)
2. Acheson, A., Bendixen, M., Blakeley, J.A., Carlin, P., Ersan, E., Fang, J., Jiang, X., Kleinerman, C., Rathakrishnan, B., Schaller, G., et al.: Hosting the .NET Runtime in Microsoft SQL server. In: Proceedings of the 2004 ACM SIGMOD international conference on Management of data, p. 865. ACM, NY (2004)
3. Amazon. Amazon web services, 2010. http://aws.amazon.com Accessed 23 Aug 2010
4. Amazon. Amazon EC2 Spot Instances, 2010. http://aws.amazon.com/ec2/spot-instances Accessed 23 Aug 2010
5. Amazon. Amazon Virtual Private Cloud *AmazonVPC*, 2010. http://aws.amazon.com/vpc Accessed 23 Aug 2010
6. Armbrust, M., Fox, A., Griffith, R., Joseph, A., Katz, R., Konwinski, A., Lee, G., Patterson, D., Rabkin, A., Stoica, I., Zaharia, M.: Above the clouds: A Berkeley view of cloud computing. Technical report, February 2009. UCB/EECS-2009-28
7. Bailey, N.J., Swinerd, G.G., Lewis, H.G., Crowther, R.: Global vulnerability to near-earth object impact. Risk Manag., 12, 31–53 (2010). http://eprints.soton.ac.uk/68976/
8. Benioff, M.: Behind the Cloud: The Untold Story of How salesforce.com Went from Idea to Billion-Dollar Company and Revolutionized an Industry. Jossey Bass, October 2009
9. Bustamante, M.: Learning WCF: A Hands-on Guide, 1st edn. O'Reilly Media, Cambridge (2007)
10. Butenhof, D.R.: Programming with POSIX threads. Addison-Wesley, MA (1997)
11. Buyya, R., Yeo, C.S., Venugopal, S.: Market-oriented cloud computing: Vision, hype, and reality for delivering it services as computing utilities. In: 10th IEEE International Conference on High Performance Computing and Communications, 2008. HPCC '08, pp. 5–13, 25–27 Sept 2008. doi: 10.1109/HPCC.2008.172
12. Calder, B., Edwards, A.: Windows azure drive. Windows Azure Platform: Whitepapers, February 2010. microsoft.com/windowsazure/whitepapers Accessed 23 Aug 2010
13. Cerebrata. Cloud storage studio, 2010. www.cerebrata.com/products/cloudstoragestudio Accessed 23 Aug 2010
14. Chaganti, P., Helms, R.: Amazon SimpleDB Developer Guide, 1st edn. Packt Publishing (2010)
15. Chang, F., Jeffrey, D., Ghemawat, S., Hsieh, W., Wallach, D., Burrows, M., Chandra, T., Fikes, A., Gruber, R.: Bigtable: A distributed storage system for structured data. ACM Trans. Comput. Syst. 26(2), 1–26 (2008). ISSN 0734-2071. doi: doi.acm.org/10.1145/1365815.1365816
16. Chappell, D.: Introducing Windows Azure, 2009. David Chappell and Associates. www.davidchappell.com
17. CodeFutures. Database sharding white paper: Cost-effective database scalability using database sharding. CodeFutures Corporation, July 2008. www.codefutures.com/database-sharding Accessed 23 Aug 2010
18. Cox, S., Chen, L., Campobasso, S., Duta, M.C., Eres, M.H., Giles, M.B., Goble, C., Jiao, Z., Keane, A.J., Pound, G.E., Roberts, A., Shadbolt, N.R., Tao, F., Wason, J.L., Xu, F.: Grid Enabled Optimisation and Design Search (GEODISE). e-Science All Hands, Sheffield (2002)
19. Date, C.J.: An Introduction to Database Systems, chapter Further Normalisation I:1NF, 2NF, 3NF, BCNF, pp. 348–379, 7th edn. Addison-Wesley, MA (2000)
20. Deelman, E., Singh, G., Livny, M., Berriman, B., Good, J.: The cost of doing science on the cloud: The montage example. In: SC '08: Proceedings of the 2008 ACM/IEEE conference on Supercomputing, pp. 1–12, Piscataway, NJ, USA, 2008. IEEE, NY. ISBN 978-1-4244-2835-9
21. DMTF: interoperable clouds: a white paper from the open cloud standards incubator, 2009. DSP-ISO101
22. Elson, J., Howell, J.: Handling flash crowds from your garage. Proceedings of the 2008 USENIX Annual Technical Conference, February 2008. Microsoft Research

23. Eres, M.H., Pound, G.E., Jiao, Z., Wason, J.L., Xu, F., Keane, A.J., Cox, S.J.: Implementation and utilisation of a grid-enabled problem solving environment in Matlab. Future Generat. Comp. Syst. 21(6), 920–929 (2005). URL http://eprints.soton.ac.uk/35492/
24. Evangelinos, C., Hill, C.N.: Cloud Computing for parallel Scientific HPC Applications: Feasibility of running Coupled Atmosphere-Ocean Climate Models on Amazons EC2. Ratio 2(2.40), 2–34 (2008)
25. Fenn, J., Raskino, M.: Mastering the Hype Cycle: How to Choose the Right Innovation at the Right Time. Harvard Business School Press, Harvard (2008)
26. Google: Google app engine, 2010. code.google.com/appengine Accessed 23 Aug 2010
27. Greenberg, A., Hamilton, J., Maltz, D., Patel, P.: The cost of a cloud: research problems in data center networks. SIGCOMM Comput. Commun. Rev. 39(1), 68–73 (2009). ISSN 0146-4833. doi: http://doi.acm.org/10.1145/1496091.1496103
28. Hay, C., Prince, B.: Azure in Action. Manning Publications, Greenwich (2010)
29. Held, G.: A Practical Guide to Content Delivery Networks. Auerbach Publications (2005)
30. Hinchcliffe, D.: An executive guide to mashups in the enterprise. Executive white paper, July 2008. jackbe.com. Accessed 23 Aug 2010
31. Huang, H.Y., Wang, B., Liu, X.X., Xu, J.M.: Identity federation broker for service cloud. In: 2010 International Conference on Service Sciences, pp. 115–120. IEEE, NY (2010)
32. Intergen and TicketDirect: Ticket seller finds ideal business solution in hosted computing platform. Microsoft Case Studies, November 2009. http://www.microsoft.com/casestudies. Accessed 23 Aug 2010
33. JBoss: Hibernate shards. hibernate.org/subprojects/shards/docs.html. Accessed 23 Aug 2010
34. Johnston, S., Takeda, K., Lewis, H., Cox, S., Swinerd, G.: Cloud computing for planetary defense. Microsoft eScience Workshop, Pittsburgh, USA, 15–17 Oct 2009
35. Johnston, S.J., Fangohr, H., Cox, S.J.: Managing large volumes of distributed scientific data. Lect. Note. Comput. Sci. 1(5103), 339–348 (2008)
36. Josefsberg, A., Belady, C., Bhandarkar, D., Costello, D., Ekram, J.: Microsofts Top 10 Business Practices for Environmentally Sustainable Data Centers, 2010. www.microsoft.com/environment. Accessed 23 Aug 2010
37. Krishnan, S.: Programming Windows Azure: Programming the Microsoft Cloud, 1st edn. O'Reilly Media (2010)
38. Lewis, H.G., Newland, R.J., Swinerd, G.G., Saunders, A.: A new analysis of debris mitigation and removal using networks. In: 59th International Astronautical Congress, September 2008. http://eprints.soton.ac.uk/68974/
39. Lewis, H.G., Swinerd, G.G., Newland, R.J.: The space debris environment: future evolution. In: CEAS 2009 European Air and Space Conference. Royal Aeronautical Society, October 2009
40. Lin, G., Fu, D., Zhu, J., Dasmalchi, G.: Cloud computing: It as a service. IT Professional 11(2), 10 –13 (2009). ISSN 1520-9202. doi: 10.1109/MITP.2009.22
41. Mendoza, A.: Guide to Utility Computing Strategies and Technologies. Artech House, March 2007
42. Microsoft: Windows Azure platform, 2010. www.microsoft.com/windowsazure. Accessed 23 Aug 2010
43. Microsoft: Windows Azure Platform Appliance , 2010. www.microsoft.com/windowsazure/appliance. Accessed 23 Aug 2010
44. Microsoft: Pinpoint homepage, 2010. pinpoint.microsoft.com. Accessed 23 Aug 2010
45. Moore, G.: Cramming more components onto integrated circuits. Electronics 38(8), 114–117 (1965)
46. Murty, J.: Programming Amazon Web Services: S3, EC2, SQS, FPS, and SimpleDB, 1st edn. O'Reilly Media, Cambridge (2008)
47. Needham, R.: Denial of service. In: CCS '93: Proceedings of the 1st ACM conference on Computer and communications security, pp. 151–153, NY, USA, 1993. ACM, NY. ISBN 0-89791-629-8. doi: http://doi.acm.org/10.1145/168588.168607

48. Nethi, D.: Scaling out with SQL Azure. Windows Azure Platform: Whitepapers, June 2010. microsoft.com/windowsazure/whitepapers. Accessed 23 Aug 2010
49. Ng, M., Johnston, S., Wu, B., Murdock, S., Tai, K., Fangohr, H., Cox, S.J., Essex, J.W., Sansom, M.S.P., Jeffreys, P.: BioSimGrid: Grid-enabled biomolecular simulation data storage and analysis. Future Generat. Comp. Syst. 22, 657–664 (2006)
50. Paluska, J., Saff, D., Yeh, T., K. Chen, K.: Footloose: A case for physical eventual consistency and selective conflict resolution. IEEE Workshop on Mobile Computing Systems and Applications, p. 170, 2003. ISBN 0-7695-1995-4
51. Peenikal, S.: Mashups and the enterprise. Strategic white paper, Sept 2009. Mphasis white paper
52. Peirce, B.: Linear Associative Algebra. D. Van Nostrand, Princeton (1882)
53. Price, A.R., Xue, G., Yool, A., Lunt, D.J., Valdes, P.J., Lenton, T.M., Wason, J.L., Pound, G.E., Cox, S.J., The GENIE team: Optimisation of integrated earth system model components using grid-enabled data management and computation. Concurrency Comput. Pract. Ex. 19(2), 153–165 (2007). URL http://eprints.soton.ac.uk/23514/
54. Redkar, T.: Windows Azure Platform, 1st edn, p. 193. Apress, New York (2010)
55. Salesforce: Salesforce homepage, 2010. www.salesforce.com. Accessed 23 Aug 2010
56. SQL Server: New T-SQL Features. Pro T-SQL 2008 Programmers Guide, pp. 525–551 (2008)
57. Smith, J.E., Nair, R.: The architecture of virtual machines. Computer 38(5), 32–38 (2005). ISSN 0018-9162. 10.1109/MC.2005.173
58. SpaceTrack: The Source For Space Surveillance. www.space-track.org. Accessed 23 Aug 2010
59. Stantchev, V., Schröpfer, C.: Negotiating and enforcing QoS and SLAs in Grid and Cloud computing. In: Advances in Grid and Pervasive Computing, vol. 5529 of Lecture Notes in Computer Science, pp. 25–35. Springer, Heidelberg (2009). 10.1007/978-3-642-01671-4_3
60. Suchi: S3Fox organizer, 2010. www.s3fox.net. Accessed 23 Aug 2010
61. Vogels, W.: Eventually consistent. Commun. ACM 52(1), 40–44 (2009)

Chapter 10
The CloudMiner

Moving Data Mining into Computational Clouds

Andrzej Goscinski, Ivan Janciak, Yuzhang Han, and Peter Brezany

Abstract Business, scientific and engineering experiments, medical studies, and governments generate huge amount of information. The problem is how to extract knowledge from all this information. Data mining provides means for at least a partial solution to this problem. However, it would be too expensive to all these areas of human activity and companies to develop their own data mining solutions, develop software, and deploy it on their private infrastructure. This chapter presents the CloudMiner that offers a cloud of data mining services (Software as a Service) running on a cloud service provider infrastructure. The architecture of the Cloud-Miner is shown and its main components are discussed: MiningCloud that contains all published data mining services, BrokerCloud which mining service providers publish services to, DataCloud that contains the collected data, and Access Point which allows users to access the Service Broker to discover mining services and supports mining service selection and their invocation. The chapter finishes with a short presentation of two use cases.

10.1 Introduction

Most of the modern scientific applications are strongly data driven [31]. Through a large number of business transactions, through sensors, experiments, and computer simulations, scientific data are growing in volume and complexity at a staggering

A. Goscinski (✉)
School of Information Technology, Deakin University, Pigdons Road, Waurn Ponds 3217, Australia
e-mail: andrzej.goscinski@deakin.edu.au

I. Janciak · Y. Han · P. Brezany
Department of Scientific Computing, University of Vienna, Nordbergstrasse 15/C/3, Vienna, Austria
e-mail: janciak@par.univie.ac.at; han@par.univie.ac.at; brezany@par.univie.ac.at

S. Fiore and G. Aloisio (eds.), *Grid and Cloud Database Management*,
DOI 10.1007/978-3-642-20045-8_10, © Springer-Verlag Berlin Heidelberg 2011

rate. The cost of the data producing and its persistent storage is very high: satellites, particle accelerators, genome sequencing, and supercomputer centers represent data generators that collectively cost billions. The situation is complicated even further by the distribution of these data; data sources, storage devices, and computation resources are located around the whole world. Without effective ways to retrieve vital information from this large mass of data that great financial expense will not yield the benefits to society that might be expected. Typically, to extract appropriate knowledge requires data mining over combinations of data from multiple data resources using different analytical tools utilizing high performance computational resources. Today, designers, diagnosticians, decision makers, and researchers who need such knowledge face difficult hurdles. To extract information from heterogeneous and distributed sites, they have to specify in much detail the sources of data, the mechanisms for integrating them, and the data mining strategies for exposing the hidden gems of information. They also have to deal with limited computational resources, which can be a bottleneck in the real time data processing. Consequently, with the current state of the art, most of that hidden knowledge remains undiscovered. Therefore, knowledge discovery in databases is still challenging research and development issue. Some attempts to tackle the problems have been made by applying grid computing. However, grid research promised but not delivered fast, interoperable, scalable, resilient, and available systems. Grids have been used mainly in research environment; grids have not been accepted by business and industry [17]. It is expected, and the expectations have been confirmed, that cloud computing would be able to address the problems of distributed and of huge amount data [10]. Cloud computing is a style of computing, strongly supported by major vendors and many IT service consumers, in which massively scalable high-throughput/performance IT-related services are provided on demand from the Internet to multiple customers. Cloud computing brings together distributed computing concepts and development outcomes, and business models. The infrastructures based on cloud computing concepts enable conducting development and use of tasks addressing large-scale data integration, preprocessing, and data mining in this challenging context. A cloud, a basis of cloud computing, exploits the concepts of services and service oriented architecture (SOA), virtualization, Web technology and standards, and computer clusters [9], although even small developers can offer specialized cloud services on small computation systems. In this chapter, the key aspects associated with efficient realization of data mining, in particular, the distributed data mining in a computational cloud based on a service-oriented environment are investigated. The overall infrastructure aiming to support on demand data processing is described in this chapter and is referred to as CloudMiner. The main objectives of the CloudMiner project are summarized as follows:

- Design a powerful, flexible, and easy to use on demand computation environment for execution of data mining tasks;
- Simplify the task of data mining service providers regarding data mining service development, deployment, and publishing;

- Support service consumers, in discovery, selection, execution, and monitoring of services that satisfy their requirements.

This chapter makes the following original contributions. First, it offers a new vision of distributed data mining by employing cloud computing. Second, it enhances the SOA by direct inclusion of data, leading toward a Data and SOA. Third, it proposes the CloudMiner, a new cloud-based distributed data mining environment. The rest of this chapter is structured as follows. Section 10.2 presents basic concepts of distributed data mining and its requirements, introduces cloud computing and clouds, and identifies cloud category that matches the requirements of distributed data mining. Section 10.3 discusses the architecture of the proposed CloudMiner and specifies its basic components: the MiningCloud, the DataCloud, the BrokerCloud, and the Access Point. Section 10.4 presents the concrete use cases and discusses some implementation issues of the CloudMiner. Section 10.5 concludes the chapter.

10.2 Data Mining and Cloud Computing

Data mining, in particular distributed data mining, allows user to extract knowledge from huge amount of data, which are very often stored in many places [14]. Cloud computing is a form of computing based on services provided in the Internet. This section presents basic concepts of distributed data mining and its requirements, introduces cloud computing and clouds, and identifies cloud category that matches the requirements of distributed data mining.

10.2.1 Distributed Data Mining

Data mining is the automated or convenient extraction of patterns representing knowledge implicitly stored in large volumes of data [20]. Typically, data mining has two high-level goals of prediction and description, which can be achieved using a variety of data mining methods, e.g., association rules, sequential patterns, classification, regression, clustering, change and deviation detection, and so forth. From the user's point of view, the execution of a data mining process and the discovery of a set of patterns can be considered either as an answer to a sophisticated database query or the result produced upon performing a set of data mining tasks. The first is called the descriptive approach, while the second is the procedural approach. To support the former approach, several data mining query languages have been developed. In the latter approach, data mining applications are viewed as complex knowledge discovery processes composed of different data processing tasks. Data mining in large amounts of data can significantly benefit from the use of parallel and distributed computational environments to improve both performance and accuracy of discovered patterns. These environments allow the compute-intensive data mining of distributed data, which is referred to as distributed data mining. In contrast to the centralized model, the distributed data mining model assumes that the data sources are distributed across multiple sites. Algorithms developed within this field address the problem of efficiently getting the mining results from all the data across these distributed sources. The application of the classical knowledge discovery process in distributed environments requires the collection of distributed data in a data warehouse for central processing. However, this is usually either ineffective or infeasible because of the storage, communication and computational costs, as well as the privacy issues involved in such an approach. Distributed data mining offers algorithms, methods, and systems that deal with the above issues to discover knowledge from distributed data in an effective and efficient way.

10.2.2 *Cloud Computing*

Cloud computing is a style of computing in which massively scalable high-throughput/performance IT-related services are provided on demand from the Internet to multiple customers. This means that computing is done in a remote unknown location (out in the Internet clouds) rather than on a local desktop. A cloud is an inexpensive Internet accessible on demand environment where clients use virtualized computing resources on a pay-as-you-go basis [11] as a utility and are freed from hardware and software provisioning issues. Companies and users are attracted to cloud computing because:

(a) Clients only pay for what they consume.
(b) Rather than spending money on buying, managing, and upgrading servers, business administrators concentrate on the management of their applications.
(c) The required service is always there – availability is very high that leads to short times from submission to the completion of execution.
(d) Cloud computing provides opportunities to small businesses by giving them access to world class systems otherwise unaffordable. On the other hand, even small companies can export their specialized services to clients.

Applications could be deployed on public, private, and hybrid clouds [9]. The decision with regard to which cloud model should be selected depends on many factors, among them cost, trust, control over data, security, and quality of service (QoS). Cloud computing means using IT infrastructure as a service over the network. The question is how different infrastructures form a basis of these services. Virtualization lays the foundation for sharable on demand infrastructure, on which three basic cloud categories [9] are offered on demand:

• Infrastructure as a Service (IaaS) – IaaS makes basic computational resources (e.g., storage, servers) available as a service over the network on demand.

- Platform as a Service (PaaS) – The PaaS platform is formed by integrating an operating system, middleware, development environment, and application software, and encapsulated such that is provided to clients, human users or another services as a service.
- Software as a Service (SaaS) – SaaS allows complete end-user applications to be deployed, managed, and delivered as a service usually through a browser over the Internet. SaaS clouds only support provider's applications on their infrastructure.

The CloudMiner primarily belongs to the category of SaaS clouds; however, it includes some features of PaaS. This implies that data mining algorithms implemented and deployed using a variety of languages and application development tools will be exposed as Data Mining Services.

10.2.3 Data Mining Services

Web services enable to achieve interoperability between heterogeneous data mining applications through platform and language-independent interfaces. As the number of available data mining services increases, it becomes more difficult to find a service that can perform a specific task at hand. Moreover, there may also be no such single data mining service capable of performing the specific task, but a combination of other existing services may provide this capability. Hence, a standardized set of interfaces and data interchange formats between services is necessary to discover suitable services as well as to enable composition of the services into complex workflows. The data mining service performs analytical tasks with certain QoS guarantees. The service wraps and exposes its functionality to the external applications or to other cooperating services through a well-defined interface. Quality of Web services has been already explored in various contexts [1,4]. In general, there are two types of QoS parameters that specify quantitative and qualitative characteristics of services. For a data mining service, the qualitative parameters may be, for example, an accuracy of a data mining model, and quantitative, for example, time it takes to train the model. In our approach, only the quantitative characteristics such as price or processing time. Semantic description of Web services aims to tackle the problem of discovering of demanded functionality by grounding the Web services to particular port types or by an extension of the interface description. In both cases, it requires semantically annotated interfaces with well-described input parameters and results produced by the services. Hence, a well described semantics is required to describe data mining tasks performed by data mining Web services. Data mining applications contain a number of generic components which interact in well-defined ways. These applications typically exhibit different levels of capability within each of these generic components. To achieve interoperability between data mining applications, a standardized set of interfaces and data interchange formats between such components is necessary. A set of requirements for data mining services arising from different application domains has been identified. These requirements are driven mainly from different user points of views such as application domain users, data mining experts, application developers, and system administrators. These kinds of users are concerned with different requirements for the data mining services presented in what follows:

- Ease of use – End users should be able to use the services without a need to know any technological details about the implementation of the service. The interfaces and the way how to interact with the service must be clearly defined.
- Scope of mining techniques – The service should be capable of accommodating a widely differing set of application domains like, for example, business, bioinformatics, health management, text mining, Web mining, spatial data mining, etc., and should support tasks on different levels of abstraction as defined by the CRISP-DM reference model [12].
- Seamless collaboration – Well-defined cooperation scenarios are needed to create processes aiming to realize more generic tasks in the hierarchy of data mining project. For example, a user should be able to define a data preprocessing task consisting of two services where the first is a data cleansing service and the second data formatting service.
- Control and monitoring – A user must have the possibility to interact with the whole process as well as with a particular service. Therefore, the user should be able to monitor the progress of the overall workflow execution and actual state of the involved services. The ability of the service such as starting, stopping, and resuming execution is required.
- Extensibility – The service's interface should be independent of any particular data mining technique or algorithm. The requirement here is that the service must be able to accept via its interfaces any parameters needed for mining algorithms. Moreover, the design of the service must be flexible enough to permit entirely new data mining functionality.
- QoS awareness – The service providers may offer either equivalent or similar services, so the services should expose the QoS guarantees. This allows a user to select the most suitable service on the basis of his personal QoS requirements.
- Manageability – System administrators should be able to easily deploy a new service as well as port the service to a different location without its modification or limiting its functionality.
- Security and privacy – The service must ensure secure communication with the clients and its peers. The security should be provided on the message as well as on transport level. Data privacy must be strongly respected and the service must not allow access to the data for the nonauthorized users.
- Fault tolerance – Support for fault tolerance should be considered a necessity for services, rather than as an additional feature. The full recovery of the failed process should be also supported.

In our previous work on the GridMiner project [5], several data mining Web services that form the initial set of CloudMiner services have been developed. Most of the above considerations were used during reimplementation of the services toward the Web Service Resource Framework. The architectural design of the data mining services is described in Reference Model for Data Mining Web Services [23]. The initial set of services includes: DecisionTreeServise, NeuralNetwork-Service, AssociationRulesService, SequenceRulesService, ClusteringService, and OLAPService.

10.2.4   Related Work

Many data mining processes feature high data intensity and high computational load. In such cases, cloud computing can provide a low-cost solution. The IBM Smart Analytics [22], like the CloudMiner, supports standard data mining algorithms, which can be applied on data resources in a cloud. This system supports real-time scoring of data records on mined data. Besides, a set of rich presentation components is offered to enable visual analysis of data mining results. In comparison, the CloudMiner presents two major differences: first, it exposes all its data mining programs as Web services; second, it allows users to develop and deploy customized mining algorithms. Another data mining oriented cloud implementation comparable to CloudMiner is the Sphere/Sector cloud [19], established on a cloud infrastructure which provides resources and/or services via the Internet, this cloud is used to archive, analyze, and mine large size distributed data sets. Analogous to the CloudMiner, it also possesses two major elements: a storage cloud providing storage services, along with a compute cloud providing compute services. The major difference resides in that, while Sphere/Sector cloud requires its applications to be programmed in a special parallel computing language, CloudMiner does not require its services to obey any special programming model. Data mining services on the CloudMiner belong to the SaaS category. Another cloud service which can be identified as in the same category is the Cloud Service for Mathematica [32]. This is a service supporting cloud-based execution of Mathematica application. Using this batch submission service, the user can upload a script file along with any input data files onto the preconfigured Amazon EC2 [2] or R Systems cloud [27]. During the job execution, the status of the job can be monitored in real time on Web pages. In comparison, CloudMiner is not constrained in using a particular calculation tool, like Mathematica. On the contrary, it allows users to develop and deploy any kind of mining tools. Although falling into a different cloud category, the PaaS category, the Distributed Computational Service Cloud (DCSC) [21] has some analogous features as CloudMiner. First, both cloud systems allow for the deployment of cloud services by the user. Second, both support the execution of service workflows. The primary distinction is that DCSC is not purely data mining oriented and can be used for any SaaS clouds. A distinctive feature of the CloudMiner is that it is based on the Resources Via Web Services (RVWS) framework [8]. This allows for the attributes of services, the exposed (cloud) resources and providers to be published to a registry by Web service WSDL documents directly. In the Cluster as a Service paradigm [9], RVWS is used to monitor and publish the state of clusters in cloud. In comparison, CloudMiner uses it for state monitoring and publishing of the data mining services.

10.3 Architecture

The basic architecture of the CloudMiner follows the SOA, and as such contains three basic components. First, services developed by service providers and published to a registry. Second, the registry that stores and manages all information about published services. The third component of the architecture is a client (end user). Normally, a SaaS cloud offers hundreds of services offered by one single service provider [29]. However, we propose here that many research institutions, companies, and even individual users could offer their mining services. Thus, it is possible that many service providers will contribute their products to the CloudMiner. The SOA architecture does not contain data that will be processed by services. Data mining is about data, huge amount of data stored in files and databases managed by file and database management systems. Therefore, the architecture of the CloudMiner contains data, the data mining service carry out computation services on. The registry contains information about published services. A cloud may have one registry to store that information or could use a number of registries for reliability and performance reasons. We also assume that with the increase of amount of data being collected around the world and different areas of applications of those data, there could be more than one cloud each supported by its registry. These two cases lead to a solution of multi-registry infrastructure. Usually, a registry contains basic information about service location and service provider. We propose to contain more information about each individual published service. The additional information contains semantics of a mining service. This implies a need for the use of the concept of a broker, which offer intelligence, which is not provided by simple registries. The user, the third basic component of the SOA architecture, should be able to learn about data mining services offered by the cloud. We propose that the CloudMiner offers through a portal a possibility to carry out operations on data and to learn about all services offered by service provider. The user accesses a registry to learn about data mining services and select a service that could solve her problem. The management of computation using the selected data mining service on specified data is be carried out directly by the end user. The registry, made more intelligent, also could carry out this task. The CloudMiner architecture offers both options. Figure 10.1 illustrates the components of CloudMiner. The SOA-based architecture involves: MiningCloud, DataCloud, BrokerCloud; and two actors, end user and service provider.

A client is a major component of the SOA architecture. In CloudMiner, the client learns about data mining services via the Access Point, which is Web-based portal. The end user as a consumer of the results provided by the services in MiningCloud uses the Access Point to select a service that satisfies her requirements, parameterizes data mining tasks, invokes mining service on specified data, and manages the overall mining process. The end user communicates with the Service Broker via Access Point so in general it simplifies interaction between a Service Broker and end-user. Via the Access Point also the third party applications and services can

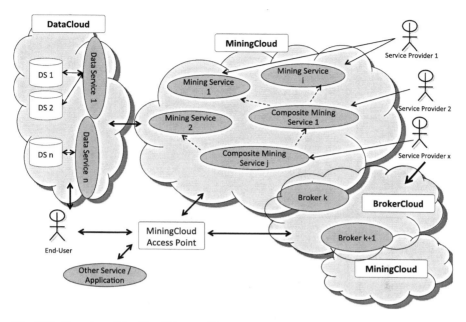

Fig. 10.1 Overview of the CloudMiner architecture

access the MiningCloud. As shown in Fig. 10.1, the service provider deploys data
mining services directly to the MiningCloud and registers them to the BrokerCloud.

10.3.1 The MiningCloud

Implementations of data mining tasks are exposed as Web services deployed into the
MiningCloud. A service provider, who developed the data mining service, publishes
it to the Service Broker. In other words, the service provider deploys a data mining
service to the MiningCloud and publishes its description in the BrokerCloud. In
summary, by the MiningCloud it is meant an infrastructure that provides a set
of data mining services over the Internet. Since the best known and frequently
used implementation of SOA-based infrastructures is Web services, the CloudMiner
exploits Web services.

10.3.1.1 Handling Service State

By their nature, Web services are stateless. This is a satisfactory solution for some
users and businesses. However, in the domain of distributed data mining there is
a need for stateful Web services for two main reasons: first, to manage resources

they expose effectively, and second, to allow users to make a "good" selection of a service that satisfies user's requirements. For example, the user should be advised by the Service Broker whether the service the user is after is running on a single-tenant virtual computer or multi-tenant virtual computer. The former implies that the cost of consuming this service could be cheaper, but the waiting time is longer because other users are in the waiting queue. The position in the queue describes the state of the Web services as it reflects the resource exposed by the services. The latter means that the user can consume the service very quickly but the cost is higher. In this case, the state of a service reflects the level of multi-tenancy and availability of virtual machines that provides the service. Here, the services of the MiningCloud are stateful Web services which expose their state either as a part of its description using RVWS [7] or as a state of a stateful resource representing an instance of the Web service. The problem is how to make services stateful and publish this information to the Service Broker to be accessed and taken advantage of by users. We propose to use one or a combination of the following approaches.

- The first approach has been proposed by the Web Services Resource Framework (WSRF) [16]. This framework offers a set of Web service specifications that provide means to model and manage state in the context of the stateless service context. However, this approach requires additional steps and operations to publish the state, in particular when it changes dynamically.
- The second approach is based on the RVWS framework. This approach does not suffer from the publishing difficulty of the WSRF-based approach. According to the RVWS framework, the state of a resource and service itself are provided in the service WSDL document. (The framework allows also passing to the WSDL document information such as price, QoS parameters, etc.) Any changes to the state are passed on procedure that updates WSDL and provides this information to the Service Broker on line. The updates performed are triggered by events or are time driven.

The MiningCloud infrastructure is implemented with a network of compute nodes, on each of which a certain number of Mining Services are deployed and maintained. The Mining Services are allowed to communicate among themselves by performing intra-services communications, in the case of multi-tenant clouds, or inter-services communications, in the case when services run on different computers, during their life cycles.

10.3.1.2 Mining Services

As can be seen in Fig. 10.1, two types of services are proposed, single mining services and composite mining services. Although both types are stand-alone and evocable Web services, they have distinct functions and uses:

- *Single mining service:* It is developed and deployed by a service provider to implement a data mining algorithm. During the execution, it can store and retrieve data resources to and from the DataCloud. Either users or other services can invoke a single mining service. We denote the set of single mining services existing in the MiningCloud as $\langle SSi \rangle, i \in 1, \ldots, I$; SS represents a single mining service.

- *Composite mining service:* The major task of this type of services is to carry out service invocations in a manner required by the service provider. By doing this, a composite mining service represents a service workflow. We denote the set of composite mining services hosted in the MiningCloud as $\langle CSj \rangle, j \in 1, \ldots, J$, with $CSj = \langle SSm, \ldots, SSn \rangle$; CS stands for a composite service, and m, n are integers, where $m, n \leq I$.

10.3.2 The DataCloud

The DataCloud is a platform that allows for the deployment and management of Web-based Data Services. The platform is mainly focused on data access and integration of the distributed data, which are the indispensable aspects of a cloud-based delivery model. The DataCloud hosts Data Services that are managed by end users due to the data privacy and security. It means that end users can deploy and configure their own data services and enable controlled access to the data resources which might be any kind of digital data. Physically, these resources may span across multiple servers.

10.3.2.1 Virtualization in the DataCloud

The DataCloud, as depicted in Fig. 10.1, hosts and virtualizes a collection of data sources. Exposed by Data Services, these data sources can be accessed and used by the Mining Services as well as by the remote users. However, some aspects of the data resources, such as their physical implementation, data formats, and locations, should be hidden from the users by data services. Also, data services need to tackle the heterogeneity in different resources and enable the user to concentrate on the data itself. Therefore, the Data Service provides a standard interface such as the one specified by WS-DAI [18].

10.3.2.2 Data Services

Data services support access and integration of data from separate data sources which are typically distributed. This allows different types of data resources including relational databases, XML, CSV files, or multimedia content to be exposed to the third-party applications. The services provide a way of querying, updating, transforming, and delivering data through a consistent, data resource independent way. This is achieved via Web services that can be combined to provide higher-level services supporting data federation and distributed query processing. Hence, the major strength of the services lies in supporting the integration of data from various data resources. Furthermore, it allows querying metadata about the data as well as resources in which this data is stored. Additionally, the Data Services might make use of data available in other "Storage Cloud." There are two types of data requests which are supported by the DataCloud, internal requests, and external request.

- *Internal requests:* These refer to the data requests issued by the services hosted in the MiningCloud. Every mining service is allowed to possess and use its data resource in the DataCloud. There are interfaces established between the Mining-Cloud and DataCloud to carry out the data requests from the MiningCloud.
- *External requests:* These are the data requests sent directly by the end-user. Users might be interested in monitoring the state of her data resource or, rather than via mining services, directly storing data to and retrieving data from the cloud. Therefore, the direct access is provided by the DataCloud.

As an example, we can use the OGSA-DAI [18] to satisfy the requirements raised above. The OGSA-DAI supports the federating and access of data resources (e.g., relational or XML databases, or files) via Web services inside or outside cloud. Typically, OGSA-DAI is accessed through a Web service presentation layer. These Web services expose data resources managed by OGSA-DAI. Clients only need to specify the resource of interest during interaction with the services. The services address the aspects, among other, such as data request execution, data resource information, and establishment of data sink and data source. To enable the utilization of OGSA-DAI in the DataCloud, two issues should be addressed:

- Adapters should be developed and attached to each service in the MiningCloud in order that requests of data resource access issued by mining services can be forwarded to the OGSA-DAI services and processed by the OGSA-DAI then.
- Proper tools and mechanism, which are similar to those used in the Service Monitor of the MiningCloud, should be setup on the DataCloud to enable the OGSA-DAI services to publish the state information of the data resources. Modules and methods need to be designed and planted into the Web service based Cloud Access Point to support direct access to the data resources via OGSA-DAI services.

10.3.3 The BrokerCloud

A registry is a basic element of any SOA-based infrastructure. Currently, simple registries (e.g., UDDI) satisfy basic requirements in the world of business [3]. However, such registries are unsatisfactory for clouds supporting scientific applications. There is a need for intelligence to support service discovery and selection. They are in particular unsatisfactory when services offered and published are stateful. Therefore, we propose to use Service Brokers in the CloudMiner. The Service

Broker is an intermediary between a service provider and the end-users mainly for locating the mining service and obtaining its description. The role of the Service Broker is to transform functional and nonfunctional requirements provided by the end-users to search for proper services in the MiningCloud. In summary, by the BrokerCloud, we mean: *an infrastructure that provides information about the Web services in the MiningCloud via a set of Service Brokers.*

10.3.3.1   Service Brokers

Service providers publish information about mining services to the Service Broker. In the case of stateless Web service, only a basic WSDL document, which provides methods to invoke services, and service URL, which provides its location in the Internet, are published. In this case, it is the responsibility of users to learn about published Web services. In the case of stateful Web services, the Service Brokers store information about the state, as presented in the description of the approaches (discussed in Sect. 10.3.1.1) to making Web service stateful. Thus, in the case of the first approach, the service providers must include additional resource properties that reflect the semantic of the published service. If the second approach to making Web services stateful was used, the information passed in the WSDL document, which contains state and additional parameters (e.g., price, QoS), is passed in the WSDL document. The information stored in the Service Broker is accessed by users. As we stated earlier, there could be more than one Service Broker that serve the MiningCloud for reliability and performance reasons or there are other Service Brokers that support other MiningClouds. There are different forms of cooperation among these Service Brokers. For example, if there is no service that satisfies user requirements in the normally served MiningCloud, the local Service Broker can transparently access one by one, or concurrently other Service Brokers, and make enquires on behalf of the end user. The local Service Broker can even make a decision to access a service that belongs to another MiningCloud. This means that a BrokerCloud is in use. Service Broker monitors and potentially limits the usage of exposed services, based on the end user's requirements and system resource usage. Figure 10.2 shows the internal components of the Service Broker.

The Service Broker tries to satisfy requirements of users in the tasks described in the following sections.

10.3.3.2   Registering a New Mining Service

Service providers are responsible for publishing newly deployed services to the Service Broker. The only mandatory information they generate automatically is a WSDL document and service's URL. The WSDL document contains enough information for invocation of the service but typically it lacks semantic description of the service functionality. Extension of WSDL, such as SAWSDL [26] or RVWS [8], might be used to provide the semantics, which can significantly help during the process of service discovery and its selection by end users. The service providers

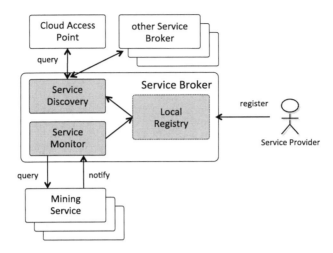

Fig. 10.2 Components of the service broker

register the service to the Service Broker, which is attached to the MiningCloud and is accessible via the Access Point. The Service Broker stores the location of the registered service in a local data store together with values of parameters being monitored by ServiceMonitor component of the Service Broker. Since each service provides a different set of parameters that can be monitored, it is up to the service provider to identify these parameters and to configure the ServiceMonitor to listen to their changes. A typical example is the monitoring of service properties attached to the stateful Web services confirming WS-Resource or resources exposed through the application of the RVWS framework in the WSDL stateful document specification.

10.3.3.3   Acquiring State of a Mining Service

The actual state of a mining service is obtained by the Service Monitoring component of the Service Broker. When a new service is registered by the service provider, the Monitoring component starts to listen to the changes in the specified parameters of the service. Each time an update is detected the Service Monitor updates the local data store of the Service Broker. The update can be detected by regular querying of the service parameters typical for RVWS approach or by Web service notification mechanism. The Web service notification mechanism requires implementation of WS-Notification and WS-BrokerNotification specifications which are companion of WSRF.

10.3.3.4   Service Discovery

The service discovery is an essential feature of any distributed system based on SOA. From the end-user point of view, it is not important if a single or composite service is selected during the discovery process; the important are the functional and QoS parameters of the service, based on which the user makes selection of the service. Therefore, a ranking of the discovered services should be also supported by the Service Broker.

10.3.3.5   Service Composition

CloudMiner allows the user to create composite mining services on demand. Let $\langle Si \rangle$ represent the services located in the discussed MiningCloud (Sect. 10.3.1.2) with $i \in 1 \ldots I$, and let $\langle Sj \rangle$ represent the services hosted in other clouds where $j \in M \ldots J$ with M being an integer and $I < M < J$. Here, S represents a single or composite mining service. Then, a composite mining service composed of $\langle Sp, \ldots Sq, Sx \ldots Sy \rangle$ can be created, p, q, x, y are integers, where $p < q \leq I$ and $M \leq x < y \leq J$. By creating and using an on-demand composite mining service, the user can execute a service workflow defined by her. Such a workflow consists of multiple service invocations, which are organized in a user-customized manner, e.g., they are arranged in a sequence or a loop. Given the functionalities of composite mining services required by the service provider, we propose to implement composite mining services based on the workflow engine from Workflow Enactment Engine Project (WEEP) [24]. The WEEP Engine is a Web service workflow engine, which supports dynamic invocation of Web services based on workflow description in WS-BPEL [30]. It can be easily encapsulated into a Web service and started by service invocation.

10.3.3.6   Gathering Information About State of a Mining Service

Given that the mining services implement different data mining tasks, it is useful to let the user acquire information regarding the data mining tasks it provides, progress of the tasks during execution, and resources they consume. The acquiring and publishing of the state information are done using the RVWS, which extends the WSDL document by inserting optional XML elements containing information about the monitored parameters. We propose that Service Broker exposes these three types of information about the mining services as follows:

- Information about data mining tasks includes parameters, which describe the nature of an implemented algorithm, nevertheless, are unrelated to the algorithm execution. Such parameter can be, for example, a description of data mining algorithm, indicating whether a decision tree algorithm uses pruning or whether a neural network algorithm supports momentum, etc. Such information can assist an end user in selecting the requested services.
- Progress monitoring of data mining tasks gives information about the state of the tasks being executed. Here, we assume that a mining service exposes the state of the task progress based on its life cycle. A typical data mining task comprises multiple phases, each of which the user becomes aware of the progress. For example, the phases of a decision tree-based algorithm can be: preprocessing input data, training model, pruning tree, evaluating model, etc. Accordingly, the

user can decide to terminate or restart the service if it has been trapped in one of the stages for an unacceptably long period of time.

- Information about consumed resources during execution of the service is important for billing of the cloud usage. The RVWS framework enables to expose the information about resources such as disk, memory, and CPU usage. The Service Broker monitors also these parameters and offers users only the mining services that have enough available resources.

The publishing of the above information is carried out by the Service Broker, to be used in the service discovery. Also, user can easily attain the published information and let it facilitate state monitoring and visualization.

10.3.4 The Access Point

The Access Point exposes the functionality of CloudMiner components to the outer world. Hence, it is a kind of a gateway to the underlying cloud infrastructure. All the communication between the end users and the Web services is done via the interface of the Access Point. Note that it is not the case for the Service Providers, who have direct access to the MiningCloud and deploy their services there. The Access Point plays an important role in the usability of the cloud since it simplifies the interaction with the CloudMiner components and hides its complexity. The Access Point navigates the end-user in the process of selecting appropriate data, data mining services, their execution, and monitoring. To prepare and configure data mining tasks, the Access Point provides an easy to use step-by-step configuration wizard. Figures 10.3 and 10.4 show examples of the configuration pages of the wizard.

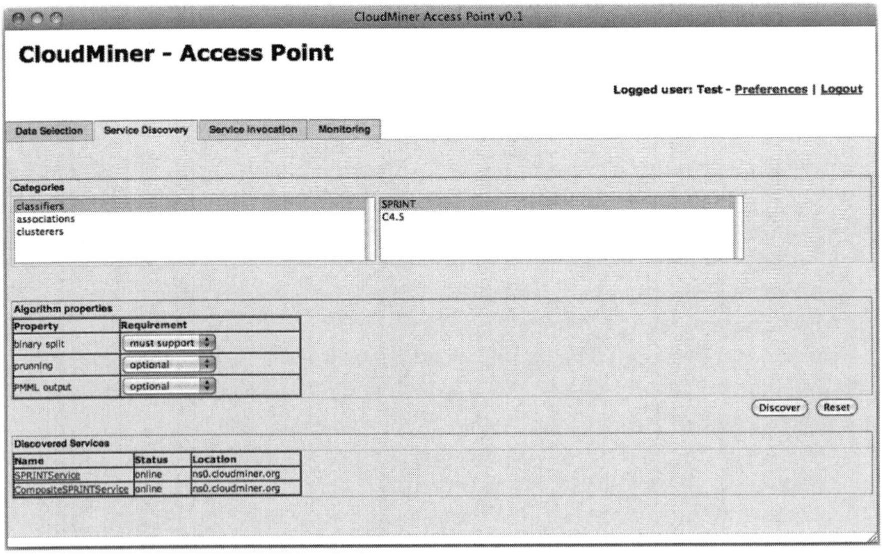

Fig. 10.3 Access point – service discovery

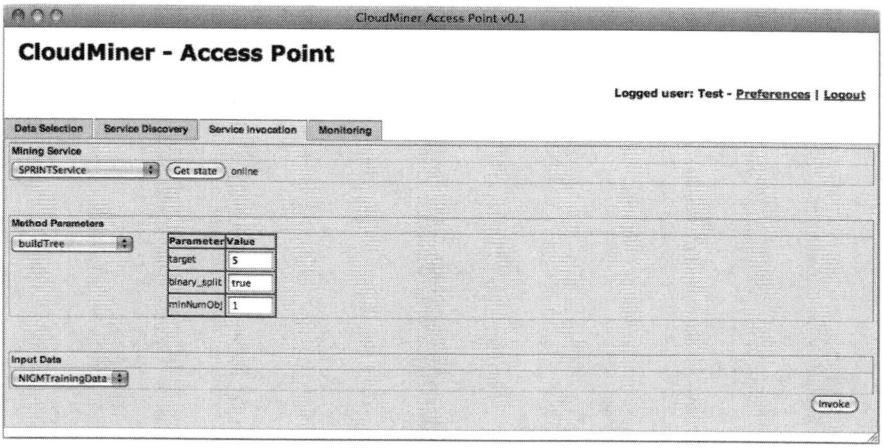

Fig. 10.4 Access point – service invocation

The first step in the wizard is data selection where the end user selects DataService, which provides input data for the data mining task. Here, the end user specifies a query and resource managed by the DataService hosted in the DataCloud. The second step deals with discovery of an appropriate mining service providing required task (see Fig. 10.3). Here, the end user selects a data mining algorithms the service should support. Based on the input parameters, the Service Broker returns a list of available mining services supporting the specified algorithm. In the next step, the end user configures and invokes the selected mining service (see Fig. 10.4). This means that the end user specifies values of the input parameters and selects input data set specified in the first step. After the parameterization of the execution task, the user can invoke the service and observe its state in the monitoring panel.

10.4   Use Cases

Our first attempt to evaluate the proposed architecture has been done by extending the implementation of the Distributed Computational Services Cloud (DCSC) [21], built up on top of the NIMBUS Toolkit [25]. In our approach, the MiningCloud deploys the scalable distributed service computing models using SPRINT algorithm [28] for building classification models. It is a decision-tree-based classification algorithm, which eliminates memory restrictions, thus it is scalable enough to evaluate our system. Furthermore, SPRINT has been designed to be easily parallelized. "Parallel" here refers to the working model where multiple processors work together to build one consistent model. This feature ensures the possibility of distributing the computation over services deployed in the computational cloud. Thus, the service is suitable for evaluation of the proposed CloudMiner

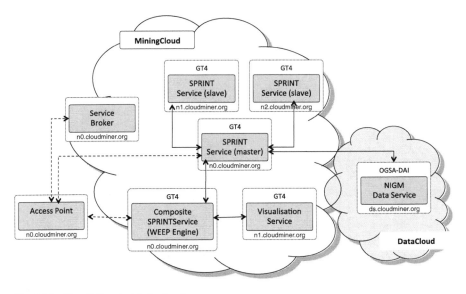

Fig. 10.5 CloudMiner test infrastructure

architecture. In SPRINT algorithm, a decision tree is constructed by recursively binary-partitioning the set of data objects until each single partition is either pure, that is all members have the same value of the class attribute, or sufficiently small, that is the size of the partition has reached a value set by the user. The implemented Web service adopts the master-slave paradigm where the master service controls the slave services processing subset of the input data. The service obtains input data from DataCloud, which hosts one DataService provided by the OGSA-DAI middleware. In our experiment, we used data taken from the NIGM Project [6]. Figure 10.5 shows the distribution of the Web services used for two use cases deployed into the CloudMiner's infrastructure.

There are two mining services deployed in the MiningCloud. The first one is the SPRINTService and the second one is the CompositeSPRINTService, which represents a composite Web service supported by a workflow engine. Both services are published to the Service Broker. In the following sections, we describe them in more details.

10.4.1   Use Case A: Single Classification Service

In this use case, there are installed three GlobusToolkit 4 (GT4) [15] containers each on a different node. Each of the containers is hosting one SPRINTService. The first one (node n0.cloudminer.org) plays role of a master service delegating the actual building of the decision tree to two other slave services. The master SPRINTService represents a mining service in the MiningCloud and is published to a Service Broker, which is installed on the same node. The service implements notification mechanism of the WSRF framework and notifies the service monitoring component of the Service Broker about changes in the resource properties being monitored, that is, CPU usage and memory usage. The input parameters of the service provided by the end user are: (1) a SQL query expression selecting training data from a DataService; (2) specification of a target attribute; (3) binary split applicability; and (4) tree pruning applicability. On the output, the SPRINTService returns decision tree model in the PMML format [13].

10.4.2   Use Case B: Composite Classification Service

This use case adds two additional Web services to the MiningCloud. The first one is VisualisationService, which transforms the decision tree model in PMML format produced by SPRINTService into a graphical representation in scalable vector graphics format. The second is CompositeSPRINTService, which invokes the SPRINTService and passes the PMML to the VisualisationService for transformation. The composite service is implemented by the WEEP Engine [24] deployed in the GT4 container. Upon its invocation, the SPRINTService and VisualisationService are executed in a sequence. This sequence of the services execution is described using WS-BPEL and deployed as composite service using capabilities of the enactment engine, which allows redirecting of incoming requests and outcoming responses of the CompositeSPRINTService to an instance of the engine, which controls the execution as specified in the deployed WS-BPEL document.

10.5   Summary

In this chapter, we presented our approach for building a SOA-based cloud data mining environment, the CloudMiner. The environment consists of three different clouds, namely MiningCloud, DataCloud, and BrokerCloud, each hosting Web services aiming to satisfy requirements for on-demand Cloud Computing. The environment also contains the Access Point, a Web-based interface, that allows users to access the Service Broker to discover mining services and supports their selection and invocation. This chapter makes the following contributions to cloud and data mining. First, it offers the architecture of a novel infrastructure for distributed data mining called CloudMiner. Second, it allows both vendors and individual programmers/users to develop, deploy, and publish dynamically changing data mining services. Third, the whole cloud data mining environment allows for the provision by vendors composite data mining services as well and makes possible for users to create composite data mining services on demand. Fourth, every broker of the BrokerCloud supports publishing and discovery and selection of data mining services based on their state and attributes (e.g., execution requirements, price). The presented use cases deployed into the presented test infrastructure show that the approach, architecture, and its components are feasible for building distributed cloud environments. In the future, we will concentrate our research on the performance study of the whole system, its security, and reliability.

References

1. Al-Ali, R., von Laszewski, G., Amin, K., Hategan, M., Rana, O., Walker, D., Zaluzec, N.: QoS support for high-performance scientific Grid applications. In: Proceedings of the 2004 IEEE International Symposium on Cluster Computing and the Grid, CCGRID '04, pp. 134–143. IEEE Computer Society, Washington, DC, USA (2004)
2. Amazon: Amazon Elastic Compute Cloud (2010). URL http://aws.amazon.com/ec2
3. Banerjee, S., Basu, S., Garg, S., Garg, S., Lee, S.J., Mullan, P., Sharma, P.: Scalable Grid Service Discovery based on UDDI. In: Proceedings of the 3rd international workshop on Middleware for grid computing, MGC '05, pp. 1–6. ACM, NY, USA (2005)
4. Benkner, S., Engelbrecht, G.: A Generic QoS Infrastructure for Grid Web Services. In: Proceedings of the Advanced Int'l Conference on Telecommunications and Int'l Conference on Internet and Web Applications and Services, AICT-ICIW '06, p. 141. IEEE Computer Society, Washington, DC, USA (2006)
5. Brezany, P., Janciak, I., Tjoa, A.M.: GridMiner: An advanced support for e-science analytics. In: Dubitzky, W. (ed.) Data Mining Techniques in Grid Computing Environments, pp. 37–55. Wiley, NY (2008)
6. Brezany, P., Elsayed, I., Han, Y., Janciak, I., Wöhrer, A., Novakova, L., Stepankova, O., Zakova, M., Han, J., Liu, T.: Inside the NIGM Grid Service: Implementation, Evaluation and Extension. In: Proceedings of the 2008 4th International Conference on Semantics, Knowledge and Grid, pp. 314–321. IEEE Computer Society, Washington, DC, USA (2008)
7. Brock, M., Goscinski, A.: State aware WSDL. In: Proceedings of the sixth Australasian workshop on Grid computing and e-research – vol. 82, AusGrid '08, pp. 35–44. Australian Computer Society, Darlinghurst, Australia (2008)
8. Brock, M., Goscinski, A.: Attributed publication and selection for web service-based distributed systems. In: Proceedings of the 2009 Congress on Services – I, pp. 732–739. IEEE Computer Society, Washington, DC, USA (2009)
9. Brock, M., Goscinski, A.: A technology to expose a cluster as a service in a cloud. In: Proceedings of the Eighth Australasian Symposium on Parallel and Distributed Computing – vol. 107, AusPDC '10, pp. 3–12. Australian Computer Society, Darlinghurst, Australia (2010)
10. Brock, M., Goscinski, A.: Toward a Framework for Cloud Security. In: ICA3PP (2), pp. 254–263 (2010)
11. Brock, M., Goscinski, A.: Toward ease of discovery, selection and use of clusters within a cloud. In: IEEE International Conference on Cloud Computing, pp. 289–296 (2010)
12. Chapman, P., Clinton, J., Kerber, R., Khabaza, T., Reinartz, T., Shearer, C., Wirth, R.: CRISP-DM 1.0 Step-by-step data mining guide. Tech. rep., The CRISP-DM consortium (2000)
13. Data Mining Group: Predictive Model Markup Language, version 4.0 (2010)
14. Demers, A., Gehrke, J.E., Riedewald, M.: Research issues in distributed mining and monitoring. In: Proceedings of the National Science Foundation Workshop on Next Generation Data Mining. Baltimore, MD (2002)
15. Foster, I.: Globus Toolkit Version 4: Software for Service-Oriented Systems. In: IFIP International Conference on Network and Parallel Computing, no. 3779 in LNCS, pp. 2–13. Springer, Berlin (2005)

16. Foster, I., Frey, J., Graham, S., Tuecke, S., Czajkowski, K., Ferguson, D., Leymann, F., Nally, M., Sedukhin, I., Snelling, D., Storey, T., Vambenepe, W., Weerawarana, S.: Modeling stateful resources with web services v.1.1. Tech. rep., Globus Alliance (2004)
17. Goscinski, A., Brock, M.: Toward dynamic and attribute based publication, discovery and selection for cloud computing. Future Gener. Comput. Syst. **26**, 947–970 (2010)
18. Grant, A., Antonioletti, M., Hume, A., Krause, A., Dobrzelecki, B., Jackson, M., Parsons, M., Atkinson, M., Theocharopoulos, E.: OGSA-DAI: Middleware for Data Integration: Selected Applications. In: IEEE Fourth International Conference on eScience '08, p. 343 (2008)
19. Grossman, R., Gu, Y.: Data mining using high performance data clouds: Experimental studies using sector and sphere. In: Proceeding of the 14th ACM SIGKDD International Conference on Knowledge Discovery and Data Mining, KDD '08, pp. 920–927. ACM, NY, USA (2008)
20. Han, J.: Data Mining: Concepts and Techniques. Morgan Kaufmann, CA (2005)
21. Han, Y., Brezany, P., Janciak, I.: Cloud-Enabled Scalable Decision Tree Construction. In: International Conference on Semantics, Knowledge and Grid, pp. 128–135. IEEE Computer Society, Los Alamitos, CA, USA (2009)
22. IBM: IBM Smart Analytics System (2010).
    URL http://www-01.ibm.com/software/data/infosphere/smart-analytics-system/data.html
23. Janciak, I., Brezany, P.: A Reference Model for Data Mining Web Services. In: International Conference on Semantics, Knowledge and Grid, pp. 251–258. IEEE Computer Society, Los Alamitos, CA, USA (2010)
24. Janciak, I., Kloner, C., Brezany, P.: Workflow enactment engine for WSRF-compliant services orchestration. In: Proceedings of the 2008 9th IEEE/ACM International Conference on Grid Computing, GRID '08, pp. 1–8. IEEE Computer Society, Washington, DC, USA (2008)
25. Keahey, K., Freeman, T.: Science Clouds: Early Experiences in Cloud Computing for Scientific Applications. In: Cloud Computing and its Applications (CCA) (2008)
26. Kopecký, J., Vitvar, T., Bournez, C., Farrell, J.: SAWSDL: Semantic Annotations for WSDL and XML Schema. IEEE Internet Comput. **11**, 60–67 (2007)
27. R Systems: (2010). URL http://www.rsystems.com/index.asp
28. Shafer, J.C., Agrawal, R., Mehta, M.: SPRINT: A Scalable Parallel Classifier for Data Mining. In: Proceedings of the 22th International Conference on Very Large Data Bases, VLDB '96, pp. 544–555. Morgan Kaufmann, CA (1996)
29. Hoch, F., Kerr, M., Griffith, A.: Software as a service: strategic backgrounder. Tech. Rep., Software Inform. Indus. Assoc. (2001)
30. Alves, A., Arkin, A., Askary, S., Bloch, B., Curbera, F., Goland, Y., Kartha, N., Sterling, Konig, D., Mehta, V., Thatte, S., van der Rijn, D., Yendluri, P., Yiu, A.: Web Services Business Process Execution Language Version 2.0. OASIS Committee Draft (2006)
31. Wang, G.: Domain-oriented data-driven data mining (3DM): Simulation of human knowledge understanding. In: Proceedings of the 1st WICI International Conference on Web Intelligence Meets Brain Informatics, WImBI'06, pp. 278–290. Springer, Heidelberg (2007)
32. Wolfram Research: Cloud services for mathematica (2010). URL http://www.nimbisservices.com/page/what-cloud-services-mathematica

Chapter 11
Provenance Support for Data-Intensive Scientific Workflows

Fakhri Alam Khan and Peter Brezany

Abstract Data-intensive workflows process and produce large volumes of data. The volume of data, number of workflow participants and activities may range from small to large numbers. The traditional way of logging experimental process is no longer valid. This has resulted in a need for techniques to automatically collect information on workflows known as provenance. Several solutions for e-Science provenance have been proposed but these are predominantly domain and application specific. In this chapter, the requirements of e-Science provenance systems are first clearly defined, and then a novel solution named the Vienna e-Science Provenance System (VePS) that satisfies these requirements is proposed. The VePS not only promises to be light weight, workflow enactment engine, domain and application independent, but it also measures the significance of workflow parameters using the Ant Colony Optimization meta-heuristic technique. Major contributions include: (1) interoperable provenance system, (2) quantification of parameters significance, and (3) generation of executable workflow documents.

11.1 Introduction

e-Science [26] is defined as computational intensive science that processes huge amount of data over large distributed network. The main theme of e-Science is to promote collaboration among researchers across their organizational boundaries and disciplines – to reduce coupleness and dependencies and encourage modular, distributed, and independent systems. This has resulted in *dry lab* experiments also known as in silico experiments. Workflows are designed and executed by scientists to make full use of the e-Science. Workflow [33] combines tasks (activities) and specifies their execution order. A task may be an autonomous activity offering a

F.A. Khan (✉) · P. Brezany
Department of Scientific Computing, University of Vienna, Nordbergstrasse 15/C/3, Austria
e-mail: khan@par.univie.ac.at; brezany@par.univie.ac.at

S. Fiore and G. Aloisio (eds.), *Grid and Cloud Database Management*,
DOI 10.1007/978-3-642-20045-8_11, © Springer-Verlag Berlin Heidelberg 2011

specific functionality while residing at some remote location and may take a data input, process it and produce a data output. Workflows may contain several hundreds of activities.

The data produced and consumed by a workflow vary both in terms of volume and variance [7]. The data variance is defined as the rate at which the structure of data changes. A typical example of data variance are the applications that collect data from a variety of distributed resources and transform it into a specific format according to participant requirements. A data-intensive workflow can be defined as *A business process logic that processes large volumes of data with highly variable characteristics* [22]. The data processed by a workflow may range from few Megabytes to Gigabytes. Furthermore, data-intensive workflows may allow fixed or unforeseen number of workflow participants (users). Data-intensive workflows have raised new issues (such as, annotation of experiments and collecting data about experiments), to make experiments reusable and reproducible. The information about workflow activities, data, and resources is known as workflow provenance [15,28]. To help keep track of workflow activities, workflow provenance describes the service invocations during its execution, information about services, input data, and data produced [27].

Moreover, workflows in scientific domains usually have exploratory nature. A scientific workflow is executed multiple times using different value combination of workflow parameters to study a scientific phenomenon. This process is best described as *hit and trial* process. This approach becomes infeasible and time consuming for real-world complex workflows, which may contain numerous parameters. It is critical to find through provenance information resolving the questions: (1) Which parameters are critical to the final result? and (2) What value combination produces optimal results? This issue is addressed via workflow parameters significance measurement using Ant Colony Optimization (ACO) [10] based approach.

In recent years, many e-Science provenance systems have been proposed, but in one way or the other they are bound to a specific domain or Scientific Workflow Management System (SWfMS). The SWfMS can be defined as an application, which interprets the process definition and interacts with workflow participants. This dependence of provenance systems is aberration from the e-Science vision, which is realized by modular, independent, and distributed infrastructures such as Grid. Keeping this in mind, in this chapter, the focus is on a domain-independent, portable and light weight e-Science provenance system. This chapter addresses not only the workflow and domain independence but also the reproducibility of experiments. A provenance system named Vienna e-Science Provenance System (VePS) having these required features is proposed. The workflow execution engine is the core component in enacting e-Science workflows and the ability of the VePS provenance framework to be independent of workflow execution engine makes it inter-operable and domain independent. The core functionalities of the VePS include: (1) provenance collection, (2) measurement of workflow parameters significance via meta-heuristic guided approach, and (3) generation of an executable workflow process from provenance information. Major contributions of the research presented in this chapter include:

- *Portable provenance system*: The VePS provenance framework is independent of workflow, workflow activities, and workflow enactment engine. This decoupling means that now the provenance system can work across multiple domains and hence be more portable.
- *Interoperable provenance information*: Along with provenance system portability, it is critical for provenance information to be usable across heterogeneous platforms as well. An XML-based approach for storing provenance information to be used by multiple users across different heterogeneous platforms is proposed and adapted.
- *Quantify parameters significance*: The information on workflow parameters significance is of paramount importance for researchers to fine tune their results and make the process of *come to optimized results* significantly faster. This contributes to higher productivity of the workflow developers. For this purpose, a methodology to quantify the impact significance of workflow parameters and integrate this information into provenance for any future use is proposed and implemented.
- *Reproducibility*: The VePS provenance framework supports reproducibility and generates an executable Business Process Execution Langauge (BPEL) [23] document describing the workflow that was executed. This results in enhanced trust and allows users to modify and compare the results of the experiments.

The rest of this chapter is organized as follows. In Sect. 11.2, introduction to the provenance in general and e-Science provenance in particular is presented. Moreover, it also describes the requirements of e-Science provenance systems. Section 11.3 details the VePS provenance framework and its main components. The provenance interceptor, parser, and transformer components of the VePS are introduced in Sect. 11.4. The ACO-based approach for estimating the workflows parameters significance is detailed in Sect. 11.5. The VePS approach to data curation and visualization is described in Sect. 11.6. The related work is discussed in Sect. 11.7. The chapter summary is provided in Sect. 11.8.

11.2 Provenance Definition and Requirements

In the following two subsections, a brief introduction to provenance from the computing and workflow perspectives is provided first. Then, e-Science provenance system requirements are detailed.

11.2.1 Provenance Definition

Workflows typically order the tasks associated with e-Science, e.g., which services will be executed and how they will be coupled together. In the Grid context,

Fox defines a workflow as *the automation of the processes, which involves the orchestration of a set of Grid services, agents and actors that must be combined together to solve a problem or to define a new service* [11]. Computational problems are complex and scientists need to collect information on workflow components and data. The information about components (activities) and data is known as provenance.

There are numerous definitions of provenance: (1) *The fact of coming from some particular source or quarter* (Oxford English Dictionary) and (2) *The history of ownership of a valued object or work of art or literature* (Merriam-Webster Online Dictionary). Provenance has been defined by computer scientists in different ways depending upon the domain, in which it is applied [34]. Buneman et al. [5] defined data provenance as the description of the origin of data and the process by which it arrived at the database. Lanter [27] defines provenance in the context of geographic information systems as information that describes materials and transformations applied to derive the data. Yogesh et al. [27] define data provenance as information that helps to determine the derivation history of a data product, starting from its original sources.

Greenwood et al. [12] have expanded Lanter's definition in the context of scientific workflows as metadata recording the process of experiment workflows, annotations, and notes about the experiments. The information about workflow activities, data, and resources is known as workflow provenance. Workflow provenance describes the workflow services invocation during its execution, information about services, input data, and data produced to help keep track of workflow activities. More fine grained workflow provenance includes information about the underlying infrastructure (processors, nodes, etc.), input and output of workflow activities, their transformations and context used.

11.2.2   e-Science Provenance System Requirements

There is a strong need to propose and build a provenance system that is in-line with e-Science core theme of modularity and de-coupleness. Characteristics of general (trivial) provenance systems are collection, storage, and dissemination. However, for a provenance system to satisfy e-Science goals, the requirements are as follows:

- *Provenance system interoperability.* The provenance system should be able to work across different workflow execution engines. An e-Science provenance system should be intelligent and independent and should collect information across multiple domains. Trivial provenance systems mainly focus on provenance data collection, whereas e-Science provenance systems need to focus on interoperability as well.
- *Provenance information interoperability.* Since e-Science encourages collaboration and dry lab experiments. The provenance information collected should be stored in a manner that it can be used by users across heterogeneous platforms.

This raises the requirement for provenance data interoperability. Compared to the trivial provenance systems collection requirements, e-Science provenance systems need to address information interoperability as well.

- *Support for parameter significance measurement.* e-Science workflows are usually complex and exploratory. To facilitate the e-Science exploratory workflow scientists, it should collect provenance information about workflow parameters and measure their significance.
- *Reproducibility.* Exploratory and scientific workflows results are questioned until they are not reproduced and re-executed by peer scientists. It is necessary to collect provenance information on workflow execution and produce executable workflow documents.
- *Visualization and report generation (dissemination).* The provenance system should have a visualization and report generation component. This will help researchers to analyze data and get an insight into the data and workflow with less effort.
- *Light weight.* The provenance collection is an ad hoc functionality. The computational overhead it exerts on workflow execution should be minimal.

One of the fundamental research issue in e-Science provenance systems is interoperability. It is the ability of the provenance system to readily work across different domains, applications, and workflow execution engines. Domain and application independence form the main criteria of our provenance research.

11.3 VePS: The Vienna e-Science Provenance System

The open provenance model (OPM) [21] has tried to standardize the provenance collection, but despite their efforts, most provenance systems for e-Science revolve around workflow enactment engine. The fundamental design of e-Science provenance system needs to be changed. It has to come out of the passive mode (passively listening to SWfMS notifications) and has to adapt a more active and intelligent mode. Keeping in mind these characteristics, a workflow execution engine independent provenance system named VePS is designed and proposed. Core functionalities of the VePS include: (1) workflow execution engine independent provenance collection, (2) measuring workflow parameters significance via a heuristic guided approach and (3) generation of executable workflow document.

The VePS shares the middleware (Apache Axis2 [14]) with the engine (i.e., both the VePS and engine are deployed onto the same middleware). The VePS records the workflow activities, whereas multiple users and clients can submit and execute their workflows. Apache Axis2 was chosen as the middleware because of its stability, popularity, maturity, healthy user community and its ability to keep evolving to cater changes in the technology and user requirements [2]. Provenance information such as service name(s), method(s) invoked, input data name(s), input data type, input data value, output data names, and values are collected by the VePS provenance framework.

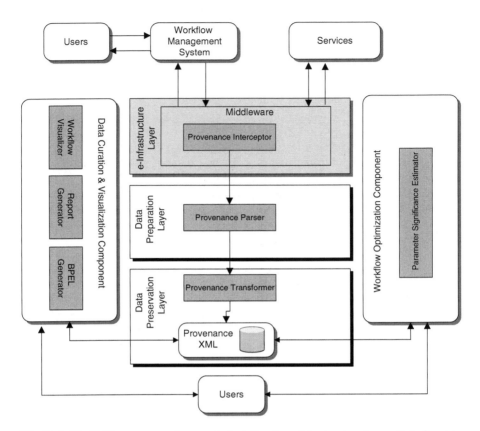

Fig. 11.1 The VePS provenance framework depicting the *e-Infrastructure Layer*, *Data Preparation Layer*, *Data Preservation Layer*, *Workflow Optimization Component*, and *Data Curation and Visualization Component*

The VePS framework consists of three layers, namely *e-Infrastructure Layer*, *Data Preparation Layer*, and *Data Preservation Layer*, and two components, namely *Workflow Optimization Component* and *Data Curation and Visualization Component* as shown in Fig. 11.1. Apart from the *e-Infrastructure Layer*, the remaining layers work outside of the underlying infrastructure, and hence put less performance burden on workflow execution. The *e-Infrastructure Layer* consists of *Provenance Interceptor* component. Its role is to catch messages passing through the middleware and asynchronously pass these intercepted messages to the *Data Preparation Layer*. This layer contains the *Provenance Parser* component, which extracts and collects the provenance information from the received messages. The collected provenance information is passed to the *Data Preservation Layer*, that transforms it into a well-defined XML structure and stores it. The *Data Curation and Visualization Component* of the VePS is used to visualize the provenance information and produce a BPEL workflow document. The *Workflow Optimization*

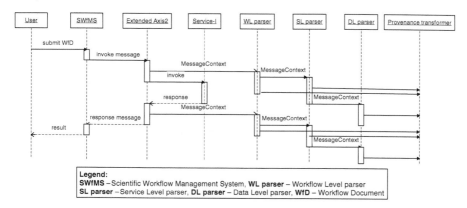

Fig. 11.2 Sequence diagram of the VePS depicting provenance collection activities when a user executes a workflow over e-Infrastructure

Component addresses the issue of complex exploratory workflows and is based on Ant Colony Optimization for Parameter Significance (ACO4PS) [17].

From the sequence diagram shown in Fig. 11.2, it can be seen that on submission of a workflow document to the SWfMS, it parses the document and generates an invoke message. Upon reception of a message by Axis2, it transforms it into SOAP message and sends an invoke request to the desired service. Before the delivery of invoke message, provenance handlers intercept this message and asynchronously deliver it to the *Provenance Parser*. This component parses workflow, activities, and data relevant provenance information from the intercepted messages. These data are then sent to the *Provenance Transformer* component, which properly structures the data and stores it in an XML document. From the VePS framework (as shown in Fig. 11.1), it is clear that it no longer depends on the workflow enactment engine, as it lies at the middleware level to intercept communication messages between the client and components (services). This ability makes it possible to collect provenance not only for one workflow enactment engine but also for any engine. This makes VePS portable and domain independent. More fine-grained details of the *e-Infrastructure Layer*, *Data Preparation Layer*, and *Data Preservation Layer* are discussed in Sect. 11.4, whereas the *Workflow Optimization Component* is introduced in detail in Sect. 11.5. The *Data Curation and Visualization Component* is described in Sect. 11.6.

11.4   Provenance Collection, Preparation, and Preservation

The provenance data collection is performed at the *e-Infrastructure Layer* level. It mainly consists of the *Provenance Interceptor* component and four subcomponents namely *Inflow*, *Outflow*, *Infaultflow*, and *Outfaultflow* provenance handlers as shown

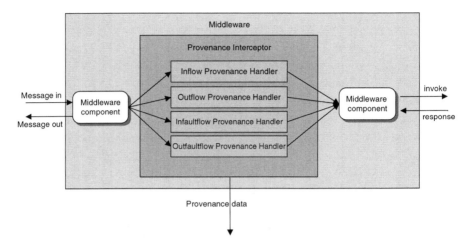

Fig. 11.3 Detailed view of *e-Infrastructure Layer* components and information flow

in Fig. 11.3. The collected data are prepared and meaningful information is collected from it at *Data Preparation Layer*, which is composed of the *Provenance Parser* component. Then, this information is properly structured, transformed, and stored at *Data Preservation Layer*, which contains the *Provenance Transformer* component. The *Provenance Interceptor*, *Provenance Parser*, and *Provenance Transformer* components of the VePS provenance framework are detailed in the following subsections.

11.4.1 Provenance Interceptor

The Apache's Axis2 architecture has defined four flows, namely *InFlow*, *outFlow*, *InFaultFlow* and *OutFaultFlow*. All communication (incoming SOAP messages, outgoing messages, incoming faulty messages, and outgoing faulty SOAP messages) pass through it. Each flow is composed of two types of phases, pre-defined and user-defined, which in turn consist of handlers. A handler is the smallest component in the Axis2 and is defined as the message interceptor. The *Provenance Interceptor* component, which lies at the *e-Infrastructure Layer* consists of four provenance handlers *InFlowProvenanceHandler*, *OutFlowProvenanceHandler*, *InFaultFlowProvenanceHandler*, and *OutFaultFlowProvenanceHandler*. The provenance handlers are integrated into Axis2 with the functionality to intercept the underlying communication between the workflow enactment engine and services passing through Axis2 as shown in Fig. 11.3.

The handlers work like "T" pipes. They catch the underlying SOAP communication messages and forward them to the *Data Preparation Layer* and leaving the Aixs2 and workflow enactment engine from dealing with the provenance collection overhead. The VePS provenance framework continuously monitors the engine as well as the workflow through the middleware. The ability of *Provenance Interceptor* to function independently of the workflow execution engine, clients, and activities makes it domain and application independent. It also means that the VePS is no longer coupled tightly to SWfMS, and hence can work across multiple engines.

11.4.2   Provenance Parser

The *Data Preparation Layer* resides outside of the Axis2 core and is composed of the *Provenance Parser* component. The provenance data parsing is performed off-line to the middleware system. This keeps the overhead to minimum and enables the VePS to exert less computational burden on the workflow enactment engine, middleware and services. *Provenance Parser* is the "go to" component for a provenance handler. The intercepted requests, responses, or faulty messages are forwarded to the provenance parsers by provenance handlers. After receiving these SOAP message objects and WSDL documents, they are parsed. The relevant provenance information such as sender, receiver, time, operation name, input, and output data type are collected. This filtered provenance data are sent to the *Provenance Transformer* component of the *Data Preservation Layer*. The *Provenance Parser* component has three sub-parsers: *Workflow Level Parser*, *Service Level Parser*, and *Data Level Parser* as shown in Fig. 11.4. These three sub-parsers collect information about workflow, services, and data exchanged, respectively:

- *Workflow level parser*: It collects workflow level provenance information such as workflow input(s), output(s), activities names, start time, and termination time of workflow. This provenance information is useful in determining the invocation sequence of workflow services (activities).

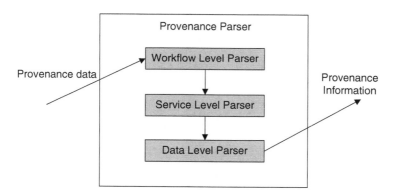

Fig. 11.4 The *Provenance Parser* component of the *Data Preparation Layer* and its subcomponents

- *Service level parser*: It gathers information about individual activities. Provenance information such as description of service parameters, input(s)/output(s) data to the activity, and invocation interface are included.
- *Data level parser*: For each workflow, multiple *Data Level Parsers* are designed. Each single data file used by a service is associated with one of them. It collects detailed provenance information on data used, such as number of rows, mean, variance, unit, and size.

11.4.3 Provenance Transformer

Data without proper structure and storage are useless. For this reason, the *Provenance Transformer* component of the *Data Preservation Layer* has been developed. XML is chosen to store and share provenance information because of its portability, platform independence, vendor independence, and system independence [4]. The main role of the *Provenance Transformer* is to identify the provenance category, transform the data into a well-designed XML format, and write it into its exact hierarchy. The *Data Preservation Layer* also lies outside the Axis2 architecture and performs all of its computation off-line to the underlying infrastructure. The provenance XML document generated is shown in Fig. 11.5.

The root element "Provenance" has an attribute "numberOfFields," which contains information about the total number of services executed by a workflow. Each activity (a service call) in a workflow document is represented by an element named

```xml
<?xml version="1.0" encoding="UTF-8"?>
<Provenance numberOfFields="5">
<!--Provenance XML document-->
<Service name="SIServiceNew" parallel="false" parallelTo="none" success="true">
<invocation methodName="executeSI" source="127.0.0.1" target="http://localhost:8080/axis2/services/SIServiceNew" time="1277392647250">
<InputData name="Input_Output" type="string" value="http://www.par.univie.ac.at/~khan/test-data/si/si-data.xml"/>
<InputData name="StandardInput" type="string" value="http://www.par.univie.ac.at/~khan/test-data/si/si-input.xml"/>
</invocation>
<response methodName="executeSI" time="1277392649484">
<OutputData name="return" type="string" value="Z:/public_html/test-data/si-generated/StandardOutput1.xml"/>
</response>
</Service>
<Service name="KFServiceNew" parallel="false" parallelTo="none" success="true">
<invocation methodName="executeKF" source="127.0.0.1" target="http://localhost:8080/axis2/services/KFServiceNew" time="1277392649781">
<InputData name="StandardOutput" type="string" value="http://www.par.univie.ac.at/~khan/data/kf/StndOutput1.xml"/>
</invocation>
<response methodName="executeKF" time="1277392650437">
<OutputData name="return" type="string" value="file://Z:/public_html/test-data/kf-generated/After-KF.xml"/>
</response>
</Service>
</Provenance>
```

Fig. 11.5 Provenance XML document generated by the VePS

"Service" in the provenance XML document. Data about service name, execution (successful/failed) are stored as attributes "name" and "success" of the Service element. The Service element has two child elements: "invocation" and "response." The invocation element contains provenance data about service execution. The time of service invocation and the name of the method invoked are represented by attributes "method-Name" and "time", respectively. Every invocation element has child element(s) named "InputData." The number of InputData elements depends on the number of parameters passed to the service. For every corresponding parameter, an "InputData" element is created. It contains information on parameter name, type, and value, while, the response element contains data about activities results and responses. Every response element has "OutputData" child element. The "OutputData" element represents result data type and value. For details of the provenance XML, refer to Fig. 11.5.

11.5   Workflow Parameters Significance Estimation

Workflows prototype a scientist experiments. In Directed Acyclic Graph (DAG) terms, a workflow is represented as $G = (N, E)$. From this representation, workflow can be defined as *Set of nodes (N) and directed edges (E), where nodes represent workflow computational tasks and edges indicate the sequence of activities*. This definition stresses that (1) a workflow can be composed of any number of computational tasks (activities); and (2) the outcome of a workflow is dependent on the order of execution of the included activities. Workflows are usually used for exploratory purposes. This means that workflows are executed multiple times in an attempt to understand a scientific phenomena or to find answers to scientific questions. Workflow and the activities in a workflow have parameters. These parameters have strong impact on the workflow outcome. Furthermore, different value combination of workflow parameters results in different final outcome. The usual practice for a scientist to come to the optimized result is a *brute force* method. This means that different or all possible combinations of parameters are tested empirically.

Consider the scenario that a workflow has N tasks. The number of tasks in a workflow may vary from small numbers to a very large number. Moreover, suppose that the workflow has n number of parameters (p). To put it mathematically, $P_n = \{p_1, p_2, p_3, \ldots, p_n\}$. Each parameter has an allowed range of values (R). The allowed range may be either discrete or continuous. The number of times the workflow need to be executed, if one wants to come to the optimized output via a *brute force* technique will be:

$$(\left|R_{(p_1)}\right| \times \left|R_{(p_2)}\right| \times \left|R_{(p_3)}\right| \times \ldots \left|R_{(p_n)}\right|) = \prod_{i=1}^{n} \left|R_{(p_i)}\right|, \qquad (11.1)$$

where, $R_{(p_i)}$ represents the value range of the i_{th} parameter. From 11.1, for achieving an optimized result, the execution pattern of the experiment increases exponentially with the increase in parameters and their ranges. In such complex exploratory workflows scenario, it is interesting and necessary to know, which parameters have significant impact on the final outcome and which have the least significance. Parameter significance can be defined as *The amount of effect a certain parameter exerts on the final outcome of a computational model* [16]. For a user of scientific workflow, it is important to know: (1) what parameter of workflow have strong effect on the final product and what have the least effect (parameter significance)? and (2) what combination of parameter values result in optimized or desired product?

Such problems, where the solution time increases significantly with the problem size are known as NP-hard problems [1]. To solve these problems with reasonable amount of time and by reaching near optimal results, they are tackled by meta-heuristic algorithms [31]. The *Workflow Optimization Component* addresses the issue of complex exploratory workflows via *parameter significant estimator (PSE)* module. It is based on ACO4PS [17]. The PSE uses information from the provenance XML to enable scientists to find near optimal solutions.

The PSE retrieves information such as parameter names, types, and their total number from the provenance XML and gets the data on the value range of individual parameters from the user. First, the workflow parameters are represented in the form of nodes and transitions. For each parameter, two new parameters *cost* and *profit* are defined, which are critical in guiding the *Parameter Significance Estimator*. The *cost* of a workflow parameter is defined as *The computation time and resources it requires*. There are numerous factors which effect the *cost* of a parameter. But to be able to numerically quantify cost of parameter, it is associated with the parameters allowed value range. The *cost* of parameter i is calculated as follows:

$$cost_i = \frac{range_i}{\sum_{k=1}^{n} range_k} \times 100, \qquad (11.2)$$

where $range_i$ represents the allowed discrete value range of parameter i. Profit of a parameter can defined as *The impact of a change in a parameter value on the known output*. Profit of a parameter is determined by knowing to which activities it is associated and how much impact it has on the workflow final product. Profit of a workflow parameter is calculated as:

$$profit_i = \frac{PoP_{(i)}}{\sum_{k=1}^{n} PoP_{(k)}} \times 100$$

$$\left\{ PoP_{(i)} = \frac{PC_{(i)}}{NF}, PC_{(i)} = \frac{r_i}{p_i}, NF = \frac{\sum_{i=1}^{n} PC_{(i)}}{n} \right\}, \qquad (11.3)$$

where r_i and p_i represent percent change in result and parameter, respectively, and NF is the normalization factor. The ACO4PS is applied on the workflow after it has been represented in the form of nodes (each node having τ, α, β, *cost*, and *profit*), the ACO parameters (such as number of ant, iterations, etc.) are setup, and the *cost* and *profit* are determined successfully. At the start, the ants are assigned randomly to different nodes. Then, each ant starts the transition process iteratively until the significance of all parameters has not been determined. Each transition from node i to j is made probabilistically according to following definition:

$$p^k_j = \frac{(\tau_j + profit_j)^\alpha + cost^\beta_j}{\sum_{c \notin mem}(\tau_c + profit_c)^\alpha + cost^\beta_c} \quad \forall \, c \notin mem \quad (11.4)$$

From (11.4), it can be seen that the transition of node j being selected by an ant k depends on the pheromone trial (τ_j), *profit*$_j$ and *cost* of parameter j. Furthermore, pheromone trial (τ_j) and *profit*$_j$ have direct effect on the probabilistic selection of parameter j, while the selection probability of node j is inversely proportional to the cost associated with parameter j. The *mem* represents the list of nodes to be visited by ant.

After each transition the *local pheromone trial update* process takes place. This means that after node j is selected as the next node, the ant moves to the selected node and the pheromone trial value on this node is updated. The *local pheromone trial update* criteria is defined as below:

$$\tau_j = (1 - \rho)\tau_j + \rho\tau_0 \quad (11.5)$$

The pheromone trial is volatile substance and it evaporates at constant value. In (11.5), the evaporation constant is represented by ρ, whereas the initial and actual pheromone trial is represented by τ_0 and τ_j, respectively. After each transition process, ant updates its memory (*mem*) and removes the selected node from the list of nodes to be visited. The transition process (next node selection process) is repeated and takes place until the desired number of parameters significance is not determined. On completion of a tour by an ant k, the cost of the trip is computed. This cost is compared to the existing best tour, and if it is found lower than the existing solution, then this trip is set as *best tour*. Furthermore, more pheromone trial is deposited on the new best tour (global pheromone trial update) according to the following criteria:

$$\tau_j = (1 - \rho)\tau_j + \rho\Delta\tau_j \quad \ldots\ldots \quad \{\Delta\tau_j = \frac{1}{L_{best}}\}, \quad (11.6)$$

where, ρ, τ_j, and L_{best} mean evaporation constant, pheromone trial, and cost of best tour, respectively. When each ant of ACO4PS (working independently and communicating via pheromone trial) completes its solution, a list of parameter significance is achieved. The resultant final best tour is a sorted list of parameters significance. The list and results of ACO4PS on estimating parameters significance are helpful for users and researchers to know:

- Out of numerous workflow parameters, which has the strongest effect on the final product of workflow? The answer to this question is satisfied by fetching the most significant parameter (first parameter from the final best tour) from the ACO4PS list.
- What parameter(s) should I consider, modifications on which will optimize the workflow product? The answer is to select a subset of most significant parameters from the list. This enable researchers to concentrate on subset of parameters, hence, considerably reducing the time and efforts.
- What parameter(s) should I ignore, modifications on which do not have high impact on the workflow result? The answer to this question lies by concentrating on the bottom half of the final sorted list.

11.6 Data Curation and Visualization

Data curation is a concept that involves data management activities like refining data through reading journal articles, finding relevant information through searching databases, and preserving data to make it usable in the future [20]. A wide range of scholarly research and scientific data resources exists including digital libraries, clinical data sets, geo-spatial data sets, and biological genes extraction data resources, to social science data resources such as census data sets, UK Border data archive to mention a few. Long-term access to such data resources is essential to enable the verification and validation of scientific discovery and provide a data platform for future research [25]. Many analytical tools and applications are used to access, manage, and visualize these data resources. However, with the rapid growth in ICT hardware and software resources, keeping data up to date and preserved to be valid and correct for future use is a challenge. The need of digital curation of data and particularly e-Research data in the scientific domain was highlighted in [6, 18, 20, 25]. Data curation can be summarized as, *communication across time with the ultimate goal to make data refined and usable in future and present* [25]. Data curation of large scientific workflows can be divided into two major types described below:

1. *Curation of workflow and scientific processes*: Workflow curation can be defined as *making the experiments available to future scientists and users*. Workflows are designed by scientists to access distributed resources, process large volumes of data sets, and are executed to derive scientific knowledge. Data-intensive scientific workflows are usually built upon earlier works by other scientists. Thus, for collaborative scientists to be able to execute, extend, and derive knowledge from these workflows, they need to know information about past workflows. The information such as what resources or activities were used in the workflow, data sets used, order of activities, etc. is needed for workflow curation. An interesting aspect of the workflow curation is to trace the workflow creation process (such as what activities were used, their relationships, order, and type of resources) to re-produce a workflow [7]. The reproducibility, which is defined as, *making*

possible the future execution is an important part of workflow curation. Workflow reproducibility is only possible if detailed information on workflow models, activities, input(s), output(s), resources, and parameters is available.

2. *Curation of data used and generated during workflow execution*: Scientific workflows often involve consumption and dissemination of a huge amount of data. Data-intensive workflows orchestrating data resources from different disciplines to support interdisciplinary research have challenges ranging from data format, semantic description of data storage methods, and future preservation of data. The data-intensive workflows highlight a number of issues related to data curation: (1) long-term preservation of data resources, (2) locating data sets that matches user requirements, and (3) collecting and preserving information about intermediate data products.

The *Data Curation and Visualization Component* of the VePS provenance framework addresses the core issue of workflow curation. It consists of three modules, namely *Workflow Visualizer*, *Report Generator*, and *BPEL Generator*. These modules are explained in the following subsections.

11.6.1 Workflow Visualizer and Report Generator

A provenance system is incomplete and ineffective without proper visualization, dissemination, and presentation. Currently, though the querying capabilities are not yet addressed, as the main focus was on collection of provenance information independent of workflow enactment engine. However, for presenting the collected provenance data, a simple validation application is developed. The *Report Generator* takes as input the provenance XML document, parses it and produces a provenance information report on workflow. The report contains information such as the total number of services in workflow, successfully invoked services, unsuccessful services, name of the least efficient service, name of the most efficient service, and names of the services having maximum and minimum parameters. The *Workflow Visualizer* module visualizes the workflow with help of boxes and lines. The rectangular boxes represent services of the workflow, and the lines show their sequence of execution.

11.6.2 Workflows Reproducibility

The ability of e-Science workflows to manage distributed complex computations has resulted in accelerating the scientific discovery process [8, 19]. To manage and automate the execution of computational tasks, workflows are provided with the specific details of resources and model locations (URI) and parameters settings. Once workflows are properly represented as an abstract process, they are executed and realized via SWfMS. The results of a workflow, if notable, are shared with peer scientists and the research community. But for reviewers and researchers to trust these results, they need to prototype and execute the experiment. This raises the requirement of scientific workflows reproducibility.

Reproducibility not only results in validation and enhanced trust but it also helps in establishing known phenomena. Workflow reproducibility can be defined as *documenting workflow activities, data, resources and parameters used so that the same results are reproduced by an independent person* [9]. To be able to reproduce experiments, rich set of provenance information is required (such as URIs of resources, data used, sequence of computational tasks, parameters used, interface of resources, etc.). For reproducibility, a provenance system must capture information on workflow intermediate inputs, outputs, and their relationship. For these reasons, the VePS provenance framework supports reproducibility. The workflows that are executed are transformed into BPEL executable processes. The BPEL is chosen as a workflow representation language because it is increasingly becoming a de facto standard for defining workflows and is widely used. The ability of BPEL to be executed on any platform (such as Java and .NET platforms) makes it attractive and suitable for e-Science infrastructures, where different users may use different platforms.

The *BPEL Generator* module of the VePS takes as input the workflow provenance XML document and parses the information (such as workflow activities names, count and URIs of WSDLs). Upon reception of provenance information, it checks the WSDL of the activities for *partnerLinkType*. For creating a communication link with the activities, the *partnerLinkType* is used by the *partnerLink* element of the BPEL. If the original target WSDL file does not contain *partnerLinkType*, then the WSDL is extended. This process is repeated for each activity and once it is completed, the *header*, *import*, and *partnerLink* elements of the resultant BPEL are created as shown in Fig. 11.6. The reproducibility component extracts data about

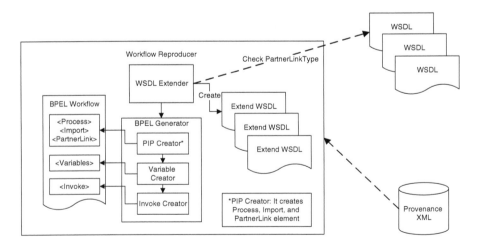

Fig. 11.6 *BPEL Generator* module of the *Data Curation and Visualization Component*

variables (such as request and response variable of an activity), and the *variables* element is created with variables declared and initialized. Finally, the *invoke* call elements are created and the BPEL is finalized.

11.7 Related Work

There are numerous existing techniques and surveys [3, 27, 32] conducted on the provenance collection and some are coming up with promising results. But all of them describe the provenance from a different perspective. The survey and taxonomy presented by Simmhan et al. [27] is a more interesting survey, but even this survey is too generic as its authors try to cover provenance issues from several different areas of computer science. The goal of this chapter is to enhance the understanding of e-Science provenance, while considering the real requirements of e-Science; that is a fully modular, portable and considering the autonomy of resource providers.

Zhao et al. [35] outline the provenance log produced in the myGrid environment by the Freefluo workflow enactment engine. In this provenance documents are linked together to browse and annotate using a Conceptual Open Hypermedia Agent (COHSE) agent. Provenance documents are exported to the COHSE environment and ontologies are prepared for annotating the documents and related web pages and finally the documents can be annotated with the concepts drawn from the myGrid ontology. Clicking on an annotation navigates to other workflow records associated semantically. Groth et al. [13] articulate the limitations of provenance generation in myGrid, e.g., it does not address general architectures or protocols for recording provenance but is more domain and project specific.

An infrastructure level recording of provenance in service-oriented architectures has been proposed in [30]. Provenance can be recorded either along-side the data as metadata or it can be stored in a separate repository made accessible as a Grid or Web service. However, the first solution required maintaining integrity of the provenance record as transformations take place and significant changes may need to be made in the existing data storage structure.

Rajbhandari and Walker [24] have presented a provenance collection and querying model in a service-oriented architecture environment. Here, the provenance can be collected synchronously or asynchronously through a provenance collection service from the workflow enactment engine. This is based on a centralized approach. It collects messages between the enactment engine and associated workflow services. However, not all interactions of services pass through an enactment engine. For example, in some workflow, service call other services directly for some of their computations, leaving the enactment engine unaware of such interactions. Moreau [29] presents a technology-independent logical architecture for a provenance system. The architecture consists of provenance store and its interfaces, libraries, application services, processing services, and different policies. In a service-oriented architecture, provenance can be categorized into actor provenance and interaction provenance. Actor provenance architecture is proposed which records provenance of the actors and their interactions with the enactment engine through provenance handlers. Since an enactment engine may invoke services from multiple domains, it may use multiple provenance handlers. However, given that the services and resources in the remote domain will unlikely be under the control of this single enactment engine, the deployment of remote provenance handlers may not be possible.

11.8 Conclusions

With the emergence of e-Science and associated data-intensive workflows, keeping log of the experimental activities during workflow execution has become a challenge. In this chapter, the VePS provenance framework is proposed, which is capable of provenance data collection, storage, reproducibility, visualization, and parameters significance measurement for e-Science applications. The VePS is workflow enactment engine independent and its framework consists of three layers, namely *e-Infrastructure Layer*, *Data Preparation Layer*, *Data Preservation Layer* and two components called *Workflow Optimization Component* and *Data Curation and Visualization Component*.

The ability of the *Provenance Interceptor* component to run independently of the workflow execution engine, clients, and activities makes it domain and application independent. It also means that the VePS is no longer coupled tightly to SWfMS and hence can work across multiple engines. The *Provenance Parser* and *Provenance Transformer* components lay outside the underlying infrastructure. This keeps the overhead to minimum and enables the VePS provenance framework to exert less computational burden on the workflow enactment engine, middleware, client(s), activities, and services. To enable proper use of provenance data and store in interoperable format, the *Provenance Transformer* is designed and implemented. It structures and stores the provenance data in the XML format.

The PSE is based on the ACO meta-heuristic algorithm and generates a sorted list of workflow parameters by their significance. An advantage of having sorted significance list is that it enhances users knowledge on parameters and guide them in choosing which parameters to consider and which to ignore. Furthermore, the PSE is capable of working with parameters of both discrete and continuous value ranges types and works well when the expected or desired results of workflows are known.

To enable the users of real-world complex workflows to re-execute and reproduce their experiments, the *Data Curation and Visualization Component* is proposed. It produces an executable workflow document specified in the BPEL language. The benefits of reproducibility include enhanced trust, comparison of results, and authentication. It also visualizes the workflow and produces an information report to give an insight into the workflow.

The VePS promises to be light weight, domain, and SWfMS-independent (portable) provenance framework in-line with the core theme of e-Science (modularity and de-coupleness). The issues identified include differentiation of workflow activities in multiple workflow execution scenario, dependence on the middleware and the limitation of PSE to work only when expected outcome is known. We believe that the VePS provenance framework is one step forward, and the future lies in fully independent provenance systems.

References

1. Hochbaum, D.S. (ed.): Approximation Algorithms for NP-Hard Problems. Course Technology, Florence (1996). ISBN: 978-0534949686
2. Azeez, A.: Axis2 popularity exponentially increasing. http://afkham.org/2008/08/axis2-popularity-exponentially.html (URL)
3. Bose, R., Frew, J.: Lineage retrieval for scientific data processing: a survey, pp. 1–28 (2005)
4. Bray, T., Paoli, J., Sperberg-McQueen, C.M., Maler, E., Yergeau, F., Cowan, J.: Extensible markup language (XML) 1.1 (2004)
5. Buneman, P., Khanna, S., Tan, W.C.: Why and where: A characterization of data provenance, pp. 316–330. LNCS, London (2001)
6. Carole, G., Robert, S., et al.: Data curation + process curation=data integration + science. Brief Bioinform. **6**, 506–517 (2008)
7. Deelman, E., Chervenak, A.: Data management challenges of data-intensive scientific workflows (2008)
8. Deelman, E., Taylor, I.: Special issue on scientific workflows. J. Grid Comput. **3–4**, 151–151 (2005)
9. Donoho, D.L., Maleki, A., et al.: Reproducible research in computational harmonic analysis, pp. 8–18 (2009)
10. Dorigo, M., Sttzle, T.: Ant colony optimization. MIT, MA (2004)
11. Fox, G., Gannon, D.: Workflow in grid systems. pp. 1009–1019 (2006)
12. Greenwood, M., Goble, C., et al.: Provenance of e-Science Experiments – Experience from Bioinformatics, pp. 223–226 (2003)
13. Groth, P., Luck, M., Moreau, L.: Formalising a protocol for recording provenance in grids, pp. 147–154 (2004)
14. Jayasinghe, D.: Quickstart Apache Axis2: A practical guide to creating quality web services. Packt Publishing (2008)
15. Khan, F.A., Han, Y., Pllana, S., Brezany, P.: Provenance support for grid-enabled scientific workflows, pp. 173–180. IEEE, Beijing, (2008)
16. Khan, F.A., Han, Y., Pllana, S., Brezany, P.: Estimation of parameters sensitivity for scientific workflows. In: Proceedings of International Conference on ICPP, Vienna, Austria. IEEE Computer Society (2009)
17. Khan, F.A., Han, Y., Pllana, S., Brezany, P.: An ant-colony-optimization based approach for determination of parameter significance of scientific workflows, pp. 1241–1248 (2010)
18. Lord, P., Macdonald, A., Lyon, L., Giaretta, D.: From data deluge to data curation, pp. 371–375 (2004)
19. Ludaescher, B., Goble, C.: Special section on scientific workflows. SIGMOD Rec. **3**, 1–2 (2005)
20. Moreau, L., Foster, I.: Provenance and annotation of data. In: International Provenance and Annotation Workshop, LNCS. Springer, Berlin (2006)

21. Moreau, L., Clifford, B., et. al. The Open Provenance Model Core Specification (v1.1). Future Generation Computer Systems, New York (2010)
22. Muehlen, M.Z.: Volume versus variance: Implications of data-intensive workflows (2009)
23. OASIS: The WS-BPEL 2.0 specification. http://www.oasis-open.org/committees/download.php/23964/wsbpel-v2.0-primer.htm(2007)
24. Rajbhandari, S., Walker, D.W.: Incorporating provenance in service oriented architecture, pp. 33–40. IEEE Computer Society, USA (2006)
25. Rusbridge, C., Burnhill, P., Ross, S. et al.: The digital curation centre: A vision for digital curation, pp. 31–41 (2005). doi: http://doi.ieeecomputersociety.org/10.1109/LGDI.2005.1612461
26. Schroeder, R.: e-Sciences as research technologies: reconfiguring disciplines, globalizing knowledge. Soc. Sci. Inf. Surles Sci. Sociales **2**, 131–157 (2008). doi: 10.1177/0539018408089075
27. Simmhan, Y., Plale, B., Gannon, D.: A survey of data provenance in e-Science, pp. 31–36 (2005)
28. Simmhan, Y.L., Plale, B., Gannon, D.: Karma2: Provenance management for data-driven workflows. Int. J. Web Service Res. **2**, 1–22 (2008)
29. Stevens, R.D., Tipney, H.J., Wroe, C.J., et al.: Exploring Williams-Beuren syndrome using myGrid. In: In Proceedings of 12th International Conference on Intelligent Systems in Molecular Biology (2003)
30. Szomszor, M., Moreau, L.: Recording and reasoning over data provenance in web and grid services, pp. 603–620 (2003)
31. Talbi, E.G.: Metaheuristics: From design to implementation (Wiley Series on Parallel and Distributed Computing). Wiley, NY (2009). http://www.amazon.com/Metaheuristics-Design-Implementation-El-Ghazali-Talbi/dp/0470278587
32. Tan, W.C.: Research problems in data provenance, pp. 45–52 (2004)
33. Taylor, I.J., Deelman, E., Gannon, D.B., Shields, M. (eds.): Workflows for e-Science: Scientific workflows for grid. Springer, Berlin (2006)
34. Uri, B., Avraham, S., Margo, S.: Securing provenance, pp. 1–5. USENIX Association, CA, (2008)
35. Zhao, J., Goble, C., Greenwood, M., Wroe, C., Stevens, R.: Annotating, linking and browsing provenance logs for e-Science, pp. 158–176 (2003)

Chapter 12
Managing Data-Intensive Workloads in a Cloud

R. Mian, P. Martin, A. Brown, and M. Zhang

Abstract The amount of data available for many areas is increasing faster than our ability to process it. The promise of "infinite" resources given by the cloud computing paradigm has led to recent interest in exploiting clouds for large-scale data intensive computing. Data-intensive computing presents new challenges for systems management in the cloud including new processing frameworks, such as MapReduce, and costs inherent with large data sets in distributed environments. Workload management, an important component of systems management, is the discipline of effectively managing, controlling and monitoring "workflow" across computing systems. This chapter examines the state-of-the-art of workload management for data-intensive computing in clouds. A taxonomy is presented for workload management of data-intensive computing in the cloud and use the taxonomy to classify and evaluate current workload management mechanisms.

12.1 Introduction

Economic and technological factors have motivated a resurgence in shared computing infrastructure with companies such as Amazon, IBM, Microsoft, and Google providing software and computing resources as services. This approach, known as *cloud computing*, gives customers the illusion of infinite resources available on demand while providing efficiencies for application providers by limiting up-front capital expenses and by reducing the cost of ownership over time [1, 2].

Cloud computing is, in turn, helping to realize the potential of large-scale data-intensive computing by providing effective scaling of resources. A growing number of companies, for example, Amazon [3] and Google [4], rely on their ability to

R. Mian (✉) · P. Martin · A. Brown · M. Zhang
School of Computing, Queen's University, Kingston, ON, K7L 3N6, Canada
e-mail: mian@cs.queensu.ca; martin@cs.queensu.ca; brown@cs.queensu.ca;
myzhang@cs.queensu.ca

S. Fiore and G. Aloisio (eds.), *Grid and Cloud Database Management*,
DOI 10.1007/978-3-642-20045-8_12, © Springer-Verlag Berlin Heidelberg 2011

process large amounts of data to drive their core business. The scientific community is also benefiting in application areas such as astronomy [5] and life sciences [6] that have very large datasets to store and process.

Data-intensive computing presents new challenges for systems management in the cloud. One challenge is that data-intensive applications may be built upon conventional frameworks, such as shared-nothing database management systems (DBMSs), or new frameworks, such as MapReduce [7], and so have very different resource requirements. A second challenge is that the parallel nature of large-scale data-intensive applications requires that scheduling and resource allocation be done so as to avoid data transfer bottlenecks. A third challenge is to support effective scaling of resources when large amounts of data are involved.

Workload management is an important component of systems management. One may define a *workload* to be a set of requests that each access and process data under some constraints. The data access performed by a request can vary from retrieval of a single record to the scan of an entire file or table. The requests in a workload share a common property, or set of properties, such as the same source application or client, type of request, priority, or performance objectives [8].

Workloads executing on the same data service compete for system resources such as processors, main memory, disk I/O, network bandwidth and various queues. If workloads are allowed to compete without any control, then some workloads may consume a large amount of the shared system resources resulting in other workloads missing their performance objectives.

Workload management is the discipline of effectively managing, controlling and monitoring "workflow" across computing systems [9]. In a cloud, the two main mechanisms used for workload management are scheduling requests and provisioning resources. Since the load on a data service in the cloud can fluctuate rapidly among its multiple workloads, it is impossible for system administrators to manually adjust the system configurations to maintain the workloads' objectives during their execution. It is therefore necessary to be able to automatically manage the workloads on a data service.

The primary objective of this chapter is to provide a systematic study of workload management of data-intensive workloads in clouds. The contributions of this chapter are the following:

- Taxonomy of workload management techniques used in clouds.
- A classification of existing mechanisms for workload management based on the taxonomy.
- An analysis of the current state-of-the-art for workload management in clouds.
- A discussion of possible directions for future research in the area.

The remainder of this chapter is structured as follows. Section 12.2 gives an overview of workload management in traditional DBMSs as background for the remainder of this chapter. Section 12.3 describes the taxonomy for workload management in clouds. Section 12.4 uses the taxonomy to survey existing systems and techniques to manage data-intensive workloads in clouds. Section 12.5 summarizes the paper and presents directions for future research.

12.2 Background

First, an overview of workload management in traditional DBMSs is provided
to highlight the main concepts in workload management. Next, the different
architectures that have been proposed to support data-intensive computing in the
cloud are identified and their impact on workload management is discussed.

12.2.1 Workload Management in DBMSs

Workload management, as defined earlier, is the discipline of effectively managing,
controlling and monitoring "workflow" across computing systems [9]. The trend
of consolidating multiple individual databases onto a single shared data resource
means that multiple types of workloads are simultaneously present on a single
data resource. These workloads may include on-line transaction processing (OLTP)
workloads, which consists of short and efficient transactions that may require only
milliseconds of CPU time and very small amounts of disk I/O to complete, as well
as on-line analytical processing (OLAP) workloads, which typically are longer,
more complex and resource-intensive queries that can require hours to complete.
Workloads submitted by different applications or initiated from distinct users may
also have performance objectives that need to be satisfied. Workload management
is necessary to ensure that different workloads meet their performance objectives,
while the DBMS maintains high utilization of its resources.

Workload management in DBMSs involves three common types of control
mechanisms, namely admission, scheduling and execution controls [10]. *Admission
control* determines whether or not newly arriving queries can be admitted into the
system. It is intended to avoid increasing the load on an already busy system.
The admission decision is based on admission control policies, the current load
on the system and estimated metrics such as the arriving query's cost, resource
usage, or execution time. *Query scheduling* determines when admitted queries are
given to the database engine for execution. Its primary goal is to decide how many
queries from different types of workloads with different business priorities can be
sent to the database engine for execution at the same time. *Execution control*, in
contrast with admission control and scheduling that are applied to queries before
their execution, is imposed while the query is executing. It involves either slowing
down or suspending a query's execution to free up shared system resources for
use by higher priority queries. As query costs estimated by the database query
optimizer may be inaccurate, some long-running and resource-intensive queries
might get the chance to enter a system when the system is experiencing a heavy load.
These problematic queries compete with others for the limited available resources
and may result in high priority queries getting insufficient resources and missing
their performance goals. Execution control manages the running of problematic
queries based on execution control policies and determines what queries should be
controlled as well as to what degree.

12.2.2 Data-Intensive Computing Architectures

Data-intensive computing in the cloud involves diverse architectures and workloads, which adds complexity for workload management compared with traditional DBMSs. Workloads in the cloud can range from ones consisting of relational queries with complex data accesses to others involving highly parallel MapReduce tasks with simple data accesses. The workloads in clouds can also differ from DBMS workloads with respect to granularity of requests, that is the amount of data accessed and processed by a request. Cloud workloads are typically coarse-grained to localize data access and limit the amount of data movement. This chapter considers the following four different architectures for data-intensive computing in the cloud:

- *MapReduce* is a popular architecture to support parallel processing of large amounts of data on clusters of commodity PCs [7], [11]. MapReduce enables expression of simple computations while hiding the details of parallelization, fault-tolerance, data distribution, and load balancing from the application developer. A MapReduce computation is composed of two phases, namely Map and Reduce phases. Each phase accepts a set of input key/value pairs and produces a set of output key/value pairs. A Map task takes a set of input pairs and produces sets of key/value pairs grouped by intermediate key values. All pairs with the same intermediate key are passed to the same Reduce task, which combines these values to form a possibly smaller set of values. Examples of MapReduce systems include Google's implementation [11] and the open-source implementation Hadoop [12].
- *Dataflow-Processing* models parallel computations in a two-dimensional graphical form [13]. Data dependencies between individual tasks are indicated by directed arcs. The edges represent data moving between tasks. Dataflow systems implement this abstract graphical model of computation. Tasks encapsulate data processing algorithms. Tasks may be custom-made by users or adhere to some formal semantics such as relational queries. Examples of dataflow processing systems in a cluster include Condor [14], Dryad [15], Clustera [16], and Cosmos [17]. Arguably, these systems can be extended to clouds with little effort.
- *Hybrid DBMS* architectures try to move relational DBMSs to the cloud. Large-scale database processing is traditionally done with shared-nothing parallel DBMSs. Queries are expressed as a dataflow graph, where the vertices are subqueries executed on individual DBMSs [18]. While shared-nothing parallel DBMSs exploit relational DBMS technology, they suffer from several limitations including poor fault tolerance, poor scaling, and a need for homogeneous platforms. There are a number of recent proposals for a hybrid approach for clouds that combines the fault tolerance, heterogeneity, and ease-of-use of MapReduce with the efficiency and performance of shared-nothing parallel DBMSs. Examples of hybrid DBMSs include Greenplum [19] and HadoopDB [20]
- *Stream-processing* is one of the most common ways in which graphics processing units and multi-core hosts are programmed [21]. In the stream-processing technique, each member of the input data array is processed independently

by the same processing function using multiple computational resources. This technique is also called Single Program, Multiple Data, a term derived from Flynn's taxonomy of CPU design [22]. Sphere is an example of a stream-processing system for data-intensive applications [21].

12.3 Workload Management Taxonomy

The taxonomy proposed for the management of data-intensive workloads in Cloud Computing provides a breakdown of techniques based on functionality. The taxonomy is used in the next section of this chapter to classify and evaluate existing workload management mechanisms. The top layer of the taxonomy, which is shown in Fig. 12.1, contains four main functions performed as part of workload management.

Workload characterization is essential for workload management as it provides the fundamental information about a workload to the management function. Workload characterization can be described as the process of identifying characteristic classes of a workload in the context of workloads' properties such as costs, resource demands, business priorities, and/or performance requirements.

Provisioning is the process of allocating resources to workloads. The ability of clouds to dynamically allocate and remove resources means that provisioning should be viewed as a workload management mechanism. Provisioning of data-intensive workloads needs to balance workload-specific concerns such as service level objectives (SLOs) and cost with system-wide concerns such as load balancing, data placement, and resource utilization.

Scheduling controls the order of execution of individual requests in the workloads according to specified objectives. The scheduling of data-intensive workloads is impacted by the presence of multiple replicas of required datasets placed at different geographical locations, which makes it different from scheduling compute-intensive workloads.

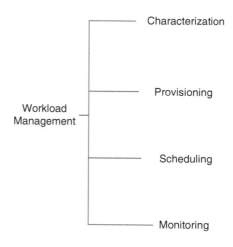

Fig. 12.1 Workload management taxonomy – top layer

Monitoring is essential to provide feedback to the scheduling and provisioning processes and can be integrated into these processes or exist as a separate autonomous process. It tracks the performance of cloud components and makes the data available to the other processes. CloudWatch [23] for Amazon's EC2 is an example of a monitoring service. It is used by the AutoScaling provisioning service [24] and provides users with information on resource utilization, operational utilization, and overall demand patterns.

The remainder of this chapter focuses on the scheduling and provisioning functions of workload management in the cloud. The categories used are explained within each of the functions and then used to categorize existing systems and work from the research literature. The workload characterization and monitoring functions are left for future work.

12.3.1 Scheduling

Figure 12.2 depicts the scheduling portion of the taxonomy. A number of key features of a scheduling approach is presented that can be used to differentiate among the approaches. Clouds are viewed as similar to Grids in a number of ways [25] and the taxonomy builds on previous studies of scheduling in Grids [26, 27].

12.3.1.1 Work Units

Scheduling policies can be classified according to the job abstraction (work unit) exposed to the scheduler for execution. The work units can range from simple queries (fine-grained data-intensive tasks) to coarser levels such as workflows of tasks. The taxonomy identifies two subclasses of work units, namely tasks and workflows.

A data-intensive *task* is defined as an arbitrary computation on data from a single node where data access is a significant portion of task execution time and so affects the scheduling decision. Examples of tasks include a relational query on a single database and a Map or Reduce task on a node.

A *workflow* represents a set of tasks involving multiple nodes that must be executed in a certain order because there are computational and/or data dependencies among the tasks. Scheduling individual tasks in a workflow therefore requires knowledge of the dependencies. Examples of a workflow are a distributed query on a set of shared-nothing DBMSs and a MapReduce program.

12.3.1.2 Objective Functions

A scheduling algorithm tries to minimize or maximize some objective function. The objective function can vary depending on the requirements of the users and the architecture of the specific cloud. Users are concerned with the performance

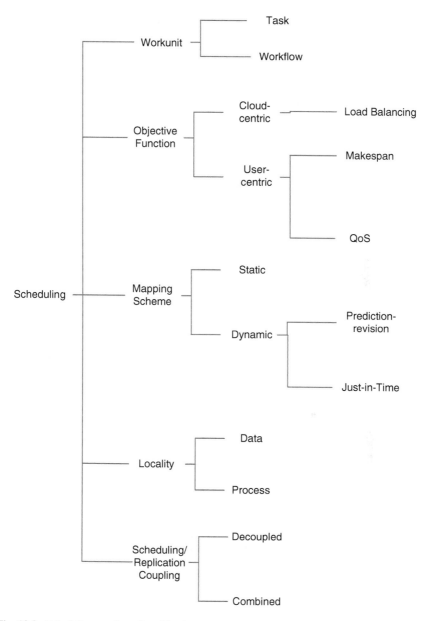

Fig. 12.2 Scheduling portion of workload management taxonomy

of their workloads, and the total cost to run their work, while cloud providers, on
the contrary, care about the utilization of their resources and revenue. Objective
functions can therefore be categorized as user-centric and cloud-centric.

User-centric objective functions aim to optimize the performance of each work unit. An objective function based on the *makespan* aims to minimize the average total completion time of a work unit. An objective function based on *Quality-of-Service (QoS)* aims to meet performance requirements specified for a work unit or workload such as a maximum cost, an average response time, or a completion deadline.

Cloud-centric objective functions are primarily concerned with maximizing revenue or resource utilization. To maximize revenue in a competitive cloud market, providers typically offer multiple levels of performance and reliability with different pricing. The aim of scheduling policies with a cloud-centric objective function is to provide predictable and reliable behavior. *Load-balancing*, which distributes load in the data center so that maximum work can be obtained out of the physical resources, is a commonly used cloud-centric objective function.

12.3.1.3 Mapping Scheme

There are two basic types of methods to map work units to resources in workload scheduling, namely static and dynamic methods. In *static* mapping schemes, it is assumed that all information about the resources and the workloads are available a priori and the complete execution schedule is determined prior to the actual execution of the work units. The mapping decisions are based on predictions of the behavior of the work units. In the case of unanticipated events such as failures, the execution schedule is recalculated and the work units are re-executed ignoring any previous progress.

Dynamic mapping schemes, however, are aware of the status of execution and adapt the schedule accordingly. Dynamic mapping schemes are further classified into prediction-revision and just-in-time schemes. *Prediction-revision* schemes create an initial execution schedule based on estimates and then dynamically revise that schedule during the execution as necessary. *Just-in-time* schemes do not make an initial schedule and delay scheduling decisions for a work unit until it is to be executed [28]. These schemes assume that planning ahead in large-scale heterogeneous environments made up of commodity resources may produce a poor schedule since it is not easy to accurately predict execution time of all workload components or account for failures in advance.

12.3.1.4 Locality

Locality is a key issue for scheduling and load-balancing in parallel programs [29, 30] and for query processing in databases [31]. It has similar importance for scheduling of data-intensive workloads in the cloud. The type of locality exploited is identified as either data or process locality. *Data* locality involves placing a work unit in the cloud such that the data it requires is available on or near the local host, whereas *process* locality involves placing data near the work units. In other words, data locality can be viewed as moving computation to data, and process locality as moving data to computation.

12.3.1.5 Scheduling/Replication Coupling

In a cloud environment, the location where the computation takes place may be separated from the location where the input data is stored, and the same input data may have multiple replicas at different locations. The differentiating criteria for scheduling policies here is whether they combine the management of scheduling and replication or keep them separate.

In each case, the policy exploits locality differently. *Decoupled scheduling* manages scheduling and replication separately. In exploiting data locality, the scheduler takes the data requirements of work units into account and schedules them close to a data source. In exploiting process locality, the scheduler brings input data to the work unit and then takes output data away. Data replicas may be created to facilitate data locality or for independent reasons such as general load balancing or fault-tolerance.

Combined scheduling manages scheduling and replication together. In exploiting data locality, the scheduler creates replicas of data, either proactively or reactively, and schedules work units to be executed on nodes with the replicas. In exploiting process locality, if the scheduler creates a replica for one work unit, then subsequent work units requiring the same data can be scheduled on that host or in its neighborhood.

12.3.2 Provisioning

Provisioning is the process of allocating resources for the execution of a task. Clouds' support for elastic resources means that provisioning should be viewed as a workload management technique since resources can be dynamically allocated or deallocated to match the demands of a workload. Provisioning for data-intensive workloads is further complicated by the need to move or copy data when the resource allocation changes. The provisioning portion of the taxonomy, which is depicted in Fig. 12.3, identifies key features of provisioning that can be used to categorize approaches.

12.3.2.1 Resource Type

Clouds currently provision two types of resources, namely virtual and physical resources. *Virtual resources* emulate the hardware with a virtual machine (VM). A VM can be packaged together with applications, libraries, and other configuration settings into a *machine image* that gives the illusion of a particular platform while

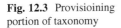

Fig. 12.3 Provisioining portion of taxonomy

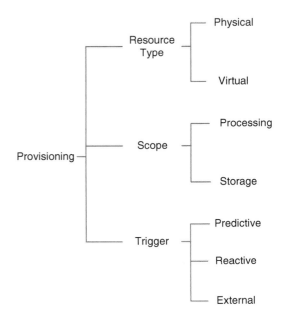

hiding the underlying hardware. In case of *physical resources*, a user either sees underlying hardware directly, or has knowledge of the hardware supporting the virtual resources.

12.3.2.2 Scope

The scope of provisioning is the kind of resources that are varied, that is processing resources or storage resources. It is clear to see how the key properties of clouds, namely short-term usage, no upfront cost, and infinite capacity on demand, apply for processing resources; it is less obvious for storage resources [1]. In fact, it is still an open research problem to create a storage system that has the ability to combine management of data with the clouds to scale on demand. *Processing resources* are typically provisioned in terms of acquiring or relinquishing images. For example, when an application's demand for computation increases, then more images can be acquired and the workload multiplexed across the increased set. Similarly, when the demand tails off then some images can be released. *Storage resources* require more complicated provisioning actions such as migrating a database [32] or varying the number of data nodes [33].

12.3.2.3 Trigger

The trigger is the method used to initiate provisioning. Two kinds of internal triggers are identified here, namely predictive and reactive triggers, which are part of the controller managing provisioning in the cloud. External triggers that can initiate the provisioning from outside the controller are also identified. *Predictive triggers* use models to forecast the need for variation in resources. They anticipate the need to provision new resources and so minimize the impact of the provisioning process on workload execution. *Reactive triggers* initiate provisioning when certain conditions are met; for example, a workload's SLOs are violated. *External triggers* are driven by decisions outside the provisioning controller; for example, from the user or from the scheduler. These conditions are set by the user. SLOs are mutual agreement on QoS between the user and the cloud provider.

The predictive and reactive approaches have well-known advantages and disadvantages. Simple predictive models are quick and inexpensive but may result in poor estimates of resource requirements. Accurate predictive models take longer to produce estimates and may not be timely enough in practice. Reactive methods, on the contrary, know when provisioning is required but may result in more disruption to workload execution while the provisioning adjusts.

12.4 Workload Management Systems

This chapter now examines workload management systems for data-intensive workloads in the cloud presented in the current research literature. The taxonomy described above is used to categorize and evaluate this work. The scheduling and provisioning aspects of the systems are explored in detail.

12.4.1 Scheduling Techniques

Scheduling, as noted earlier, has been examined with respect to other areas such as grids. This chapter focuses on strategies that explicitly deal with data during processing and on features such as adapting to environments with varied data sources and scheduling work units to minimize the movement of data. In presenting the survey of scheduling the systems are organized according to the four data-intensive computing architectures discussed in the background section of this chapter.

12.4.1.1 MapReduce

In the *Google MapReduce implementation (GoogleMR)* [7], Map and Reduce functions are encapsulated as tasks that perform some computation on data sets. Tasks are grouped into a workflow (MR-workflow) in which Map tasks are executed to produce intermediate data sets for Reduce tasks. Data is managed by the Google File System (GFS) [34]. GFS uses replication to provide resiliency, and these replicas are exploited by GoogleMR to perform decoupled-scheduling.

The scheduler exploits data locality by taking the location information of the input files into account and scheduling Map tasks on or near a host that contains a replica of its input data. The scheduler uses a dynamic mapping scheme to address execution skew and failures, and is likely to be just-in-time mapping. The objective function of the scheduler is to reduce the makespan of the MR-workflow.

Hadoop [35] is an open source implementation of MapReduce that closely follows the GoogleMR model. Hadoop consists of two layers, namely the Hadoop Distributed File System (HDFS) [36] and a data processing layer based on MapReduce Framework [20]. The MapReduce Framework follows a simple master–slave architecture.

The master is a single JobTracker and the slaves are TaskTrackers. The JobTracker handles the runtime scheduling of an MR-workflow and maintains information on each TaskTracker's load and available data hosts. The JobTracker exploits data locality by matching a TaskTracker to Map tasks that process data local to the TaskTracker. It load balances by ensuring all available TaskTrackers are assigned tasks. TaskTrackers regularly update the JobTracker with their status through heartbeat messages. Hadoop's built-in scheduler runs tasks in FIFO order with five priority levels [37] and has a just-in-time mapping. When a task slot becomes free, the scheduler scans through MR-workflows in order of priority and submit time to find a task of the required type.

12.4.1.2 Dataflow-Processing

Condor is a high-throughput distributed batch computing system that provides a task management mechanism, scheduling policy, priority scheme, resource monitoring and resource management [14]. Directed Acyclic Graph Manager (DAGMan) is a service built on top of Condor for executing multiple tasks with dependencies. Condor uses combined scheduling since coordination among data components and tasks can be achieved at a high level using DAGMan to dispatch both ready tasks and data placement requests. It also provides a hybrid of data and process locality since both tasks and data are being dispatched. DAGMan ensures that tasks are executed in the right order and presents the tasks to the Condor scheduler, which maps tasks to hosts at execution time employing a dynamic just-in-time mapping scheme. With a workflow of tasks, a user is probably interested in reducing the makespan of his or her workflow rather than throughput of individual tasks in the workflow.

Dryad, which draws from cluster management systems such as Condor, MapReduce implementations, and parallel database systems, is a general-purpose framework for developing and executing coarse-grain data parallel applications [15]. Dryad applications consist of a dataflow graph (which is really a workflow of tasks) where each vertex is a program or a task and edges represent data channels implemented via sockets, shared-memory message queues, or files. It is a logical computation graph that is automatically mapped onto data hosts by the runtime assuming dynamic mapping. Dryad provides support for scheduling the vertices on the data hosts of a cluster, establishing communication channels between computations, and dealing with software and hardware failures.

Dryad uses a distributed storage system in which, like GFS, large files can be broken into small pieces that are replicated and distributed across the local disks of the cluster computers. The computer on which a graph vertex or task is scheduled is therefore, in general, nondeterministic and the amount of data written in intermediate computation stages is typically not known before a computation begins. A dynamic just-in-time decoupled scheduling approach exploiting data locality is thus used for scheduling and a makespan objective function is used.

Clustera [16] shares many of the same goals as Dryad and is similarly categorized with the taxonomy. Both are targeted toward handling a wide range of work units from fine-grained data-intensive tasks (SQL queries) to coarse-grained data-intensive tasks and workflows. The two systems, however, use radically different implementation methods. Dryad uses techniques similar to those first pioneered by the Condor project based on the use of daemon processes running on each host in the cluster to which the scheduler pushes tasks for execution. In contrast, Clustera uses a pull model where a data host is implemented as a web-service client that requests work from the server. If a suitable data host cannot be found, then the scheduler will try to minimize the amount of data to transfer.

Cosmos is a distributed computing platform for storing and analyzing massive data sets [17]. The Cosmos Storage System, like GFS, supports data distribution and replication. It is optimized for large sequential I/O and all writes are append-only. SCOPE is a declarative language that allows users to focus on the data transformations required to solve the problem while hiding the complexity of the underlying platform.

The SCOPE compiler and optimizer generate an efficient workflow and the Cosmos runtime engine executes the workflow so SCOPE/Cosmos provide a prediction-revision mapping. The Cosmos runtime scheduler uses a decoupled scheduling approach and tries to schedule tasks to execute on the same data host as their input data or at least within the same rack as the data in an attempt to exploit data locality. Cosmos schedules a workflow onto the hosts when all the inputs are ready, monitors progress, and, on failure, re-executes parts of the workflow. SCOPE/Cosmos also uses a makespan objective function.

12.4.1.3 Stream-Processing

Sector [21] is a distributed storage system that can be deployed over a wide area. It allows users to obtain large datasets from any location but assumes a high-speed network connection. In addition, Sector automatically replicates files for better reliability, availability and access throughout the WAN.

Sphere [21] is a compute service built on top of Sector that allows users to write distributed data-intensive applications using a stream abstraction. A Sphere stream consists of multiple data segments and the segments are processed by Sphere Processing Engines (SPEs). An SPE can process a single data record from a segment, a group of data records or the complete segment. User-defined functions (UDFs) are supported by the Sphere Cloud over data both within and across data centers.

Parallelism is achieved in two ways. First, a Sector dataset consists of one or more physical files and these files can be processed in parallel. Second, Sector is typically configured to create replicas of files for archival purposes, and these replicas can be processed in parallel. Sphere achieves data locality because often data can be processed in place without moving it.

The SPE is the major Sphere service or task. Each SPE is based on a user-defined function. Usually, there are many more segments than SPEs, which provides a simple mechanism for load balancing. SPEs periodically report the progress of the processing to the user. If an SPE does not report any progress before a timeout occurs, then the user abandons the SPE. The segment being handled by the abandoned SPE is assigned to another SPE and processing of that segment is started again. Sphere therefore provides a just-in-time mapping.

Unlike the systems discussed so far, the user is responsible for orchestrating the complete running of each Sphere task. One of the design principles of the Sector/Sphere system is to leave most of the decision making to the user, so that the Sector master can be quite simple. The objective function is therefore user-centric and makespan is used as an example [21]. Sector independently replicates for parallelism, which conforms to decoupled scheduling. Gu and Grossman argue that both stream-processing and MapReduce are ways to simplify parallel programming and that MapReduce-style programming can be implemented in Sphere by using a Map UDF followed by a Reduce UDF.

12.4.1.4   Hybrid DBMS

The *Pig* project at Yahoo [38] and the open source *Hive* project [39] integrate declarative query constructs from the database community into MapReduce software to allow greater data independence, code reusability, and automatic query optimization. Pig and Hive both use Hadoop as the underlying MapReduce framework and so their workload management is the same as Hadoop.

HadoopDB provides such a hybrid structure at the systems-level [20]. It uses MapReduce as the communication layer above multiple data nodes running single-node DBMS instances. Queries are expressed in SQL, translated into MapReduce (Hadoop) tasks by extending existing tools (Hive), and as much work as possible is pushed into the DBMSs at the nodes. HadoopDB inherits the workload management characteristics of Hadoop. The objective function is to minimize the makespan, which is accomplished by pushing as much of the work as possible into the DBMSs. Scheduling and replication are managed separately, which conforms to decoupled scheduling and scheduling exploits data locality by dispatching tasks to DBMSs containing the required data.

12.4.1.5 Discussion

A summary of the evaluation using the scheduling taxonomy is given in Table 12.1. Moving large amounts of data is expensive and causes significant delays in processing. As a result, almost all of the surveyed systems exploit data locality by bringing computations to the data source or near it. Arguably, this is the right direction for data-intensive workload management.

Note that all the schedulers use a decoupled approach and try to place tasks close to data. They do not, however, consider the need to create replicas in the face of increased workload demand and so may overload data resources [40]. Therefore, there appears to be a need for research to explore different replication strategies that are independent (decoupled scheduling) and that work in concert with the scheduler (combined scheduling).

Most of the systems surveyed use workflow as a unit of execution and employ just-in-time mapping. This mapping approach is scalable and adapts to resource heterogeneity and failures. Nevertheless, systems can still benefit from prediction-revision mapping techniques that incorporate some pre-execution planning, work-flow optimization, heuristics or history analysis. The additional analysis can help in creating the appropriate number of replicas or determining the appropriate amount of resources required for a computation.

Makespan is the prevalent objective function in the survey. Clouds, however, are competitive and dynamic market systems in which users and providers have their own objectives. Therefore, objective functions related to cost and revenue, or participants' utilities, are appropriate and require further study. Because the economic cost and revenue are considered by cloud users and cloud providers, respectively, objective functions and scheduling policies based on them need to be developed.

12.4.2 Provisioning Techniques

Let us consider provisioning at the infrastructure level and identify three provisioning techniques currently in use, namely scaling, migration and surge computing. The presentation of provisioning in clouds for data-intensive workloads is organized based on the technique used.

12.4.2.1 Scaling

Scaling involves increasing or decreasing the amount of resources allocated depending on demand. Scaling is presently the most prevalent mechanism for dealing with variations in the workload. Commercial clouds typically offer customers the choice of a small number of fixed configuration VM types that differ in their computational or data capacity [41].

Table 12.1 Summary of the scheduling in large-scale data processing systems

System	Architecture	Objective function	Mapping	Scheduling/replication coupling	Locality	Work unit
GoogleMR	MapReduce	User→Makespan; Cloud→LoadBalancing	Just-in-time	Decoupled	Data	Workflow
Hadoop	MapReduce	User→Makespan; Cloud→LoadBalancing	Just-in-time	Decoupled	Data	Workflow
DAGMan/condor	Dataflow processing	User→Makespan	Just-in-time	Combined	Hybrid	Task
Dryad	Dataflow processing	User→Makespan	Just-in-time	Decoupled	Data	Workflow
Clustera	Dataflow processing	User→Makespan	Just-in-time	Decoupled	Data	Workflow
SCOPE/cosmos	Dataflow processing	User→Makespan	Prediction-revision	Decoupled	Data	Workflow
Sector/sphere	Stream processing	User→Makespan; Cloud→LoadBalancing	Just-in-time	Decoupled	Data	Task
Pig/hive	Hybrid DBMS	User→Makespan; Cloud→LoadBalancing	Just-in-time	Decoupled	Data	Workflow
HadoopDB	Hybrid DBMS	User→ Makespan	Just-in-time	Decoupled	Data	Workflow

Amazon EC2 provides scaling of virtual processing resources called instances. AutoScaling [24] uses a reactive trigger mechanism and allows a user to automatically scale instances up or down according to user-defined conditions based on CloudWatch metrics.

Elastic storage addresses elastic control for multi-tier application services that acquire and release resources in discrete units, such as VMs of predetermined configuration [33]. It focuses on elastic control of the storage tier where adding or removing a data node (which consists of VMs) requires rebalancing stored data across the data nodes. The storage tier presents new challenges for elastic control, namely delays due to data rebalancing, interference with applications and sensor measurements, and the need to synchronize variation in resources with data rebalancing. They use a reactive trigger mechanism and use an integral control technique called *proportional thresholding* to regulate the number of discrete data nodes in a cluster.

Google AppEngine scales a user's applications automatically for both processing resources and storage. The scaling is completely transparent to the user. The system simply replicates the application enough times to meet the current workload demand. A reactive trigger is likely used to initiate the scaling. The amount of scaling is capped in that resource usage of the application is monitored and cannot exceed its quota. There is a base level of usage available for free with a payment system available to pay for higher quotas. The monitored resources include incoming and outgoing bandwidth, CPU time, stored data and e-mail recipients. [4]

Microsoft Windows Azure does not offer automatic scaling but it is the primary tool for provisioning. Users can provision how many instances they wish to have available for their application. Like Amazon, the instances are virtual processing resources [42].

12.4.2.2 Migration

Migration is a workload management technique used in clouds where an application execution is moved to a more appropriate host. Live migration of virtual machines has been shown to have performance advantages in the case of computation-intensive workloads [43] as well as fault tolerance benefits [44]. Migration with data-intensive workloads, however, faces problems with high overhead and long delays because large data sets may have to be moved [32].

Elmore et al. [32] analyze various database multi-tenancy models and relate them to the different cloud abstractions to determine the tradeoffs in supporting multi-tenancy. So, the scope is the storage tier. At one end of the spectrum is the shared hardware model, which uses virtualization to multiplex multiple data nodes on the same host with strong isolation. In this case, each data node has only a single database process with the database of a single tenant. At the other end of the spectrum is the shared table model, which stores multiple tenants' data in shared tables providing the finest level of granularity.

They provide a preliminary investigation and experimental results for various multi-tenancy models and forms of migration. For shared hardware migration, using a VM abstracts the complexity of managing memory state, file migration and networking configuration. Live migration only requires Xen be configured to accept migrations from a specified host. Using Xen and a 1 Gbps network switch, the authors were able to migrate an Ubuntu image running MySQL with a 1 GB TPC-C database between hosts on average in only 20 s. Running the TPC-C benchmark in a standard OS vs. a virtual OS, the authors observed an average increase of response times by 5–10%. In this case, the resource type is virtual.

On the contrary, shared table migration is extremely challenging and any potential method is coupled to the implementation. Isolation constructs must be available to prevent demanding tenants from degrading system wide performance in systems without elastic migration. Some shared table models utilize tenant identifiers or entity keys as a natural partition to manage physical data placement [45]. Finally, using a "single" heap storage for all tenants [46] makes isolating a cell for migration extremely difficult. Without the ability to isolate a cell leaves efficient migration of shared tables an open problem.

12.4.2.3 Surge Computing

Surge computing is a provisioning technique applicable in Hybrid (private/public) clouds. The resources of the private cloud are augmented on-demand with resources from the public cloud [47]. In these scenarios, the clouds are typically connected by a WAN so that there are latency implications with moving data to the public cloud.

Zhang et al. present a comprehensive workload management framework for web applications called *resilient workload manager (ROM)* [47]. ROM includes components for load balancing and dispatching, offline capacity planning for resources, and enforcing desired QoS (e.g., response time). It features a fast workload classification algorithm for classifying incoming workload between a base workload (executing on a private cloud) and trespassing workload (executing on a public cloud). So, the scope of ROM is to vary processing resources. Resource planning and sophisticated request dispatching schemes for efficient resource utilization are only performed for base workload. The private cloud runs a small number of dedicated hosts for the base workload, while images (VMs + applications etc.) in the public cloud are used for servicing the trespassing workload. So, the resource type is hybrid. The data storage in the private cloud is decoupled from that in the public cloud so shared or replicated data is not needed.

In the ROM architecture, there are two load balancers, one for each type of workload. The base load balancer makes predictions on the base workload and uses integrated offline planning and online dispatching schemes to deliver the guaranteed QoS. The prediction may also trigger an overflow alarm. Then, workload classification algorithm sends some workload to the public cloud for processing. ROM operates an integrated controller and load balancer in the public cloud. The controller reacts to the external alarm and provisions images and the load balancer services trespassing workload on the provisioned images using round-robin policy.

Moreno-Vozmediano et al. [48] analyze the deployment of generic clustered services on top of a virtualized infrastructure layer that combines the *OpenNebula* [49] VM manager and Amazon EC2. The separation of resource provisioning, managed by OpenNebula, from workload management, provides elastic cluster capacity. The capacity can be varied by deploying (or shutting down) VMs on demand, either in local hosts or in remote EC2 instances. So, the resource type is virtual.

The variation in the number VMs in OpenNebula is requested by an external provisioning module. For example, a provisioning policy limits the number of VMs per host to a given threshold.

Two experiments operating over the Hybrid cloud are reported by Moreno-Vozmediano et al. [48]. One shares insights in executing a typical High Throughput Computing application, and the other at latencies in a clustered web server. For the experiments, the scope is variation in processing resources.

12.4.2.4 Discussion

A summary of the classification of provisioning techniques is given in Table 12.2. It can be observed that most of the current work on provisioning in clouds involves scaling and is applied to web applications that do not involve large-scale data processing, so scaling data storage is not an issue. In the ROM system, the data storage in the private cloud is decoupled from that in the public cloud so that the latter is not tied to the former through shared or replicated data resources. This seems a reasonable approach for large read-only data.

Note that the reactive techniques examined involve the user defining (condition, action) pairs to control the reaction. With multiple rules, many questions arise such as: can multiple rules be defined on the same metrics, can rules (conditions,

Table 12.2 Summary of provisioning for large-scale data processing

Technique	System	Scope	Trigger	Resource type
Scaling	Amazon			
	AutoScaling	Processing	Reactive	Virtual
	Elastic storage	Storage	Reactive	Virtual
	Google			
	AppEngine	Processing/storage	Reactive	Physical
	Microsoft windows azure	Processing	External	Virtual
Migration	Multi-tenant			
	DB migration	Storage	External	Virtual
Surge computing	Resilient workload			
	manager (ROM)	Processing	External	Hybrid
	OpenNebula	Processing	External	Virtual

actions) overlap and can rules contradict. Armbrust et al. point out that there is a need to create a storage system that can harness the cloud's advantage of elastic resources while still meeting existing expectations on storage systems in terms of data consistency, data persistence and performance [1].

Systems that jointly use Scheduling and Provisioning have been explored in Grids. The Falkon [50] scheduler triggers a provisioner component for host increase/decrease. This host variation has also been explored during the execution of a workload hence providing dynamic provisioning. Presently, tasks stage data from a shared data repository. Since this can become a bottleneck as data scales, scheduling exploiting data locality is suggested as a solution. The MyCluster project [51] similarly allows Condor or SGE clusters to be overlaid on top of TeraGrid resources to provide users with personal clusters. Various provisioning policies with different trade-offs are explored including dynamic provisioning. The underlying motivation is to minimize the wastage of resources. However, MyCluster is aimed at compute-intensive tasks. Given the similarities between Grids and Clouds, the joint techniques for Scheduling and Provisioning in these systems and related work are worth exploring for their relevance in clouds.

12.5   Conclusions and Future Research

The amount of data available for many areas is increasing faster than our ability to process it. The promise of "infinite" resources given by the cloud computing paradigm has led to recent interest in exploiting clouds for large-scale data-intensive computing. Data-intensive computing presents new challenges for systems management in the cloud including new processing frameworks, such as MapReduce, and costs inherent with large data sets in distributed environments. Workload management, an important component of systems management, is the discipline of effectively managing, controlling and monitoring "workflow" across computing systems. This chapter examines the state-of-the-art of workload management for data-intensive computing in clouds.

A taxonomy is presented for workload management of data-intensive computing in the cloud. The top level of the taxonomy identifies four main functions: workload characterization, provisioning, scheduling and monitoring. The focus is on the scheduling and provisioning functions in this chapter. The scheduling portion of the taxonomy categorizes scheduling methods in terms of several key properties: the work unit scheduled; the objective function optimized by the scheduling; the scheme used to map work units to resources; the type of locality exploited in the scheduling, if any, and whether data replica management is integrated with scheduling or not. In examining current scheduling approaches, note that most consider entire workflows, optimize a user-centric objective like the makespan of a workflow, use simple dynamic mapping and exploit knowledge of replicas in placing tasks. Arguably, there is a need to better integrate scheduling and replica management and to balance global metrics such as cost with user metrics in scheduling.

The provisioning portion of the taxonomy categorizes provisioning methods in terms of the kind of resource provisioned (physical or virtual), the kind of trigger used to initiate provisioning (Predictive, Reactive or Trigger) and the scope of the provisioning (processing or storage). Note that systems use three methods to provision resources, namely scaling, migration, and surge computing. Scaling is the primary method used in public clouds such as Amazon's EC2 where virtual processing images are automatically scaled using a reactive trigger based on user-defined conditions. Surge computing is used in the case of a hybrid private–public cloud combination. There is little work so far on provisioning storage resources.

There are numerous opportunities for research into workload management of large-scale data computing in the cloud. First, provisioning of storage resources in a dynamic manner involves a number of problems including effective partitioning and replication of data, minimizing the impact of dynamic reallocation of resources on executing work, and finding new definitions of consistency appropriate for the cloud environment. Second, workload management methods that integrate scheduling and provisioning should be explored. Third, data-intensive workloads can be diverse ranging from coarse-grained MapReduce workflows to fine-grained dataflow graphs of distributed queries. It is unclear what techniques are appropriate and should be chosen and applied to be most effective for a particular workload executing on the clouds under certain particular circumstances, or how the multiple techniques can be coordinated to ensure that all running workloads meet their required performance goals. Fourth, predictive models are valuable tool for workload management; however, in many instances they rely on being able to estimate system capacity. The models will need to be dynamic to fit into the elastic resource model of the cloud.

References

1. Armbrust, M., Fox, A., Griffith, R., Joseph, A.D., Katz, R., Konwinski, A., Lee, G., Patterson, D., Rabkin, A., Stoica, I., Zaharia, M.: A view of cloud computing. Commun. ACM **53**(4), 50–58 (2010). doi:10.1145/1721654.1721672
2. Armbrust, M., Fox, A., Griffith, R., Joseph, A.D., Katz, R.H., Konwinski, A., Lee, G., Patterson, D.A., Rabkin, A., Stoica, I., Zaharia, M.: Above the clouds: A berkeley view of cloud computing. Technical Report No. UCB/EECS-2009–28. University of California at Berkeley (2009)
3. Amazon Elastic Compute Cloud (amazon ec2). http://aws.amazon.com/ec2/ (2010). Accessed 19 May 2010
4. Google App engine. http://code.google.com/intl/de-DE/appengine/ (2010). Accessed 19 May 2010
5. Raicu, I., Foster, I., Szalay, A., Turcu, G.: Astroportal: A science gateway for large-scale astronomy data analysis. In: TeraGrid Conference, 12–15 June 2006
6. Desprez, F., Vernois, A.: Simultaneous scheduling of replication and computation for data-intensive applications on the grid. J. Grid Comput. **4**(1), 19–31 (2006)
7. Dean, J., Ghemawat, S.: Mapreduce: Simplified data processing on large clusters. Commun. ACM **51**(1), 107–113 (2008)
8. Ahmad, M., Aboulnaga, A., Babu, S., Munagala, K.: Modeling and exploiting query interactions in database systems. Paper presented at the proceeding of the 17th ACM conference on information and knowledge management, Napa Valley, CA, USA (2008)

9. Niu, B., Martin, P., Powley, W.: Towards autonomic workload management in DBMSs. J. Database Manag. **20**(3), 1–17 (2009)
10. Krompass, S., Kuno, H., Wiene, J.L., Wilkinson, K., Dayal, U., Kemper, A.: Managing long-running queries. In: Proceedings of the 12th International Conference on Extending Database Technology: Advances in Database Technology, EDBT'09, Saint Petersburg, Russia, 2009. Association for Computing Machinery, pp. 132–143
11. Dean, J., Sanjay, G.: Mapreduce: Simplified data processing on large clusters. In: Proceedings of the Sixth Symposium on Operating Systems Design and Implementation (OSDI'04), Berkeley, CA, USA, 2004. USENIX Assoc, pp. 137–149
12. Apache Hadoop. http://hadoop.apache.org/ (2010). Accessed 19 Aug 2010
13. Gurd, J.R., Kirkham, C.C., Watson, I.: The manchester prototype dataflow computer. Commun. ACM **28**(1), 34–52 (1985)
14. Thain, D., Tannenbaum, T., Livny, M.: Distributed computing in practice: The condor experience. Concurr. Comput-Pract. Exp. **17**(2–4), 323–356 (2005)
15. Isard, M., Budiu, M., Yu, Y., Birrell, A., Fetterly, D.: Dryad: Distributed data-parallel programs from sequential building blocks. Paper presented at the Proceedings of the 2nd ACM SIGOPS/EuroSys European Conference on Computer Systems 2007, Lisbon, Portugal, 2007
16. DeWitt, D.J., Paulson, E., Robinson, E., Naughton, J., Royalty, J., Shankar, S., Krioukov, A. Clustera: An integrated computation and data management system. Proc. VLDB Endow. **1**(1), 28–41 (2008). doi:10.1145/1453856.1453865
17. Chaiken, R., Jenkins, B., Larson, P., Ramsey, B., Shakib, D., Weaver, S., Zhou, J.: Scope: Easy and efficient parallel processing of massive data sets. Proc. VLDB Endow. **1**(2), 1265–1276 (2008). doi:10.1145/1454159.1454166
18. Dewitt, D., Gray, J.: Parallel database systems. The future of high performance database systems. Commun. ACM **35**(6), 85–98 (1992)
19. GreenPlum. Greenplum database architecture. http://www.greenplum.com/technology/architecture/ (2010). Accessed 19 Aug 2010
20. Abouzeid, A., Bajda-Pawlikowski, K., Abadi, D., Silberschatz, A., Rasin, S.A.: Hadoopdb: An architectural hybrid of mapreduce and dbms technologies for analytical workloads. Proc. VLDB Endow. **2**(1), 922–933 (2009)
21. Gu, Y., Grossman, R.L. Sector and sphere: The design and implementation of a high-performance data cloud. Phil. Trans. Roy. Soc. A: Math. Phys. Eng. Sci. **367**(1897), 2429–2445 (2009). doi:10.1098/rsta.2009.0053
22. Duncan, R.: Survey of parallel computer architectures. Computer **23**(2), 5–16 (1990)
23. Amazon Cloudwatch. http://aws.amazon.com/cloudwatch/ (2010). Accessed 18 May 2010
24. Amazon Auto scaling. http://aws.amazon.com/autoscaling/ (2010). Accessed 18 May 2010
25. Foster, I., Yong, Z., Raicu, I., Lu, S., Cloud computing and grid computing 360-degree compared. In: Grid Computing Environments Workshop, 2008. GCE '08, 2008, pp. 1–10
26. Dong, F.: Workflow scheduling algorithms in the grid. PhD, Queen's University, Kingston (2009)
27. Venugopal, S., Buyya, R., Ramamohanarao, K. A taxonomy of data grids for distributed data sharing, management, and processing. ACM Comput. Surv. **38**(1), 123–175 (2006). doi:http://doi.acm.org/10.1145/1132952.1132955
28. Yu, J., Buyya, R.: A taxonomy of scientific workflow systems for grid computing. Sigmod. Rec. **34**(3), 44–49 (2005)
29. Hockauf, R., Karl, W., Leberecht, M., Oberhuber, M., Wagner, M.: Exploiting spatial and temporal locality of accesses: A new hardware-based monitoring approach for dsm systems. In: Euro-par'98 parallel processing, pp. 206–215 (1998)
30. McKinley, K.S., Carr, S., Tseng, C.-W. Improving data locality with loop transformations. ACM Trans. Program Lang. Syst. **18**(4), 424–453 (1996). doi:http://doi.acm.org/10.1145/233561.233564
31. Shatdal, A., Kant, C., Naughton, J.F.: Cache conscious algorithms for relational query processing. In: International Conference Proceedings on Very Large Data Bases, Santiago, Chile, pp. 510–521. Morgan Kaufmann, CA (1994)

32. Elmore, A., Das, S., Agrawal, D., Abbadi, A.E.: Who's driving this cloud? Towards efficient migration for elastic and autonomic multitenant databases. Tecnical Report 2010–05. UCSB CS (2010)
33. Lim, H.C., Babu, S., Chase, J.S. Automated control for elastic storage. Paper presented at the Proceeding of the 7th International Conference on Autonomic Computing, Washington, DC, USA, pp. 1–10 (2010)
34. Sanjay, G., Howard, G., Shun-Tak, L.: The google file system. SIGOPS Oper. Syst. Rev. **37**(5), 29–43 (2003). doi:10.1145/1165389.945450
35. Apache Hadoop. http://hadoop.apache.org/ (2010). Accessed 3 Jun 2010
36. Apache Hadoop distribtued file system. http://hadoop.apache.org/common/docs/current/hdfs_design.html (2010). Accessed 3 Jun 2010
37. Zaharia, M., Borthakur, D., Sarma, J.S., Elmeleegy, K., Shenker, S., Stoica, I.: Job scheduling for multi-user mapreduce clusters. Technical Report No. UCB/EECS-2009–28. University of California at Berkeley (2009)
38. Olston, C., Reed, B., Srivastava, U., Kumar, R., Tomkins, A. Pig latin: A not-so-foreign language for data processing. Paper presented at the Proceedings of the 2008 ACM SIGMOD International Conference on Management of data, Vancouver, Canada (2008)
39. Thusoo, A., Sarma, J.S., Jain, N., Shao, Z., Chakka, P., Anthony, S., Liu, H., Wyckoff, P., Murthy, R. Hive: A warehousing solution over a map-reduce framework. Proc. VLDB Endow. **2**(2), 1626–1629 (2009)
40. Ranganathan, K., Foster, I.: Decoupling computation and data scheduling in distributed data-intensive applications. In: Proceedings 11th IEEE International Symposium on High Performance Distributed Computing, Piscataway, NJ, USA, 2002. IEEE Comput. Soc., pp. 352–358
41. Quiroz, A., Kim, H., Parashar, M., Gnanasambandam, N., Sharma, N.: Towards autonomic workload provisioning for enterprise grids and clouds. In: 2009 10th IEEE/ACM International Conference on Grid Computing (GRID), Banff, AB, Canada, 2009. IEEE Computer Society, pp. 50–57
42. Chappell, D.: Introducing windows azure. David Chappell & Associates. http://download.microsoft.com/documents/uk/mediumbusiness/products/cloudonlinesoftware/IntroducingWindowsAzure.pdf (2009). Accessed 24 Aug 2010
43. Voorsluys, W., Broberg, J., Venugopal, S., Buyya, R.: Cost of virtual machine live migration in clouds: A performance evaluation. In: 1st International Conference on Cloud Computing, Beijing, China, 2009. Lecture Notes in Computer Science (including subseries Lecture Notes in Artificial Intelligence and Lecture Notes in Bioinformatics). Springer, Berlin, pp. 254–265
44. Prodan, R., Ostermann, S.: A survey and taxonomy of infrastructure as a service and web hosting cloud providers. In: 2009 10th IEEE/ACM International Conference on Grid Computing, 13–15 Oct 2009, pp. 17–25
45. Chang, F., Dean, J., Ghemawat, S., Hsieh, W.C., Wallach, D.A., Burrows, M., Chandra, T., Fikes, A., Gruber, R.E.: Bigtable: a distributed storage system for structured data. ACM Trans. Comput. Syst. **26**(2), 1–26 (2008). doi:10.1145/1365815.1365816
46. Weissman, C.D., Bobrowski, S. The design of the force.Com multitenant internet application development platform. Paper presented at the proceedings of the 35th SIGMOD international conference on Management of data, Providence, RI, USA (2009)
47. Zhang, H., Jiang, G., Yoshihira, K., Chen, H., Saxena, A.: Resilient workload manager: Taming bursty workload of scaling internet applications. In: 6th International Conference on Autonomic Computing, ICAC'09, Barcelona, Spain, 2009. Proceedings of the 6th International Conference Industry Session on Autonomic Computing and Communications Industry Session, ICAC-INDST'09. Association for Computing Machinery, pp. 19–28
48. Moreno-Vozmediano, R., Montero, R.S., Llorente, I.M.: Elastic management of cluster-based services in the cloud. Paper presented at the proceedings of the 1st workshop on Automated control for datacenters and clouds, Barcelona, Spain (2009)
49. Sotomayor, B., Montero, R.S., Llorente, I.M., Foster, I. Virtual infrastructure management in private and hybrid clouds. IEEE Internet Comput. **13**(5), 14–22 (2009)

50. Raicu, I., Zhao, Y., Dumitrescu, C., Foster, I., Wilde, M.: Falkon: A fast and light-weight task execution framework. Paper presented at the proceedings of the 2007 ACM/IEEE conference on Supercomputing, Reno, Nevada (2007)
51. Walker, E., Gardner, J.P., Litvin, V., Turner, E.L.: Creating personal adaptive clusters for managing scientific jobs in a distributed computing environment. In: Challenges of Large Applications in Distributed Environments, 2006 IEEE, 2006, pp. 95–103

Part IV
Scientific Case Studies

Chapter 13
Managing and Analysing Genomic Data Using HPC and Clouds

Bartosz Dobrzelecki, Amrey Krause, Michal Piotrowski, and Neil Chue Hong

13.1 Background

Database management techniques using distributed processing services have evolved to address the issues of distributed, heterogeneous data collections held across dynamic, virtual organisations [1–3]. These techniques, originally developed for data grids in domains such as high-energy particle physics [4], have been adapted to make use of the emerging cloud infrastructures [5].

In parallel, a new database management movement, NoSQL, has emerged which attempts to address the difficulties with scaling out relational database systems. In contrast to relational databases, these data stores often do not require fixed table schemas, avoid the use of join operations, have relaxed transactional properties and are designed to scale horizontally. Such data stores suit the distributed nature of cloud computing infrastructures well and a number of systems have been developed, often on top of cloud distributed file systems.

One example of a NoSQL system which falls into the class of Wide Column Store systems is HBase, a part of the Apache Software Foundation's Hadoop project [6] which runs on top of the Hadoop Distributed File System (HDFS). These new massively parallel databases often support expression of data processing tasks using MapReduce [7], a programming model which simplifies development of parallel data processing programs.

However, in the field of genome biology, most experimental data are managed using structured folder hierarchies on the file system or in spreadsheets. Experimental setup is captured in free form files. Therefore, "database management" is being done at the file system level. One of the reasons for not using databases to manage

B. Dobrzelecki (✉) · A. Krause · M. Piotrowski · N.C. Hong
EPCC, The University of Edinburgh, James Clark Maxwell Building, The Kings Buildings, Mayfield Road, Edinburgh EH9 3JZ, UK
e-mail: bartosz.dobrzelecki@googlemail.com; a.krause@epcc.ed.ac.uk; m.piotrowski@epcc.ed. ac.uk; N.ChueHong@epcc.ed.ac.uk

S. Fiore and G. Aloisio (eds.), *Grid and Cloud Database Management*,
DOI 10.1007/978-3-642-20045-8_13, © Springer-Verlag Berlin Heidelberg 2011

raw experimental data is the lack of database support for distributed storage of large arrays and difficulty in expressing algorithms that process arrays. The database community is aware of this and recent efforts of the SciDB [8] community may change the state of affairs.

Cloud infrastructures bring new possibilities for storing, sharing and parallel processing in file-based databases. One of the technologies enabling large scale data management is a distributed file system. Processing data in such system is non-trivial as it requires tracking file parts and taking care that computation is done close to the data. This chapter explores an implementation of an algorithm, often used to analyse microarray data, on top of an intelligent runtime which abstracts away the hard parts of file tracking and scheduling in a distributed system. This novel formulation is compared with a traditional method of expressing data parallel computations in a distributed environment which uses explicit message passing.

13.2 Analysis of Microarray Data

Microarray analysis allows the simultaneous processing of thousands to millions of genes or sequences across tens to thousands of different samples. Expression profiles of genes or samples are compared using various data processing methods to identify patterns.

Microarray experiments include many stages. There are a variety of microarray platforms using the same basic method: usually, a glass slide or membrane is spotted or "arrayed" with DNA fragments that represent specific gene coding regions. Purified RNA is then radioactively labelled and hybridised to the array. During hybridisation, complementary nucleic acid sequences pair specifically with each other. After thorough washing, only strongly paired strands remain hybridised and the raw data is obtained by laser scanning.

In the image analysis, the light intensity on the laser scans of a sample is converted to an array of real numbers. The resulting data is represented by a real matrix containing the genes and samples as rows and columns. The obtained numerical data is pre-processed by applying background correction, normalisation and various filters in preparation for subsequent data mining analysis.

There are many data processing packages available for analysing genomic data. They provide methods that include clustering, classification, network analysis or per gene analysis. Such analyses are often computationally very expensive, and because of this, scientists are forced to sacrifice the robustness of the results to be able to extract information in reasonable time. The size of the data sets may also become a limiting factor. Many large-scale analyses are too taxing even for powerful workstations.

The most popular software packages not only provide microarray data-specific algorithms like MAS5 [9] or RMA [10] but also allow expression of arbitrary statistical processing. Microarray data are usually analysed in a series of steps forming a pipeline. The R Project for Statistical Computing [11] develops an open source solution used widely by the scientific community. It provides a general purpose, scriptable runtime system and a rich repository of third party libraries which includes the Bioconductor package [12] – an aggregation of extensions dedicated to bioinformatics, with strong support for microarray data-related algorithms.

As previously mentioned, the scale of some analyses is often too large for individual workstations. The SPRINT project [13], which aims to tackle this problem, carried out a user requirements survey across the bioinformatics community using R, to identify, which analysis functions are causing bottlenecks when processing genomic data. Among the most frequently mentioned functions were data clustering algorithms. The following chapters describe two parallelisation techniques which enable large-scale data clustering.

13.3 Parallel Data Processing Architectures and Approaches

The multicore revolution brought increased interest in parallel computation. Solving problems in parallel is usually based on some kind of divide and conquer approach, where a large dataset is split into small chunks processed independently in parallel with occasional collective communication between processing elements that contribute to a global result.

When attempting to solve a problem in parallel, one needs to make a choice of which programming model to use. This choice is often limited by external factors. For example, there may be an existing serial code implemented within some constrained runtime environment, or the target machine architecture may be specialised and fixed. Implementation strategies used for a shared memory machine will be different from strategies used when targeting a shared-nothing compute cluster. Additionally, distributed memory systems are diverse. They can be highly reliable and tightly coupled, as found in high performance computing (HPC) systems, or decoupled, susceptible to frequent failures and heterogeneous.

However, in data processing problems, the first decision needs to be made on a slightly higher level of abstraction. One needs to decide if a computation will be code-driven or data-driven. The imperative, code-driven approach defines computation as a sequence of statements that manipulate data. The data-driven approach, or dataflow programming, models a program as a directed graph of transformational operations applied to a flow of data. Another way of looking at it is to realise that in code-driven data processing a statement triggers a request for data and in the data-driven approach it is the availability of data that triggers the processing code.

The solutions to the problem of microarray data clustering discussed in this chapter explore different parallel implementations of the same algorithm within two distinct sets of constraints. The first set constrains the problem to a specific runtime environment – the R runtime and tightly coupled high performance compute clusters. The second set puts no constraints on the programming environment, but assumes the execution environment to be distributed and with limited reliability (a compute cloud).

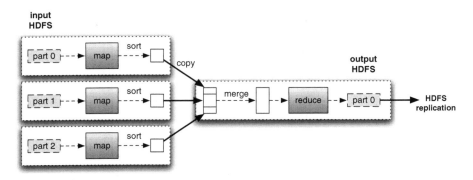

Fig. 13.1 MapReduce data flow in Hadoop. HDFS stands for Hadoop distributed file system

The solution targeting HPC systems uses an imperative approach based on explicit message passing. To exploit such systems, the R runtime must be able to utilise the multiple processors available on these systems. There are existing modules that enable R to use multiple processors, but these are either difficult to use for the HPC novice or cannot be used to solve certain classes of problems. A method of exploiting HPC systems, using R, but without recourse to mastering parallel programming paradigms is, therefore, necessary to analyse genomic data to its fullest. The SPRINT library simplifies employment of HPC resources by providing parallel implementations of commonly used analysis functions.

The second set of constraints is met by refactoring the data analysis workflow using the data centric approach. This is achieved by implementing the analysis within the MapReduce [7] framework. The MapReduce dataflow as implemented by Apache Hadoop [6] is presented in Fig. 13.1.

The Hadoop MapReduce framework relies on the HDFS. When data are put into HDFS, it is automatically split into blocks (size 64MB by default) and distributed among data nodes. File blocks are also being replicated to ensure robustness (3 replicas by default). To process the data, a programmer needs to define *Map* and *Reduce* operations. The *Map* operator receives input data from a partial dataset as (*key, value*)-pairs from one domain and maps these to a possibly empty list of (*key, value*)-pairs from another domain. The framework executes mappers in parallel taking into account data locality and takes care of sorting map output by *key*-value and grouping. The *Reduce* operator receives (*key, value*)-pairs where *value* is a list of values, produced in the map stage with the same *key*. The result of the reduce phase is put to HDFS.

Parallel processing in a MapReduce framework is simplified as the programmer code is automatically parallelised and scheduled in a distributed cluster of compute nodes. This approach trades off some performance aspects with the ability to scale the number of analyses as required. Many MapReduce frameworks automatically deal with computations on datasets that exceed the collective memory of the compute nodes.

Ability to deal with large datasets is important as emerging whole genome associative studies and clinical projects will require from several hundreds to several thousands of microarray experiments. The complexity increases even further when considering the meta-analysis of combined data from several experiments. Microarray experiment repositories such as ArrayExpress are constantly growing in size, and this trend is set to continue as advances in technology and reduction in cost are constantly contributing to an increase in the amount of data to be analysed.

The choice of runtime systems requires some explanation. Both of our target machines, i.e., a tightly coupled HPC machine programmed with MPI and loosely coupled virtual instances spawned on a cloud running a MapReduce framework, are clusters of some sort. It is important to remember that both of these execution environments are general purpose, and it is equally possible to run MPI programs on cloud resources as it is possible to use a MapReduce framework to harness a supercomputer. In a similar fashion, both of these approaches can be successfully applied to program a shared memory machine.

The most important operational difference between the MPI and MapReduce runtimes is their fault tolerance. Systems like Hadoop assume an unreliable environment and have mechanisms that deal efficiently with node failures, whereas a failure of a single node while executing an MPI program will cause an unrecoverable error. Still, if the execution environment is reasonably reliable, then the choice of paradigm should be made based on other factors, like ease of programming, maintainability or performance.

13.4   Related Work

There are several projects that introduce parallel computing to the R environment. These libraries are either low level wrappers to the MPI library or frameworks that allow expression of decoupled many-task computations. The main difference of these approaches compared to the SPRINT library is that they do not provide actual parallel implementations of R functions leaving the burden of finding and expressing parallelism with the user. The SPRINT framework allows a parallel programmer to make use of the full power of MPI to develop scalable implementations of common functions that can be used transparently by the end user. A survey of parallel R projects is presented in [14]. A more detailed comparison of these projects to the SPRINT framework is available in [13].

Examples of projects, where cloud infrastructure has been successfully applied to solve bioinformatics problems, include the Cloudburst parallel sequence mapping algorithm [15] and the Crossbow software pipeline for whole genome resequencing analysis [16]. The MapReduce paradigm has been applied to develop multicore phylogenic applications for analysing large scale collections of evolutionary trees [17].

MapReduce formulations of machine learning algorithms are discussed in [18]. It has been demonstrated that a large class of algorithms that fit the Statistical Query

Model may be mapped to some "summation form" which represents the exact alternative implementation of a given algorithm. A summation form can then be easily expressed in the MapReduce framework. Among the discussed algorithms are the k-means clustering algorithm, locally weighted linear regression and a support vector machine. All of these have applications in microarray analysis.

As far as we know, there have been no previous attempts to apply MapReduce specifically to analyse microarray data. However, some algorithms that can be used for this purpose have been implemented in the Mahout project [19].

Although we are discussing MapReduce formulations outside the R environment, readers should be aware that the two technologies can easily be integrated as exemplified by the RHIPE package [20].

13.5 Identifying Gene Expression Patterns

Once microarray data have been transformed into a gene expression matrix and preprocessed, it can be analysed to identify gene expression patterns. The matrix encodes genes as well as experiments (samples), hence the data can be analysed in two ways:

- Gene expression profiles can be compared by analysing the rows of the matrix
- Sample expression profiles can be compared by analysing the columns of the matrix

Clustering is a popular analysis method that groups together objects – genes or samples – with similar properties. It is typically used to detect groups of co-regulated genes in a dataset.

Many clustering algorithms have been proposed. They differ in computational complexity and robustness. One of the best known partitional clustering methods is the k-means algorithm [21]. In practice, biostatisticans often make use of a more robust clustering method called Partitioning Around Medoids (PAM) [22].

PAM groups objects into clusters by means of identifying a characteristic object (called medoid) for each cluster. The medoid is the object which has minimal dissimilarities – measured by a cost function, for example the euclidean distance or a correlation coefficient – to all other objects in the cluster. The use of a sum of dissimilarities as the objective function makes the PAM algorithm more robust to noise and outliers as compared to k-means.

The PAM algorithm has two stages: *build* and *swap*. During the build phase, the initial set of clusters, represented by their medoids, is computed along with the total cost of this clustering. In the swap phase, other objects are swapped with the existing medoids with the goal of minimising the value of the total cost function. The swap phase terminates when the value of the total cost function cannot be improved.

13.6   HPC Approach: SPRINT Framework

In this section, we describe a parallelisation of the PAM algorithm inside the R programming environment. As it was previously mentioned, core R does not provide any support for parallel execution. The algorithm is parallelised within the SPRINT framework.

13.6.1   The SPRINT Architecture

The architecture of the SPRINT runtime is presented in Fig. 13.2. It distinguishes between the master and worker nodes, which follow different execution paths. All nodes start from instantiating the R runtime and loading the SPRINT library. During this phase, the MPI subsystem is initialised. If this step is successful, worker nodes enter a waiting loop expecting coordination messages from the master node. The master node starts evaluating the user's R program. At some point in the script, a parallel function provided by the SPRINT library is invoked. Here, the execution is handed over to the SPRINT framework, which notifies all the worker nodes about the function that is going to be evaluated in parallel. After distributing the data, all of the nodes start to collectively evaluate a function. The local evaluation can happen inside the C runtime or may be delegated to the R runtime. On completion, the accumulated result is returned back to the R runtime on the master node. The worker nodes go back to the wait loop ready for the next function. The user script on the master node is responsible for shutting down the SPRINT framework before exiting the R runtime.

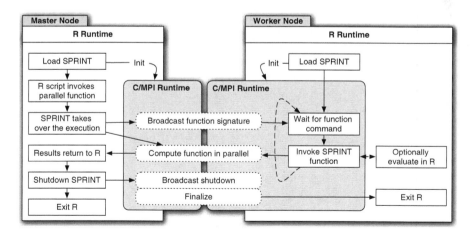

Fig. 13.2   The SPRINT framework architecture

The goal of the SPRINT library is to shield the R user from any details of the parallel implementation. The signatures of parallelised functions are as close as possible to their serial counterparts. For example, a call to a serial correlation function cor(distMtx) becomes simply pcor(distMtx).

13.6.2 Parallelisation

The SPRINT library aims to deliver parallel versions of functions currently used by the R community. A serial version of the PAM algorithm is included in the CLUSTER package (Maechler, M., Rousseeuw, P., Struyf, A., Hubert, M.: Cluster analysis basics and extensions. Unpublished (2005)) and can be obtained from CRAN [23]. The SPRINT implementation of the pam() function provides a parallelised kernel of the serial algorithm and introduces a few serial optimisations.

An important enhancement was achieved by storing the distance matrix in a more expensive, full representation which is optimised for the contiguous row-wise access that the algorithm requires. Storing a symmetric distance matrix in its full square form uses more memory, but greatly improves cache efficiency. On a single processor, the optimised serial version performs up to 20 times faster than the original implementation.

A parallel implementation of the PAM algorithm is straightforward and relies on the fact that the calculations performed inside the ForEach loop are independent of each other (see pseudo code in Fig. 13.3). In a C programming language, such a loop is implemented as interchangeable nested for loops and one could either parallelise over the set of medoids or over the set of non-medoids. To achieve better scalability, the SPRINT implementation parallelises the outer non-medoid loop.

Splitting the iteration space across the nodes results in a block distribution of the distance matrix. Each row of the distance matrix contains all the information needed to compute the cost of a swap for a single data point. Worker nodes calculate swap costs for a subset of data points, and then the best swap is selected by comparing results from all the nodes. As there is no need to redistribute data after each iteration and because the collective communication is minimal, the parallelised algorithm is very efficient.

```
Input: Distance matrix, Initial set of medoids
Output: Optimised set of medoids

Repeat until no change in medoids
  Calculate the current value of the objective function
    ForEach pair of medoid and non-medoid
      If swapping medoid with non-medoid reduces the objective
        Save as possible swap
  Apply the best swap
```

Fig. 13.3 Pseudo code for the swap phase of the PAM algorithm

The SPRINT implementation of PAM not only reduces the execution time but also enables analyses of large data sets. A sample data set that we use in the performance experiments contains 22,347 genes and yields a distance matrix of size $22,347 \times 22,347$ which uses about 3.7GB of memory. This size is beyond the capabilities of many workstations (especially since R is eager to create copies of this matrix). The solution to this problem is to store the distance matrix as a binary file on disk if it exceeds the available memory. This file is then memory mapped and the data are accessed as if it were resident in the memory. This increases the size limit of the distance matrix to the available file sizes on the underlying file system, enabling users to process datasets that are significantly larger than the available physical memory.

13.7 Cloud Approach: MapReduce

A parallel implementation described in the previous section used explicit message passing. Data distribution, splitting of the iteration space, collective communication all had to be managed by the programmer. The result was a highly efficient parallel implementation targeted at reliable parallel systems.

In a distributed environment node, failures are common. One way of achieving fault tolerance is to define computation in terms of independent functions with no side effects. In such a setup, it is possible to resubmit the failed partial calculation. Of course, the data that the failed function operated on still need to be accessible. MapReduce frameworks deal with this by relying on a distributed file system which provides some level of replication. The first step, however, is to find a MapReduce formulation of the PAM algorithm.

13.7.1 Algorithm

Both stages of the PAM algorithm are iterative. The original MapReduce framework and its Hadoop implementation do not provide support for iterations. It is possible to chain mappers and reducers but only in a static manner at compile time. This is not sufficient for iterative algorithms where a termination criteria is evaluated during runtime. Therefore, one is forced to express iterations as a loop over separate MapReduce jobs. A possible formulation of the PAM algorithm in terms of map and reduce operations follows. The input to the MapReduce PAM is the distance matrix $d(i, j)$ and the number of final clusters k.

13.7.1.1 The Build Stage

The initial set of medoids M is empty. Repeat the following steps k times where k is the number of clusters:

1. The map function receives a row $(d_i(j))$ of the distance matrix as input and the current set of medoids M.
2. For each j find the medoid m in $M \cup \{j\}$ representing the cluster to which i belongs. Note the cost (distance to the medoid m) as $cost_j(i)$ and output key j and value $cost_j(i)$ for all j.
3. The reduce function receives parameters j and a list of $cost_j(i)$ for all i. It outputs j as a key, and the total cost $\sum_i cost_j(i)$ as a value. This value is the total cost for medoid set $M \cup \{j\}$.
4. Find j with the minimum total cost $c = \sum_i cost_j(i)$ and define the new set of medoids as $M \cup \{j\}$.

At the end of the build stage, a set of k medoids has been built.

13.7.1.2 The Swap Stage

Start from the set of cluster medoids M with a total cost of c.

1. The map function receives the ith row $(d_i(j))$ of the distance matrix as input and the current set of medoids M.
2. For each j and for each $m \in M$ swap j with medoid m, i.e., consider the set of medoids $M' = M - \{m\} \cup \{j\}$. Find the new medoid $n \in M'$ representing the cluster to which i now belongs. Note the cost (distance to the medoid n) as $cost_{jm}(i)$ and output key j and value $(cost_{jm}(i))$.
3. The reduce function receives parameters j and a list of $cost_{jm}(i)$ for all i and all $m \in M$. It sums up the list of values, therefore, computing the total cost for medoid set $M' = M - \{m\} \cup \{j\}$ as $\sum_i cost_{jm}(i)$. The reduce output is the key j and the total cost as value.
4. Find j and m with minimum value $c' = \sum_i cost_j m(i)$ and swap j and m, i.e., the new medoid set is $M - \{m\} \cup \{j\}$.
5. Repeat these steps until the total cost cannot be improved, that is $c' >= c$.

13.7.2 Runtime

The MapReduce implementation of the PAM algorithm is built on top of Apache Hadoop. A Hadoop cluster has a master node running the main job tracker and a set of client nodes each of which is running a task tracker. All nodes are data nodes and part of the HDFS. Files stored in HDFS are automatically split and distributed across the data nodes, with a certain degree of replication. The split size and the replication factor are configurable.

When a MapReduce job is submitted, the master node's job tracker retrieves a list of input data splits and creates a map task for each split. The number of reduce tasks can be configured. The job tracker submits each map task to a task tracker that is close to the input splits's network location. The output produced by the map tasks is transferred to the reducer tasks. Map output is written to disk when it reaches a threshold size.

The input to MapReduce PAM is the distance matrix $d(i, j)$ and the number k of clusters to construct. The distance matrix is stored on the HDFS and distributed across the available data nodes. The size of the distance matrix determines the number of input splits and therefore the number of map tasks that are created.

Hadoop defines an operation called *Combine*, which is a local *Reduce* operation performed on the output of the mappers from a single data node prior to global reduction. Using a combiner usually results in increased efficiency. The *Reduce* functions in both stages of the MapReduce PAM algorithm are distributive and therefore our implementation makes use of a *Combiner*.

13.8 Testing and Scaling Results

Implementations of the PAM algorithms were tested on their target platforms. Testing of the MapReduce implementation was carried out on the Amazon Elastic Compute Cloud (Amazon EC2) with varying numbers of Amazon EC2 Small Instances. The MPI based implementation was tested on the HECToR supercomputer. The comparison of machine specifications is presented in Table 13.1.

The experiment clustered points using a distance matrix generated from a series of microarray experiments with 22,347 genes and 25 samples. The resulting distance matrix is of size $22,347 \times 22,347$ and occupies about 3.7GB of memory.

The Hadoop runtime used the default settings. The input distance matrix is resident in the HDFS filesystem. The time taken to calculate and store the input matrix is not taken into consideration.

Both approaches showed excellent speedup as illustrated in Fig. 13.4. The scaling degradation seen for the MPI implementation is caused by the fact that the per node problem size decreases with increasing number of cores put to the task. This affects the compute to communication time ratio putting more pressure on the interconnect. Similar behaviour can be observed for the MapReduce implementation. However, the increased communication has more dramatic influence as the performance of

Table 13.1 Hardware details of machines used for the performance experiments

	HECToR	Cluster of EC2 small instances
CPU	Quad core 2.3GHz Opteron	1 EC2 compute unit (single core equivalent to 1.0–1.2 GHz 2007 Opteron or 2007 Xeon processor)
Memory	2GB per core	1.7 GB
I/O	High performance I/O nodes with Lustre parallel filesystem	Moderate performance (virtualised)
Interconnect	High performance Cray SeaStar2 interconnect	Moderate performance (virtualised)

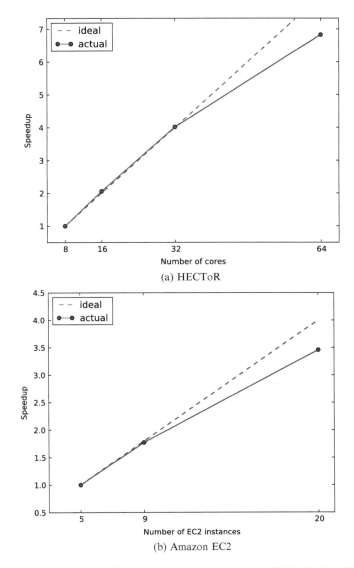

Fig. 13.4 Scaling plots for: (**a**) MPI PAM relative to 8 core run and (**b**) MapReduce PAM relative to 5 core run

the virtualised I/O system provided by the Amazon cloud is only moderate with no performance guarantees. In addition, the message exchange in Hadoop is much more expensive than in MPI mainly due to serialisation.

The actual execution times are plotted in Fig. 13.5. Comparing the actual execution times, the MPI implementation performs 300 times faster than the MapReduce implementation (MPI on 8 cores takes 53 s and MapReduce on 9 cores takes

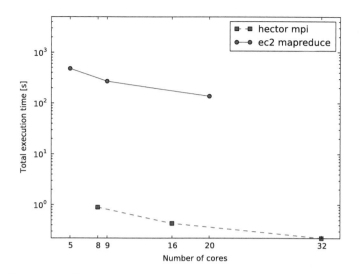

Fig. 13.5 Comparison of the total execution time for the same PAM experiment and MPI and MapReduce implementations

269 min). Even if we take into consideration the differences in architecture of the compute nodes, the MapReduce implementation is two orders of magnitude slower than the MPI implementation. In fact, even a single core run of the MPI implementation which uses external memory is faster then any of the Hadoop results. The next section includes a more in-depth discussion of this behaviour. The same experiment was run as on the Amazon Elastic MapReduce environment with 20 Amazon Elastic MapReduce Small Instances. This uses the Hadoop framework so that the same implementation could be used. The test run was cancelled when the build phase of the PAM algorithm had not completed after five hours of runtime, therefore, being very slow in comparison to the tests executed on 20 Amazon EC2 instances. Hence, Amazon Elastic MapReduce was not explored any further.

13.9   Comparison of Approaches

The most important difference of the two approaches is their total execution time. There are several reasons why the Hadoop MapReduce approach performs badly compared to the MPI approach. The lack of support for iterative computations in Hadoop is one of them. Each iteration of MapReduce PAM is a separate job, and it is impossible for mappers to store state between iterations. The lack of state and side effects allows MapReduce to automatically paralellise computation and freely reschedule map and reduce tasks. However, this feature in iterative algorithms forces the framework to reread static input data at each iteration. In case of the MPI

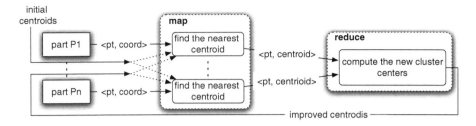

Fig. 13.6 MapReduce formulation of the k-means clustering algorithm (pt – point, coord – coordinates)

implementation once the data are read before the first iteration, it stays in memory until the algorithm terminates.

Another feature of Hadoop is its ability to seamlessly deal with out-of-core data processing. As a result of this, output from the mappers is often spilt to disk. This, together with repeated reading of the entire dataset, quickly turns a clustering calculation into an I/O intensive problem. This obviously incurs huge overhead in comparison to the MPI implementation.

Similarly, high overheads have been reported for the Hadoop MapReduce implementation of k-means clustering algorithm in [24]. The k-means algorithm (see Fig. 13.6) is different from PAM as it takes a set of point and a set of centroids as inputs and associates each point with the nearest centroid in the map phase. The reduce phase calculates new center for each cluster and the iterative map-reduce process continues until the error is sufficiently small. Each point only needs to know the set of current centroids and does not require any knowledge about other points in the dataset. This makes calculations more independent as compared to PAM and makes the overall calculation more suitable to the MapReduce approach.

The main runtime overhead for k-means algorithm is caused, as for PAM, by its iterative character. An alternative, iterative MapReduce framework proposed in [24] distinguishes between fixed and variable data in map-reduce iterations and achieves performance comparable with MPI implementations. This framework has been further extended and is now available as Twister [25].

Hadoop implementations of both PAM and k-means are very cache inefficient. Input and output data represented as (*key, value*) pairs will often be fragmented in memory, and therefore it will be impossible to achieve cache friendly memory reads enjoyed by the MPI implementations. On top of this, values that are represented as efficient primitive types in C language become wrapped in objects in Java implementations.

The MapReduce implementation based on the Hadoop framework requires very little knowledge of distributed systems and parallel programming. Hadoop comes with sensible default settings that allow the user to set up a test environment with minimal effort. Input data is automatically split and assigned to tasks on available nodes, taking data locality into account. The programmer works on a higher level of abstraction, and this usually shortens the development time.

Predicting the performance for MapReduce formulations is problematic. As yet, there is no explicit model of communication for the MapReduce framework, also the mixture of local and HDFS filesystem makes it hard to predict the load on the I/O subsystem. One needs to rely on the profiling capabilities provided by the runtime. Explicit messages used in the MPI implementation allow for exact performance modelling if exchange patterns are regular (as it is the case for the PAM algorithm). A high number of configurable runtime parameters may also be held against the Hadoop runtime. It is very hard for the user to reason about optimal settings for a specific problem.

13.10 Conclusions and Future Work

The main conclusion is that whilst programmers are usually willing to sacrifice some performance in return for a simplified programming model which enhances their productivity, the overheads of the runtime system must not be overwhelming. Our experiments show that the overheads introduced by Hadoop are prohibitive when it comes to iterative clustering algorithms. Of course, this does not mean that one should discard the Hadoop approach all together. As it has been mentioned in the related work section, several projects demonstrated benefits of using the MapReduce framework to tackle different classes of bioinformatics problems. Hadoop may be more suitable for larger datasets in the size of terabytes whilst the dataset analysed in this study did not exceed a few gigabytes.

The fact that efficient MapReduce formulations can be found only for a subset of problems raises the question of maintainability of solutions. A typical microarray workflow will use many different algorithms. The most convenient approach would be a methodology that is generic enough to provide for scalable implementations within a single framework. Given the diversity of possible microarray data analyses, investing in MPI implementations would probably lead to more sustainable libraries.

Although the MPI runtime can be deployed on cloud infrastructures, its lack of fault tolerance may become problematic. Investigating MapReduce runtime systems supporting iterative computations is an important future direction for the community.

The character of the overheads incurred by the Hadoop framework could be better understod by implementing the same PAM algorithm in other non-Java MapReduce frameworks. A shared memory MapReduce framework like [26], whilst not being applicable for very large datasets and distributed clusters, might shed light on the issues caused by serialising and transferring large datasets between nodes that are produced in the map step.

Another option to investigate is to write map and reduce operations using non-Java languages and coordinate their parallel execution using Hadoop Streams. The Malstone benchmark [27] reports significant reduction of runtime when using mappers and reducers implemented in Python.

Acknowledgements The authors of this chapter would like to thank Kostas Kavoussanakis and Radek Ostrowski of EPCC for their support and valuable comments and the SPRINT Project Team: Terry Sloan (EPCC), and Thorsten Forster and Muriel Mewissen (Division of Pathway Medicine). This work was supported by the Wellcome Trust and The Centre for Numerical Algorithms and Intelligent Software.

References

1. Finkelstein, A., Gryce, C., Lewis-Bowen, J.: Relating requirements and architectures: A study of data-grids. J. Grid Comput. **2**(3), 207–222 (2004) doi: 10.1007/s10723-004-6745-6
2. Dobrzelecki, B., Krause, A., Hume, A., Grant, A., Antonioletti, M., Alemu, T., Atkinson, M., Jackson, M., Theocharopoulos, E.: Integrating distributed data sources with ogsa-dai dqp and views. Phil. Trans. Roy. Soc. A, **368**(1926), 4133–4145 (2010) doi: 10.1098/rsta.2010.0166
3. Taniar, D., Leung, C.H.C., Rahayu, W., Goel, S.: High Performance Parallel Database Processing and Grid Databases. Wiley, NY (2008). ISBN 978-0-470-10762-1
4. Laure, E., Stockinger, H., Stockinger, K.: Performance engineering in data grids. Concurrency Comput. Pract. Ex. **17**, 171–191 (2005) doi: 10.1002/cpe.923
5. Special issue on data management on cloud computing platforms. Technical Report 1, March 2009
6. The Apache Software Foundation. The apache hadoop project. http://hadoop.apache.org (2011). Accessed 22 June 2011
7. Dean, J., Ghemawat, S.: Mapreduce: Simplified data processing on large clusters. In: Symposium on Operating System Design and Implementation (OSDI), pp. 137–150 (2004)
8. Stonebraker, M., Becla, J., Dewitt, D., Lim, K.-T., Maier, D., Ratzesberger, O., Zdonik, S.: Requirements for science data bases and scidb. In: Conference on Innovative Data Systems Research (CIDR), January 2009
9. Hubbell, E., Liu, W.-M., Mei, R.: Robust estimators for expression analysis. Bioinformatics **18**(12), 1585–1592 (2002). doi:10.1093/bioinformatics/18.12.1585
10. Irizarry, R.A., Bolstad, B.M., Collin, F., Cope, L.M., Hobbs, B., Speed, T.P.: Summaries of affymetrix GeneChip probe level data. Nucl. Acids Res. **31**(4), e15 (2003). doi:10.1093/nar/gng015
11. R Development Core Team. R: A Language and Environment for Statistical Computing. R Foundation for Statistical Computing, Vienna, Austria, 2010. URL http://www.R-project.org.
12. Gentleman, R.C., Carey, V.J., Bates, D.M., Bolstad, B., Dettling, M., Dudoit, S., Ellis, B., Gautier, L., Ge, Y., Gentry, J., Hornik, K., Hothorn, T., Huber, W., Iacus, S., Irizarry, R., Leisch, F., Li, C., Maechler, M., Rossini, A.J., Sawitzki, G., Smith, C., Smyth, G., Tierney, L., Yang, J.Y., Zhang, J.: Bioconductor: open software development for computational biology and bioinformatics. Genome Biol. **5**(10), R80 (2004). doi: 10.1186/gb-2004-5-10-r80
13. Hill, J., Hambley, M., Forster, T., Mewissen, M., Sloan, T.M., Scharinger, F., Trew, A., Ghazal, P.: SPRINT: a new parallel framework for R. BMC Bioinform. **9**(1), 558 (2008). doi:10.1186/1471-2105-9-558
14. Schmidberger, M., Morgan, M., Eddelbuettel, D., Yu, H., Tierney, L., Mansmann, U.: State of the art in parallel computing with R. J. Stat. Software **31**(1), 1–27 (2009)
15. Schatz, M.C.: CloudBurst: highly sensitive read mapping with MapReduce. Bioinformatics **25**(11), 1363–1369 (2009). doi:10.1093/bioinformatics/btp236
16. Langmead, B., Schatz, M., Lin, J., Pop, M., Salzberg, S.: Searching for snps with cloud computing. Genome Biol. **10**(11) (2009) doi: 10.1186/gb-2009-10-11-r134
17. Matthews, S., Williams, T.: Mrsrf: An efficient mapreduce algorithm for analyzing large collections of evolutionary trees. BMC Bioinformatics **11**(Suppl 1) (2010) doi: 10.1186/1471-2105-11-S1-S15

18. Chu, C.T., Kim, S.K., Lin, Y.A., Yu, Y., Bradski, G.R., Ng, A.Y., Olukotun, K.: Map-reduce for machine learning on multicore. In: Schölkopf, B., Platt, J.C., Hoffman, T. (eds.) NIPS. MIT, MA (2006)
19. The apache mahout project. http://mahout.apache.org
20. Guha, S.: Rhipe – R and hadoop integrated processing environment. http://www.stat.purdue.edu/~sguha/rhipe
21. Jain, A.K., Dubes, R.C.: Algorithms for Clustering Data. Prentice-Hall, CA (1988)
22. van der Laan, M., Pollard, K., Bryan, J.: A new partitioning around medoids algorithm. J. Stat. Comput. Simul. **73**(8), 575–584 (2003). doi:10.1080/0094965031000136012
23. Cran: The comprehensive r archive network. URL http://cran.r-project.org
24. Ekanayake, J., Pallickara, S., Fox, G.: MapReduce for data intensive scientific analyses. In: eScience, 2008. eScience '08. IEEE Computer Society, Los Alamitos, CA, USA, 2008, pp. 277–284. doi:10.1109/eScience.2008.59
25. Twister: A runtime for iterative mapreduce. URL http://www.iterativemapreduce.org
26. Ranger, C., Raghuraman, R., Penmetsa, A., Bradski, G., Kozyrakis, C.: Evaluating mapreduce for multi-core and multiprocessor systems. In: HPCA '07: Proceedings of the 2007 IEEE 13th International Symposium on High Performance Computer Architecture. IEEE Computer Society, February 2007. ISBN 1-4244-0804-0. doi: 10.1109/HPCA.2007.346181
27. Malstone: A stylized benchmark for data intensive computing. URL http://code.google.com/p/malgen/wiki/Malstone

Chapter 14
Grid Technologies for Satellite Data Processing and Management Within International Disaster Monitoring Projects

Nataliia Kussul, Andrii Shelestov, and Sergii Skakun

Abstract This chapter describes the use of Grid technologies for satellite data processing and management within international disaster monitoring projects carried out by the Space Research Institute NASU-NSAU, Ukraine (SRI NASU-NSAU). This includes the integration of the Ukrainian and Russian satellite monitoring systems at the data level, and the development of the InterGrid infrastructure that integrates several regional and national Grid systems. A problem of Grid and Sensor Web integration is discussed with several solutions and case-studies given. This study also focuses on workflow automation and management in Grid environment, and provides an example of workflow automation for generating flood maps from images acquired by the Advanced Synthetic Aperture Radar (ASAR) instrument aboard the Envisat satellite.

14.1 Introduction

Nowadays, satellite monitoring systems are widely used for the solution of complex applied problems such as climate change monitoring, rational land use, environmental, and natural disasters monitoring. To provide solutions to these problems not only on a regional scale but also on a global scale, a "system of systems" approach that is already being implemented within the Global Earth Observation System of Systems[1] (GEOSS) and Global Monitoring for Environment and Security[2]

---

[1] http://www.earthobservations.org.

[2] http://www.gmes.info.

N. Kussul (✉) · A. Shelestov · S. Skakun
Space Research Institute NASU-NSAU, Glushkov Prospekt 40, building 4/1, Kyiv 03680, Ukraine
e-mail: inform@ikd.kiev.ua; andrii.shelestov@gmail.com; serhiy.skakun@ikd.kiev.ua

S. Fiore and G. Aloisio (eds.), *Grid and Cloud Database Management*,
DOI 10.1007/978-3-642-20045-8_14, © Springer-Verlag Berlin Heidelberg 2011

(GMES) is required. This approach envisages the integrated use of satellite data and corresponding products and services, and integration of existing regional and international satellite monitoring systems.

In this chapter, existing approaches and solutions to satellite monitoring systems integration with an emphasis on practical issues in this area are discussed. The following levels of system integration are considered: *data integration level* and *task management level*. Two examples of system integration that use these approaches are discussed in detail. The first one refers to the integration of the Ukrainian (SRI NASU-NSAU) and Russian (Space Research Institute RAN, IKI RAN) systems at the data level. The second example refers to the development of an InterGrid infrastructure that integrates several regional and national Grid systems: the Ukrainian Academician Grid (with satellite data processing Grid segment, UASpaceGrid) and the Center for Earth Observation and Digital Earth of the Chinese Academy of Sciences (CEODE-CAS) Grid segment.

Different practical issues regarding the integration of the emerging Sensor Web technology with Grids are discussed in the study. We show how the Sensor Web can benefit from using Grids and vice versa. A flood application example is given to demonstrate the benefits of such integration.

A problem of workflow automation and management in Grid environment is reviewed in this chapter, and a practical example of workflow automation of Envisat/Advanced Synthetic Aperture Radar (ASAR) data processing to support flood mapping is given.

14.2   Levels of Integration: Main Problems and Possible Solutions

At present, there is a strong trend for globalization of monitoring systems with a purpose of solving complex problems on global and regional scale. Earth observation (EO) data from space are naturally distributed over many organizations involved in data acquisition, processing, and delivery of dedicated applied services. The GEOSS system is aimed at working with and building upon existing national, regional, and international systems to provide comprehensive, coordinated Earth observations from thousands of instruments worldwide, transforming the data collected into vital information for society. Therefore, a considerable need exists to support integration of existing systems for solving applied domain problems on a global and coordinated basis.

With the regard to satellite monitoring systems, integration can be done at the following levels: data integration level and task management level. The data integration approach aims to provide an infrastructure for sharing data and products. Such an infrastructure allows data integration where different entities provide various kinds of data to enable joint problem solving (Fig. 14.1). The integration at data integration level could be done using common standards for EO data exchange, user interfaces, application programming interfaces (APIs), and data and metadata catalogues.

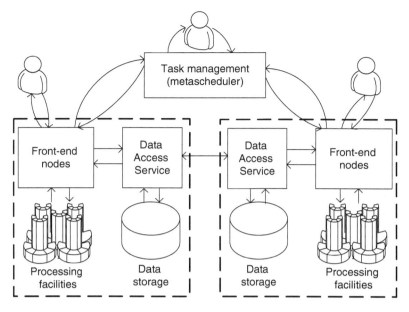

Fig. 14.1 The integration of monitoring systems at data level and at task management level

The task management level approach aims at running applications on distributed computing resources provided by different entities (Fig. 14.1). Since many of the existing satellite monitoring systems heavily rely on the use of Grid technologies, appropriate approaches and technologies should be evaluated and developed to enable the Grid system integration (we define it as *InterGrid*). In such a case, the following problems should be tackled: the use of shared computational infrastructure, development of algorithms for efficient jobs submission and scheduling, load monitoring enabling, and security policy enforcement.

14.2.1 *Data Integration Level*

At present, the most appropriate standards for data integration are the Open Geospatial Consortium[3] (OGC) standards. The following set of standards could be used to address data visualization issues: Web Map Service (WMS), Style Layer Descriptors (SLD), and Web Map Context (WMC). The OGC Web Feature Service (WFS) and Web Coverage Service (WCS) standards provide a uniform way to data delivery. To provide interoperability at the level of catalogues, a Catalogue Service for Web (CSW) standard can be used.

Since the data are usually stored at geographically distributed sites, there are issues regarding optimization of different visualization schemes. In general, there

---

[3]http://www.opengeospatial.org.

are two possible ways to do visualization of distributed data: a centralized visualization scheme and a distributed visualization scheme. Advantages and shortcomings of each of the schemes and experimental results are discussed in detail in [1].

14.2.2 Task Management Level

In this section, the main issues and possible solutions to Grid systems integration are given. The main prerequisite of such integration is enabling certificates trust. This can be done, for example, through the EGEE infrastructure that at present brings together the resources of different organizations from more than 50 countries. Other problems that should be addressed within the integration are as follows: data transfer, high-level access to geospatial data, development of common catalogues, enabling jobs submission and monitoring, and information exchange.

14.2.2.1   Security Issues

To enable security trust between different parties in the Grid system, a Public Key Infrastructure (PKI) is traditionally applied. X.509 is the most widely used format which is supported by most of the existing software.

To get access to resources of the Grid system, a user should make a request to a Certificate Authority (CA) which is always a known third party. The CA validates the information about the user and then signs the user certificate by the CA's private key. The certificate can thus be used to authenticate the user to grant access to the system. To provide a single sign on and delegation capabilities, the user can use the certificate and his private key to create a *proxy certificate*. This certificate is signed not by CA but rather the user himself. The proxy certificate contains information about the user's identity and a special time stamp after which the certificate will no longer be accepted.

To enable Grid system integration with different middleware installed and security mechanisms and policies used, the following solutions were tested:

1. To create our own CAs and to enable the trust between them
2. To obtain certificates from a well-known CA, for example, the European Policy Management Authority for Grid Authentication[4] (EUGridPMA)
3. To use a combined approach in which some of the Grid nodes accept only certificates from the local CA and others accept certificates from a well-known third party CAs.

Within the integration of the UASpaceGrid and the CEODE-CAS Grid, the second and the third approaches were verified. In such a case, the UASpaceGrid accepted

---

[4]http://www.eugridpma.org.

the certificates issued by the local CA that was established using the TinyCA, and certificates issued by the UGRID CA.[5]

It is worth mentioning that Globus Toolkit v.4[6] and gLite v.3[7] middleware implement the same standard for the certificates, but different standards for describing the certificate policies. That is why it is necessary to use two different standards for describing the CA's identity in a policy description file.

14.2.2.2 Enabling Data Transfer Between Grid Platforms

GridFTP is recognized as a standard protocol for transferring data between Grid resources [2]. GridFTP is an extension of the standard File Transfer Protocol (FTP) with the following additional capabilities:

- Integration with the Grid Security Infrastructure (GSI) [3] enabling the support of various security mechanisms
- Improved performance of data transfer using parallel streams to minimize bottlenecks
- Multicasting by doing one-source-to-many-destinations transfers

The Globus Toolkit 4 also provides the OGSI-compliant Reliable Transfer Service (RFT) to enable reliable transfer of data between the GridFTP servers. In this context, reliability means that problems arisen during the transfer are managed automatically to some extent defined by the user.

Some difficulties in using GridFTP exist in networks with a complex architecture. The bunch of these problems originates from the use of the Network Address Translation (NAT) mechanism. To overcome these problems, the appropriate configurations to the network routers and GridFTP servers should be made.

The gLite 3 middleware provides two GridFTP servers with different authorization mechanisms:

1. The GridFTP server with the Virtual Organization Membership System (VOMS) [4] authorization
2. GridFTP server with the Grid Mapfile authorization mechanism

These two servers can work simultaneously under the condition they will use different TCP ports. To transfer files between gLite and GT platforms, both versions of GridFTP servers can be applied. But the server with the VOMS authorization requires all clients to be authorized using the VOMS server. In such a case, this may pose some limitations. In contrast, the GridFTP server with the Grid Mapfile authorization mechanism does not pose such a limitation, and thus can be used with any other authorization system.

---

[5]https://ca.ugrid.org.

[6]http://www.globus.org/toolkit/.

[7]http://glite.web.cern.ch/.

To test file transfers between different platforms used at the UASpaceGrid and the CEODE-CAS Grid, the GridFTP version with the Grid Mapfile authorization was used. File transfers were successfully completed in both directions between two Grids with configured client and server roles.

14.2.2.3   Enabling Access to Geospatial Data

In a Grid system that is used for satellite data processing, corresponding services should be developed to enable access to geospatial data. In such a case, the data may be of different nature, and different formats may be used for storing them.

Two solutions can be used to enable a high-level access to geospatial data in Grids: the Web Services Resource Framework (WSRF) services or the Open Grid Services Architecture–Database Access and Integration[8] (OGSA–DAI) container. Each of these two approaches has its own advantages and shortcomings. A basic functionality for the WSRF-based services can be easily implemented, packed, and deployed using proper software tools, but enabling advanced functionality such as security delegation, third-party transfers, and indexing becomes much more complicated. The difficulties also arise if the WSRF-based services are to be integrated with other data-oriented software. A basic architecture for enabling access to geospatial data in Grids via the WSRF-based services is shown in Fig. 14.2.

The OGSA–DAI framework provides uniform interfaces to heterogeneous data. This framework allows the creation of high-level interfaces to data abstraction layer hiding the details of data formats and representation schemas. Most of the problems such as delegation, reliable file transfer, and data flow between different sources are handled automatically in the OGSA–DAI framework. The OGSA–DAI containers are easily extendable and embeddable. But comparing to the WSRF basic functionality, the implementation of an OGSA–DAI extension is much more complicated. Moreover, the OGSA–DAI framework requires a preliminary deployment of additional software components. A basic architecture for enabling access to geospatial data in Grids via the OGSA–DAI container is shown in Fig. 14.2.

14.2.2.4   Job Submission and Monitoring

Different approaches were evaluated to enable job submission and monitoring in the InterGrid composed of Grid systems that use different middleware. In particular:

1. To use a Grid portal that supports job submission mechanism for different middleware (Fig. 14.3). The GridSphere and P-GRADE are among possible solutions.

---

[8] http://www.ogsadai.org.uk.

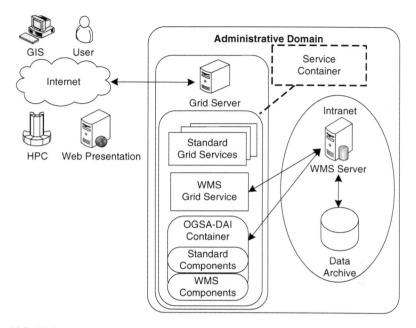

Fig. 14.2 High-level access to geospatial data via the WSRF-based services and the OGSA–DAI container

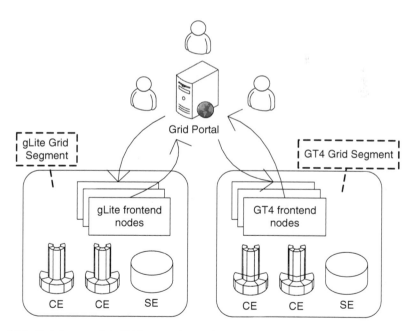

Fig. 14.3 Portal approach to Grid system integration

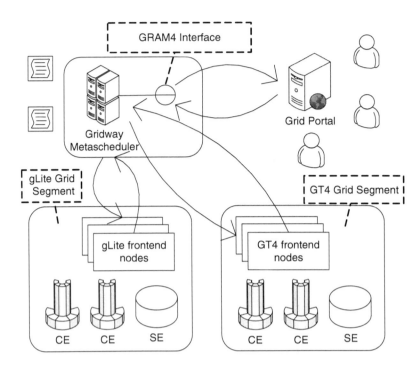

Fig. 14.4 Metascheduler approach

2. To develop a high-level Grid scheduler – *metascheduler* – that will support different middleware by providing standard interfaces (Fig. 14.4).

The Grid portal is an integrated platform to end users that enables access to Grid services and resources via a standard Web browser. The Grid portal solution is easy to deploy and maintain, but it does not provide APIs and scheduling capabilities.

On the contrary, a metascheduler interacts with low-level schedulers used in different Grid systems enabling system interoperability. The metascheduler approach is much more difficult to maintain comparing to the portal; however, it provides necessary APIs with advanced scheduling and load-balancing capabilities. At present, the most comprehensive implementation for the metascheduler is a GridWay system. The GridWay metascheduler is compatibility with both Globus and gLite middleware. Beginning from Globus Toolkit v4.0.5, GridWay becomes a standard part of its distribution. The GridWay system provides comprehensive documentation for both users and developers that is an important point for implementing new features.

A combination of these two approaches will provide advanced capabilities for the implementation of interfaces to get access to the resources of the Grid system, while a Grid portal will provide a suitable user interface.

To integrate the resources of the UASpaceGrid and the CEODE-CAS Grid, a GridSphere-based Grid-portal was deployed.[9] The portal allows the submission and monitoring of jobs on the computing resources of the Grid systems and provides access to the data available at the storage elements of both systems.

14.3 Implementation Issues: Lessons Learned

In this section, two real-world examples of system integration at the data level and task management level are given. The first example describes the integration of the Ukrainian satellite monitoring system operated at the SRI NASU-NSAU and the Russian satellite monitoring system operated at the IKI RAN at the data level. The second example refers to the development of the InterGrid infrastructure that integrates several regional and national Grid systems: the Ukrainian Academician Grid and Chinese CEODE-CAS Grid.

14.3.1 Integration of Satellite Monitoring Systems at Data Level

Figure 14.5 shows the overall architecture for integrating the satellite monitoring systems at data level. The satellite data and corresponding products, modeling data, and in situ observations are provided in a distributive way by applying the OGC standards. In particular, the SRI NSAU-NSAU provides OGC/WMS-compliant interfaces to the following data sets:

- Meteorological forecasts derived from the Weather Research and Forecast (WRF) numerical weather prediction (NWP) model [5]
- In situ observations from a network of weather stations in Ukraine
- Earth land parameters such as temperature, vegetation indices, and soil moisture derived from NASA's Moderate resolution Imaging Spectro-radiometer (MODIS) instrument onboard Terra and Aqua satellites

The IKI RAN provides OGC/WMS-compliant interfaces to the following satellite-derived products:

- Land parameters that are primarily used for agriculture applications
- Fire risk and burnt area estimation for disaster monitoring applications

The products provided by the IKI RAN cover both Russia and Ukraine countries. Coupling these products with modeling data and in situ observations provided by the SRI NASU-NSAU allows information of a new quality to be acquired in almost near-real time. Such integration would never be possible without the use of standardize OGC interfaces. The proposed approach is used for the solution of applied problems of agriculture resources monitoring and crop yield prediction.

---

[9]http://gridportal.ikd.kiev.ua:8080/gridsphere.

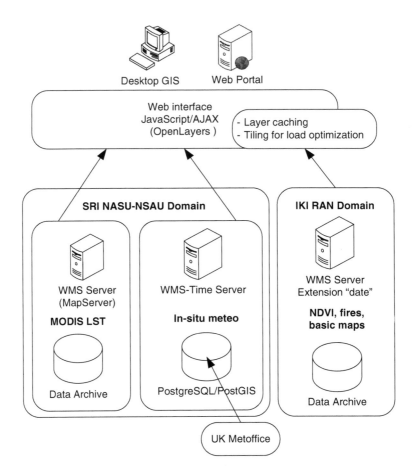

Fig. 14.5 Architecture of satellite monitoring systems integration

To provide a user interface that will enable integration of data coming from multiple sources an open-source OpenLayers[10] framework is used. OpenLayers is a thick client software based on JavaScript/AJAX technology and fully operational on a client side. Main OpenLayers features also include:

- Support of several WMS servers
- Support of different OGC standards (WMS, WFS)
- Caching and tiling support to optimize visualization
- Support of both raster and vector data

The data and satellite-based products provided by the SRI NASU-NSAU and IKI RAN are available at http://land.ikd.kiev.ua. Figure 14.6 shows a screenshot of OpenLayers interface in which data from multiple sources are being integrated.

---

[10]http://www.openlayers.org.

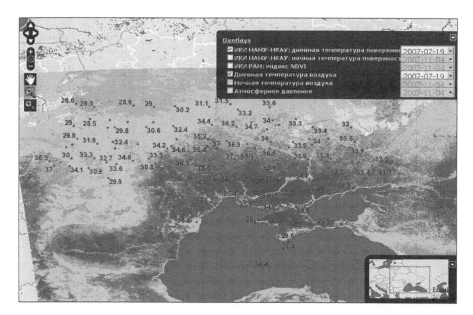

Fig. 14.6 OpenLayers interface to heterogeneous data integration from multiple distributed sources

14.3.2   The InterGrid Testbed Development

The second case study refers to the development of the InterGrid aimed at solving applications of environment and natural disasters monitoring. The InterGrid integrates the Ukrainian Academician Grid with a satellite data processing Grid segment UASpaceGrid and the CEODE-CAS Grid. This InterGrid is considered as a testbed for the Wide Area Grid (WAG) – a project initiated within the CEOS Working Group on Information Systems and Services[11] (WGISS).

An important application that is being solved within the InterGrid environment is flood monitoring and prediction. This task requires the adaptation and tuning of existing meteorological, hydrological and hydraulic models for corresponding territories [5], and the use of heterogeneous data stored at multiple sites. The following data sets are used within the flood application:

- NWP modelling data provided within the UASpaceGrid
- *Satellite data:* synthetic aperture radar (SAR) imagery acquired by ESA's Envisat/ASAR and ERS-2/SAR satellites, optical imagery acquired by Terra, Aqua and EO-1 satellites
- Products derived from optical and microwave satellite data such as surface temperature, vegetation indices, soil moisture, and precipitation

---

[11] http://www.ceos.org/wgiss.

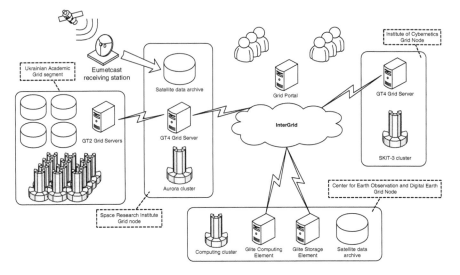

Fig. 14.7 Architecture of InterGrid

- In situ observations from weather stations
- Topographical data such as digital elevation model (DEM)

The process of model adaptation can be viewed as a complex workflow and requires the solution of optimization problems (so-called parametric study) [5]. The processing of satellite data and generation of corresponding products is also a complex workflow and requires intensive computations [6, 7]. All these factors lead to the need of using computing and storage resources of different organizations and their integration into a common InterGrid infrastructure. Figure 14.7 shows the architecture of the proposed InterGrid.

Currently, the InterGrid infrastructure integrates the resources of several geographically distributed organisations, in particular:

- SRI NASU-NSAU (Ukraine) with deployed computing and storage nodes based on the Globus Toolkit 4 and gLite 3 middleware, access to geospatial data and a Grid portal
- Institute of Cybernetics of NASU (IC NASU, Ukraine) with deployed computing and storage nodes based on Globus Toolkit 4 middleware and access to computing resources (SCIT-1/2/3 clusters,[12] more than 650 processors)
- CEODE-CAS (China) with deployed computing nodes based on gLite 3 middleware and access to geospatial data (approximately 16 processors)

In all cases, the Grid Resource Allocation and Management (GRAM) service [8] is used to execute jobs on the Grid resources.

It is also worth mentioning that satellite data are distributed over the Grid environment. For example, the Envisat/ASAR data (that are used within the flood

---

[12]http://icybcluster.org.ua.

application) are stored on the ESA's rolling archive and routinely downloaded for the Ukrainian territory. Then, they are stored at the SRI NASU-NSAU archive that is accessible via the Grid. MODIS data from Terra and Aqua satellites that are used in flood and agriculture applications are routinely downloaded from the USGS archives and stored at the SRI NASU-NSAU and IC NASU.

The GridFTP protocol was chosen to provide data transfer between the Grid systems. Access to the resources of the InterGrid is organized via a high-level Grid portal that has been deployed using a GridSphere framework.[13] Through the portal, a user can access the required data and submit jobs to the computing resources of the InterGrid. The portal also provides facilities to monitor the resources state such as CPU load and memory usage. The workflow of the data processing steps in the InterGrid is managed by a Karajan engine.[14]

14.4 Integration of Grid and Sensor Web

Decision makers in an emergency response situation (e.g., floods, droughts) need rapid access to the existing data, the ability to request and process data specific to the emergency, and tools to rapidly integrate the various information services into a basis for decisions. The flood prediction and monitoring scenario presented here is being implemented within the GEOSS Architecture Implementation Pilot[15] (AIP). It uses precipitation data from the Global Forecasting System (GFS) model and NASA's Tropical Rainfall Measuring Mission[16] (TRMM) to identify the potential flood areas. Once the areas have been identified, we can request satellite imagery for the specific territory for flood assessment. These data can be both optical (like EO-1, MODIS, SPOT) and microwave (Envisat, ERS-2, ALOS, RADARSAT-1/2).

This scenario is implemented using the Sensor Web [9, 10] and Grid [6, 7, 11, 12] technologies. The integration of sensor networks with Grid computing brings out dual benefits [13]:

- Sensor networks can off-load heavy processing activities to the Grid.
- Grid-based sensor applications can provide advance services for smart-sensing by deploying scenario-specific operators at runtime.

14.4.1 Sensor Web Paradigm

Sensor Web is an emerging paradigm and technology stack for integration of heterogeneous sensors into a common informational infrastructure. The basic

---

[13]http://www.gridsphere.org.

[14]http://www.gridworkflow.org/snips/gridworkflow/space/Karajan.

[15]http://www.ogcnetwork.net/AIpilot.

[16]http://trmm.gsfc.nasa.gov.

functionality required from such infrastructure is remote data access with filtering capabilities, sensors discovery, and triggering of events by sensors conditions.

Sensor Web is governed by the set of standards developed by OGC [14]. At present, the following standards are available and approved by consortium:

- OGC Observations and Measurements[17] – common terms and definition for Sensor Web domain
- Sensor Model Language[18] – XML-based language for describing different kinds of sensors
- Transducer Model Language[19] – XML-based language for describing the response characteristics of a transducer
- Sensor Observations Service[20] (SOS) – an interface for providing remote access to sensors data
- Sensor Planning Service[21] (SPS) – an interface for submitting tasks to sensors

There are also standards drafts that are available from the Sensor Web working group but not yet approved as official OpenGIS standards:

- Sensor Alert Service – service for triggering different kinds of events basing of sensors data
- Web Notification Services – notification framework for sensor events

The Sensor Web paradigm assumes that sensors could belong to different organizations with different access policies or, in a broader sense, to different administrative domains. However, existing standards does not provide any means for enforcing data access policies leaving it to underlying technologies. One possible way for handling informational security issues in Sensor Web is presented in the next sections.

14.4.2   Sensor Web Flood Use Case

One of the most challenging problems for Sensor Web technology implementation is global ecological monitoring in the framework of GEOSS. In this section, we consider the problem of flood monitoring using satellite remote sensing data, in situ data, and results of simulations.

Flood monitoring requires the integrated analysis of data from multiple heterogeneous sources such as remote sensing satellites and in situ observations. Flood prediction is adding the complexity of physical simulation to the task. Figure 14.8 shows the Sensor Web architecture for this case-study. It presents the integrated use

---

[17]http://www.opengeospatial.org/standards/om.

[18]http://www.opengeospatial.org/standards/sensorml.

[19]http://www.opengeospatial.org/standards/tml.

[20]http://www.opengeospatial.org/standards/sos.

[21]http://www.opengeospatial.org/standards/sps.

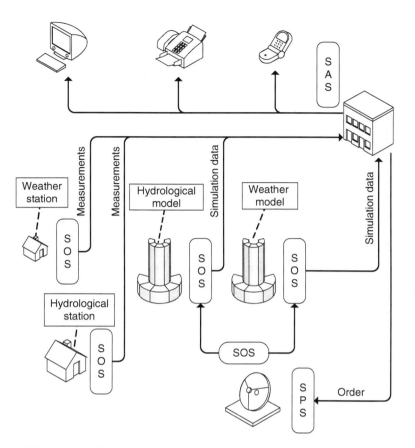

Fig. 14.8   The sensor web architecture of the flooding test case

of different OpenGIS® specifications for the Sensor Web. The data from multiple sources (numerical models, remote sensing, in situ observations) are accessed through the Sensor Observation Service (SOS). An aggregator site is running the Sensor Alert Service to notify interested organization about potential flood event using different communication means. The aggregator site is also sending orders to satellite receiving facilities using the SPS service to acquire new satellite imagery.

14.4.3   Sensor Web SOS Gridification

The Sensor Web services such as SOS, SPS, and SAS can benefit from the integration with the Grid platform like Globus Toolkit. Many Sensor Web features can take advantage of the Grid services, namely:

• Sensor discovery could be performed through the combination of the Index Service and Trigger Service.

- High-level access to XML description of the sensors and services could be made through queries to the Index Service.
- Grid platform provides a convenient way for the implementation of notifications and event triggering using corresponding platform components [15].
- The RFT service [2] provides reliable data transfer for large volumes of data.
- The GSI infrastructure provides enforcement of data and services access policies in a very flexible way allowing implementation of desired security policy.

To exploit these benefits, an SOS testbed service using Globus Toolkit as a platform has been developed. Currently, this service works as a proxy translating and redirecting user requests to the standard HTTP SOS server. The current version uses client-side libraries for interacting with the SOS server provided by the 52North in their OX-Framework. The next version will also include in-service implementation of the SOS server functionality.

The Grid service implementing SOS provides an interface specified in the SOS reference document. The key difference between the standard interfaces and the Grid-based implementations of the SOS lies in the encoding of service requests. The standard implementation uses custom serialization for the requests and responses, and the Grid-based implementation uses the Simple Object Access Protocol (SOAP) encoding.

To get advantage of Globus features, the SOS service should export service capabilities and sensor descriptions as WSRF resource property [16]. Traditionally, the implementation of such a property requires the translation between XML Schema and Java code. However, the XML Schema of the SOS service and related standards, in particular GML [15], is a very complex one, and there are no available program tools able to generate Java classes from it. This problem was solved by storing service capabilities and sensor description data as the Document Object Model (DOM) element object and using a custom serialization for this class provided by the Axis framework that is used by the Globus Toolkit. Within this approach, particular elements of the XML document cannot be accessed in an object-oriented style. However, the SOS Grid service is acting as a proxy between the user and the SOS implementation, so it does not have to modify the XML document directly. With resource properties defined in this way, they can be accessed by using a standard Globus Toolkit API or command line utilities.

14.5 Grid Workflow Management for Satellite Data Processing Within UN-SPIDER Program

One of the most important problems associated with satellite data processing for disaster management is a timely delivery of information to end users. To enable such capabilities, an appropriate infrastructure is required to allow for rapid and efficient access to processing and delivery of geospatial information that is further used for damage assessment and risk management. In this section, the use of Grid

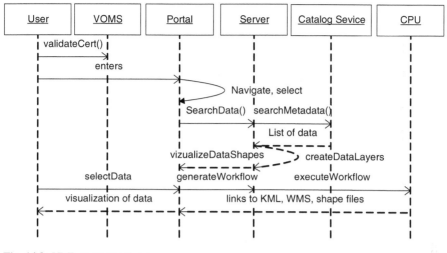

Fig. 14.9 UML sequence diagram

technologies for automated acquisition, processing and visualization of satellite SAR, and optical data for rapid flood mapping is presented. The developed services are used within the United Nations Platform for Space-based Information for Disaster Management and Emergency Response[22] (UN-SPIDER) Regional Support Office (RSO) in Ukraine that was established in February 2010.

14.5.1 Overall Architecture

Within the infrastructure, an automated workflow of satellite SAR data acquisition, processing and visualization, and corresponding geospatial services for flood mapping from satellite SAR imagery were developed. The data are automatically downloaded from the ESA rolling archives where satellite images are available within 2–4 h after their acquisition. Both programming and graphical interfaces were developed to enable search, discovery, and acquisition of data. Through the portal, a user can perform a search for the SAR image file based on geographical region and a time range. A list of available SAR imagery is returned and the user can select a file to generate a flood map. The file is transferred to the resources of the Grid system at the SRI NASU-NSAU, and a workflow is automatically generated and executed on the resources of the Grid infrastructure. The corresponding UML sequence diagram is shown in Fig. 14.9.

To enable execution of the workflow in the Grid system, a set of services has been implemented (Fig. 14.10). We followed the approach used in the Earth System Grid [17]. The four major components of the system are as follows:

---

[22]http://www.un-spider.org.

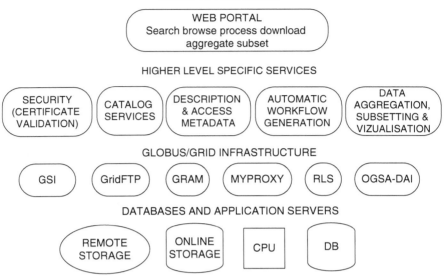

Fig. 14.10 System architecture

1. *Client applications.* Web portal is a main entry point, and provides interfaces to communicate with system services.
2. *High-level services.* This level includes security subsystem, catalogue services, metadata services (description and access), automatic workflow generation services, and data aggregation, subsetting and visualization services. These services are connected to the Grid services at the lower level.
3. *Grid services.* These services provide access to the shared resources of the Grid system, access to credentials, file transfer, job submission, and management.
4. *Database and application services.* This level provides physical data and computational resources of the system.

14.5.2 Workflow of Flood Extent Extraction from Satellite SAR Imagery

A neural network approach to SAR image segmentation and classification was developed [6]. The workflow of data processing is as follows (Fig. 14.11):

1. *Data calibration.* Transformation of pixel values (in digital numbers) to backscatter coefficient (in dB).
2. *Orthorectification and geocoding.* This step is intended for a geometrical and radiometric correction associated with the SAR imaging technology, and to provide a precise georeferencing of data.

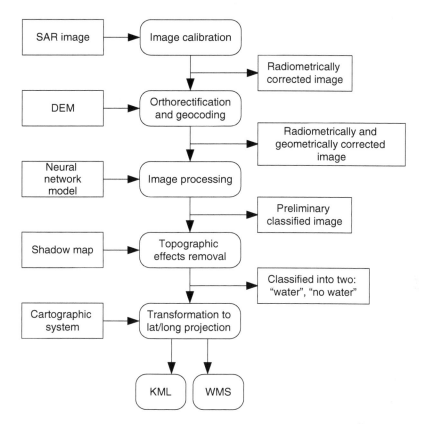

Fig. 14.11 Workflow of flood extent extraction from SAR satellite imagery

3. *Image processing.* Segmentation and classification of the image using a neural network.
4. *Topographic effects removal.* Using digital elevation model (DEM), such effects as shadows are removed from the image. The output of this step is a binary image classified into two classes: "Water" and "No water."
5. *Transformation to geographic projection.* The image is transformed to the projection for further visualization via Internet using the OGC-compliant standards (KML or WMS) or desktop Geographic Information Systems (GIS) using shape file.

14.5.3   China–Ukrainian Service-Oriented System for Disaster Management

To benefit from data of different nature (e.g., optical and radar) and provide integration of different products in case of emergency, our flood mapping service was integrated with the flood mapping services provided by the CEODE-CAS.

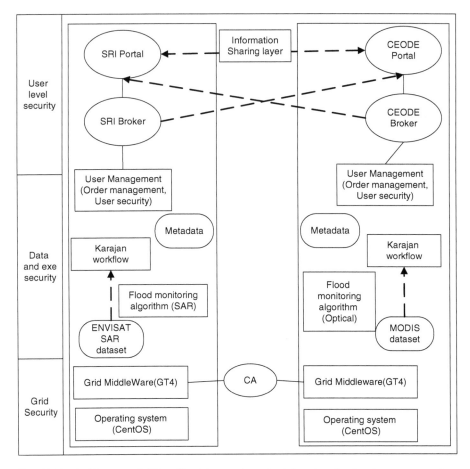

Fig. 14.12 Architecture of China–Ukrainian service-oriented system for disaster management

This service is based on the use of optical data acquired by MODIS instrument onboard Terra and Aqua satellites. Figure 14.12 shows the architecture of the China–Ukrainian service-oriented system for disaster management.

The integration of the Ukrainian and Chinese systems is done at the level of services. The portals of SRI NASU-NSAU and CEODE are operated independently and communicate with corresponding brokers that provide interfaces to the flood mapping services. These brokers process requests from both local and trusted remote sites. For example, to provide a flood mapping product using SAR data, the CEODE portal generates a corresponding search request to the broker at the SRI NASU-NSAU side based on user search parameters. This request is processed by the broker and the search results are displayed at the CEODE portal. The user selects the SAR image file to be processed, and the request is submitted to the SRI NASU-NSAU broker which generates and executes workflow, and delivers the flood maps to the CEODE portal. The same applies to the broker operated at the CEODE side that provides flood mapping services using optical satellite data. To get access to the portal, the user should have a valid certificate. The SRI NASU-NSAU runs the VOMS server to manage with this issue.

14.6 Experimental Results

14.6.1 Numerical Weather Modeling in Grid

The forecasts of meteorological parameters derived from numerical weather modeling are vital for a number of applications including floods, droughts, and agriculture. Currently, we run the WRF model in operational mode for the territory of Ukraine. The meteorological forecasts for the next 72 h are generated every 6 h with a spatial resolution of 10 km. The size of horizontal grid is 200 × 200 points with 31 vertical levels. The forecasts derived from the National Centers for Environmental Prediction Global Forecasting System (NCEP GFS) are used as boundary conditions for the WRF model. These data are acquired via Internet through the National Operational Model Archive and Distribution System (NOMADS).

The WRF model workflow to produce forecasts is composed of the following steps [18]: data acquisition, data preprocessing, generation of forecasts using WRF application, data postprocessing, and visualization of the results through a portal.

Experiments were run to evaluate the performance of the WRF model with respect to the number of computating nodes of the Grid system resources. For this purpose, we used a parallel version of the WRF model (version 2.2) with a model domain identical to those used in operational NWP service (200 × 200 × 31 grid points with a horizontal spatial resolution of 10 km). The model parallelization was implemented using the Message Passing Interface (MPI). We observed almost a linear productivity growth against the increasing number of computation nodes. For example, the use of eight nodes of the SCIT-3 cluster of the Grid infrastructure gave the performance increase 7.09 times (of 8.0 theoretically possible) comparing to a single node. The use of 64 nodes of the SCIT-3 cluster increased the performance 43.6 times. Since a single iteration of the WRF model run corresponds to the forecast of meteorological parameters for the next 1 min, the completion of 4,320 iterations is required for a 3 day forecast. That is, it takes approximately 5.16 h to generate a 3-day forecast on a single node of the SCIT-3 cluster of the Grid infrastructure. In turn, the use of 64 nodes of the SCIT-3 cluster allowed us to reduce the overall computing time to approximately 7.1 min.

14.6.2 Implementation of SOS Service for Meteorological Observations: Database Issues

To provide access to meteorological data, we implemented the Sensor Web SOS service (see Sect. 14.4.1). As a case study, we implemented an SOS service for retrieving surface temperature measured at the weather stations distributed over Ukraine.

The SOS service output is an XML document in a special scheme specified by the SOS reference document. The standard describes two possible ways for retrieving results, namely "Measurement" and "Observation." The first form is more suitable to situations when the service returns a small amount of heterogeneous data. The second form is more suitable for a long time-series of homogeneous data. Table 14.1 gives an example of the SOS service output for both cases.

The 52North software was used for the implementation of the SOS service. Since the 52North has a complex relational database scheme, we had to adapt the existing database structure using a number of SQL views and synthetic tables. From 2005 through 2008 there were nearly two million records with observations derived at the weather stations. The PostgreSQL database with the PostGIS spatial extension was used to store the data. Most of the data records were contained within a single table "observations" with indices built over fields with observation time and a station identifier. The tables of such a volume require a special handling; so the index for a time field was clusterized thus reordering data on the disks and reducing the need for I/O operations. The clusterization of the time index reduced a typical query time from 8,000 to 250 ms.

To adapt this database to the requirements of the 52North server, a number of auxiliary tables with reference values related to the SOS service such as phenomena names, sensor names, and region parameters, and a set of views that transform the underlying database structure into the 52North scheme were created. The most important view that binds all the values of synthetic tables together with observation data has the following definition:

```
SELECT observations.''time'' AS time_stamp, ''procedure''.
procedure_id, feature_of_interest. feature_of_interest_id,
phenomenon.phenomenon_id, offering.offering_id,
'' AS text_value, observations.t AS numeric_value,''
AS mime_type, observations.oid AS observation_id
FROM observations, ''procedure'', proc_foi, feature_of_interest,
proc_off, offering_strings offering, foi_off, phenomenon,
proc_phen, phen_off
WHERE ''procedure''.procedure_id::text = proc_foi.
procedure_id::text AND proc_foi.feature_of_interest_id::text =
feature_of_interest.feature_of_interest_id AND ''procedure''.
procedure_id::text = proc_off.procedure_id::text AND proc_off.
offering_id::text = offering.offering_id::text AND foi_off.
offering_id::text = offering.offering_id::text AND foi_off.
feature_of_interest_id::text = feature_of_interest.
feature_of_interest_id AND proc_phen.procedure_id::text =
''procedure''.procedure_id::text AND proc_phen.phenomenon_id::
text = phenomenon.phenomenon_id::text AND phen_off.
phenomenon_id::text = phenomenon.phenomenon_id::text AND
phen_off.offering_id::text = offering.offering_id::text AND
observations.wmoid::text = feature_of_interest.
feature_of_interest_id;
```

Table 14.1 The two different forms of the SOS service output

Measurement	Observation
`<om:Measurement gml:id="o255136">`	`<om:result>`
`<om:samplingTime>`	`2005-03-14T21:00:00+03,33506,-5@@`
`<TimeInstant xsi:type=''gml:TimeInstantType''>`	`2005-03-15T00:00:00+03,33506,-5.2@@`
`<timePosition>`	`2005-03-15T03:00:00+03,33506,-5.5@@`
`2005-04-14T04:00:00+04`	`2005-03-15T06:00:00+03,33506,-4.6@@`
`</timePosition>`	`2005-03-15T09:00:00+03,33506,-2.2@@`
`</TimeInstant>`	`2005-03-15T12:00:00+03,33506,1.7@@`
`</om:samplingTime>`	`2005-03-15T15:00:00+03,33506,1.7@@`
`<om:procedure xlink:href=`	`2005-03-15T18:00:00+03,33506,2.4@@`
`''urn:ogc:object:feature:Sensor:WMO:33506''/>`	`2005-03-15T21:00:00+03,33506,-0.7@@`
`<om:observedProperty xlink:href=`	`2005-03-16T00:00:00+03,33506,-1.4@@`
`''urn:ogc:def:phenomenon:OGC:temperature''/>`	`2005-03-16T03:00:00+03,33506,-1.1@@`
`<om:featureOfInterest>`	`2005-03-16T06:00:00+03,33506,-1.1@@`
`<sa:Station gml:id=''33506''>`	`2005-03-16T09:00:00+03,33506,-1.3@@`
`<name>WMO33506</name>`	`2005-03-16T12:00:00+03,33506,0.5@@`
`<sa:sampledFeature xlink:href=''''/>`	`2005-03-16T15:00:00+03,33506,1.7@@`
`<sa:position>`	`2005-03-16T18:00:00+03,33506,1.5@@`
`<Point>`	`</om:result>`
`<pos srsName=''urn:crs:epsg:4326''>`	
`34.55 49.6`	
`</pos>`	
`</Point>`	
`</sa:position>`	
`</sa:Station>`	
`</om:featureOfInterest>`	
`<om:result uom=''celsius''>10.9</om:result>`	
`</om:Measurement>`	

The 52North's database scheme uses a string as a primary key for auxiliary tables instead of a synthetic numerical one, and is far from being optimal in the sense of performance. It might cause problems in a large-scale SOS-enabled data warehouse. A typical SQL query from the 52North service is quite complex. Here is an example:

```
SELECT observation.time_stamp, observation.text_value,
observation.observation_id, observation.numeric_value,
observation.mime_type, observation.offering_id, phenomenon.
phenomenon_id, phenomenon.phenomenon_description,
phenomenon.unit,phenomenon.valuetype,observation.procedure_id,
feature_of_interest.feature_of_interest_name, feature_of_interest.
feature_of_interest_id, feature_of_interest.feature_type,
SRID(feature_of_interest.geom), AsText(feature_of_interest.geom)
AS geom FROM phenomenon NATURAL INNER JOIN observation NATURAL
INNER JOIN feature_of_interest WHERE (feature_of_interest.
feature_of_interest_id = '33506') AND (observation.phenomenon_id
='urn:ogc:def:phenomenon:OGC:1.0.30:temperature') AND
(observation.procedure_id = 'urn:ogc:object:feature:Sensor:
WMO:33506') AND (observation.time_stamp >= '2006-01-01 02:00:00
+0300'AND observation.time_stamp <= '2006-02-26 01:00:00+0300')
```

An average response time for such a query (assuming a 1-month time period) is about 250 ms with the PostgreSQL server running in a virtual environment on a 4 CPUs server with 8GB of RAM and 5 SCSI 10k rpm disks in RAID5 array. The increase in a query depth results in a linear increase of response time with an estimate of 50 ms per month (see Fig. 14.13).

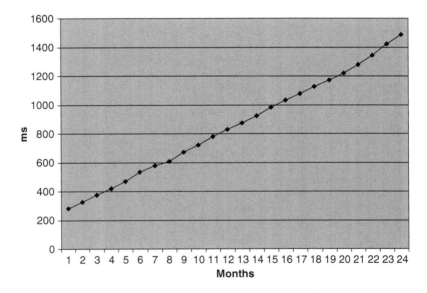

Fig. 14.13 The dependence of query response time against the depth of query

14.6.3 Rapid Flood Mapping from Satellite Imagery in Grid

Within the developed Grid infrastructure, a set of services for rapid flood mapping from satellite imagery is delivered. The use of the Sensor Web services enables automated planning and tasking of satellite (where available) and data delivery, while the Grid services are used for workflow orchestration, data processing, and geospatial services delivery to end users through the portal.

To benefit from the use of the Grid, a parallel version of the method for flood mapping from satellite SAR imagery has been developed (see Sect. 14.5.2). The parallelization of the image processing was implemented in the following way: an SAR image is split into the uniform parts that are processed on different nodes using the OpenMP Application Program Interface (www.openmp.org). The use of the Grids allowed us to considerably reduce the time required for image processing and service delivery. In particular, it took approximately more than 1.5 h (depending on image size) to execute the whole workflow on a single workstation. The use of Grid computing resources allowed us to reduce the computational time to less than 20 min.

Another case study refers to the use of the Sensor Web for tasking the EO-1 satellite through the SPS service [19]. Through the UN-SPIDER RSO in Ukraine, a request was made from local authorities to acquire satellite images over the Kyiv city area due to a high risk of a flood in spring 2010. The use of the Sensor Web and the Grid ensured a timely delivery of products to end users. In particular, Table 14.2 gives a sequence of events starting from the notification of satellite tasking and ending with generation of final products.

It took less than 12 h after image acquisition to generate geospatial products that were delivered to the Ukrainian Ministry of Emergency Situations, the Council of National Security and Defence, and the Ukrainian Hydrometeorological Centre. The information on river extent that was derived from the EO-1 image was used to calibrate and validate hydrological models to produce various scenarios of water extent for flood risk assessment.

Table 14.2 The timeline of tasking the EO-1 satellite and generating the final geospatial product during the potential flood in Ukraine in spring 2010

Date and time[a]	Event
Mon Apr 12 2010 @ 10:33 PM	Notification on EO-1 tasking through SPS
Tue Apr 13 2010 @ 11:33 AM	Image taken
Tue Apr 13 2010 @ 04:30 PM	Image available at the NASA ftp server and automatically transferred to the Grid system resources
Tue Apr 13 2010 @ 11:30 PM	Generation of geospatial products using Grid computing resources

[a]Time is local Ukrainian.

14.6.4 Discussion

Summarizing, we may point out the following benefits of using Grid technologies and Sensor Web for the case studies described in this section. Within the meteorological modeling application, the use of the Grid system resources made it possible to considerably reduce the time required to run the WRF model (up to 43.6 times). It is especially important for the cases when it is necessary to tune the model and adapt it to a specific region and thus to run the model multiple times to find an optimal configuration and parameterization [5]. For the flood application, Grids also allowed us to reduce the overall computing time required for satellite image processing, and made possible the fast response within international programs and initiatives related to disaster management. The Sensor Web standards ensured automated tasking of remote-sensing satellite and a timely delivery of information and corresponding products in case of emergency. Although a successful use case of using the Sensor Web was demonstrated in this section, it is not always the case. Moreover, the case study of the SOS service for surface temperature retrieval from weather stations showed that a lot of database issues still exist that should be properly addressed within the future implementations.

14.7 Conclusions

This chapter was devoted to the description of different approaches to integration of satellite monitoring systems using such technologies as Grid and Sensor Web. We considered integration at the following levels: *data integration level* and *task management level*. Several real-world examples were given to demonstrate such integration. The first example referred to the integration of the Ukrainian satellite monitoring system operated at the SRI NASU-NSAU and the Russian satellite monitoring system operated at the IK RAN at the data level. The second example referred to the development of the InterGrid infrastructure that integrates several regional and national Grid systems: the Ukrainian Academician Grid with a satellite data processing Grid segment UASpaceGrid and the CEODE-CAS Grid. Different issues regarding the integration of the emerging Sensor Web technology with Grids were discussed in the study. We showed how the Sensor Web can benefit from using Grids and vice versa. A flood monitoring and prediction application was used as an example to demonstrate the advantages of integration of these technologies. An important problem of Grid workflow management for satellite data processing was discussed, and automation of the workflow for flood mapping from satellite SAR imagery was described. To benefit from using data from multiple sources, integration of the Ukrainian and Chinese flood mapping services that use radar and optical satellite data was carried out.

References

1. Shelestov, A., Kravchenko, O., Ilin, M.: Distributed visualization systems in remote sensing data processing GRID. Int. J. Inf. Tech. Knowl. **2**(1), 76–82 (2008)
2. Allcock, W., Bresnahan, J., Kettimuthu, R., Link, M.: The Globus Striped GridFTP Framework and Server. In: Proceedings of ACM/IEEE SC 2005 Conf on Supercomputing. (2005). doi: 10.1109/SC.2005.72
3. Butler, R., Engert, D., Foster, I., Kesselman, C., Tuecke, S., Volmer, J., Welch, V.: A national-scale authentication infrastructure. IEEE Comp. **33**(12), 60–66 (2000)
4. Alfieri, R., Cecchini, R., Ciaschini, V., dell'Agnello, L., Frohner, Á., Gianoli, A., Lõrentey, K., Spataro, F.: VOMS, an authorization system for virtual organizations. Lect. Notes Comp. Sci. **2970**, 33–40 (2004)
5. Kussul, N., Shelestov, A., Skakun, S., Kravchenko, O.: Data assimilation technique for flood monitoring and prediction. Int. J. Inf. Theor. Appl. **15**(1), 76–84 (2008)
6. Kussul, N., Shelestov, A., Skakun, S.: Grid system for flood extent extraction from satellite images. Earth Sci. Informatics **1**(3–4), 105–117 (2008)
7. Fusco, L., Cossu, R., Retscher, C.: Open grid services for Envisat and Earth observation applications. In: Plaza, A.J., Chang, C.-I. (eds.) High Performance Computing in Remote Sensing, pp 237–280, 1st edn. Taylor & Francis, New York (2007)
8. Feller, M., Foster, I., Martin, S.: GT4 GRAM: A functionality and performance Study. http://www.globus.org/alliance/publications/papers/TG07-GRAM-comparison-final.pdf (2007). Accessed 30 Aug 2010
9. Moe, K., Smith, S., Prescott, G., Sherwood, R.: Sensor web technologies for NASA Earth science. In: Proceedings of 2008 IEEE Aerospace Conference, pp. 1–7 (2008). doi:10.1109/AERO.2008.4526458
10. Mandl, D., Frye, S.W., Goldberg, M.D., Habib, S., Talabac, S.: Sensor webs: Where they are today and what are the future needs? In: Proceedings of Second IEEE Workshop on Dependability and Security in Sensor Networks and Systems (DSSNS 2006), pp. 65–70 (2006). doi: 10.1109/DSSNS.2006.16
11. Foster, I.: The Grid: A new infrastructure for 21st century science. Phys. Today **55**(2), 42–47 (2002)
12. Shelestov, A., Kussul, N., Skakun, S: Grid technologies in monitoring systems based on satellite data. J. Automation Inf. Sci. **38**(3), 69–80 (2006)
13. Chu, X., Kobialka, T., Durnota, B., Buyya, R.: Open sensor web architecture: Core services. In: Proceedings of the 4th International Conference on Intelligent Sensing and Information Processing (ICISIP), pp. 98–103. IEEE, New Jersey (2006)
14. Botts, M., Percivall, G., Reed, C., Davidson, J.: OGC sensor web enablement: Overview and high level architecture (OGC 07–165) Accessible via http://portal.opengeospatial.org/files/?artifact_id=25562 (2007) Accessed 30 Aug 2010
15. Humphrey, M., Wasson, G., Jackson, K., Boverhof, J., Rodriguez, M., Bester, J., Gawor, J., Lang, S., Foster, I., Meder, S., Pickles, S., McKeown, M.: State and events for web services: A comparison of five WS-resource framework and WS-notification implementations. In: Proceedings of 4th IEEE International Symposium on High Performance Distributed Computing (HPDC-14), Research Triangle Park, NC (2005)
16. Foster, I.: Globus Toolkit Version 4: Software for Service-Oriented Systems. In: IFIP International Conference on Network and Parallel Computing, LNCS, vol. 3779, pp. 2–13. Springer, Heidelberg (2005)
17. Williams, D.N., et al.: Data management and analysis for the Earth System Grid. J. Phys. Conf. Ser. **125**, 012072 (2008). doi: 10.1088/1742–6596/125/1/012072
18. Kussul, N., Shelestov, A., Skakun, S.: Grid and sensor web technologies for environmental monitoring. Earth Sci. Informatics **2**(1–2), 37–51 (2009)
19. Mandl, D.: Experimenting with sensor webs using Earth observing 1. IEEE Aerospace Conference, Big Sky, MT (2004)

Chapter 15
Transparent Data Cube for Spatiotemporal Data Mining and Visualization

Mikhail Zhizhin, Dmitry Medvedev, Dmitry Mishin, Alexei Poyda, and Alexander Novikov

Abstract Data mining and visualization in very large spatiotemporal databases requires three kinds of computing parallelism: file system, data processor, and visualization or rendering farm. Transparent data cube combines on the same hardware a database cluster for active storage of spatiotemporal data with an MPI compute cluster for data processing and rendering on a tiled-display video wall. This approach results in a scalable and inexpensive architecture for interactive analysis and high-resolution mapping of environmental and remote sensing data which we use for comparative study of the climate and vegetation change.

15.1 Introduction

Data mining and visualization in very large spatiotemporal databases requires three kinds of computing parallelism: parallel file system, distributed data processor, and rendering farm for visualization. In high-performance computing, these tasks are usually solved by different types of hardware, such as a network-attached storage, a diskless computational cluster with low-latency interconnect, and a tiled-display video wall attached to a rendering farm. This architecture requires high bandwidth interconnection between the specialized hardware, which is mainly used to move data to computation nodes.

Transparent data cube is a new technology which reduces both the number of CPUs and the network load for data intensive applications by performing computations at the distributed data storage nodes. The idea is to send parallel data processing and visualization tasks to the CPUs which are located at the same

M. Zhizhin (✉) · D. Medvedev · D. Mishin · A. Poyda · A. Novikov
Space Research Institute and Geophysical Center, Russian Academy of Sciences, Moscow, Russia
e-mail: mikhail.zhizhin@gmail.com; dmedv@wdcb.ru; dmitry.mishin@gmail.com; poyda@wdcb.ru; novikov@wdcb.ru

S. Fiore and G. Aloisio (eds.), *Grid and Cloud Database Management*,
DOI 10.1007/978-3-642-20045-8_15, © Springer-Verlag Berlin Heidelberg 2011

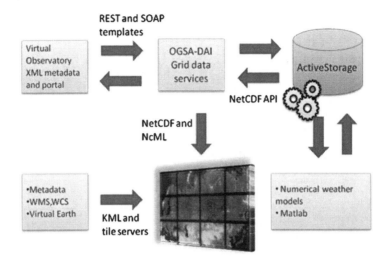

Fig. 15.1 Transparent data cube concept

cluster nodes as the hard drives with the chunks of data needed for the processing. Such transparent data cube combines several software layers on the same hardware platform: (1) a database cluster for active storage of multidimensional data arrays; (2) a computing cluster for data processing, and (3) a tiled-display video wall for data visualization. The software implementations of the storage, processor and display components communicate with each other using web or grid services (Fig. 15.1).

This "general purpose" hardware in the data cube cluster can be a commodity desktop computer or a virtual machine running in a private or public cloud, such as Eucalyptus or Amazon. In the cloud, the number of virtual machine instances in the data cube can be instantly increased to scale according to the storage and data processing tasks.

In the following sections, we describe CDM ActiveStorage architecture, provide its performance specs and API, describe environmental event data mining algorithms based on fuzzy logic and parallel video wall visualization clients.

15.2 Background

The term "active storage" was coined in the late 1990s by Riedel et al. [1] to take advantage of the processing power on individual disk drives to run application-level code. This idea was later transformed into the "object-based storage device" (OSD) concept [2] implemented in the distributed parallel file system Luster [3]. Recently, Felix et al. [4, 5] presented an implementation of active storage for the Luster file system which provides a kernel-space solution with the processing component parts implemented in the user space.

There is a lot of ongoing research and software development in this area. For web search and log analysis, Google combines distributed file system GFS [6] with the Big Table database [7] and the Map-Reduce data processing pattern [8] which provides a framework for programming and running atomic processing jobs on a CPU attached to a hard drive with some input data. Hadoop is an alternative open-source Java implementation of the distributed file system and Map-Reduce framework used by Yahoo [9]. A similar Dryad technology is being developed by Microsoft [10].

Very large database (VLDB) community uses database clusters for the similar task of distributed parallel querying of large databases [11]. The difference between the "active storage" and the "database cluster" approaches reflects mainly the difference in the object data models: files or relational tables. Instead of running "arbitrary" user-mode binary data processing code in the active storage or by the Map-Reduce steps, VLDB relies on very efficient interpretation of a limited number (20+ according to Jim Gray) of SQL programs and stored procedures by a parallel cluster of database servers [12]. The database paradigm results in a "classical" multi-tier software architecture, which includes a web client, a web and application server, a database server, and a disk storage. To achieve horizontal scalability and to overcome potential bottlenecks at the database tier, distributed databases can be designed in several ways, including partitioning, replication, or distributed control [13].

Several relatively large data management and mining projects for Earth Sciences including seismology [14], climatology [15], remote sensing [16], and space physics [17] have resulted in the same conclusions as mentioned by the authors of *The Fourth Paradigm* [12]:

1. Efficient management of very large scientific datasets requires the data to be organized in a distributed database;
2. Distributed collections of data files do not scale well enough to be an alternative for the database approach.

However, the typical Earth Science data model as a multidimensional numeric array is not directly supported by the "classical" relational database management systems. In many cases, the data array dimensions correspond to spatial coordinates and time, with possible addition of altitude, etc. Let us call this type of numeric arrays simply as spatiotemporal data.

Spatiotemporal data can be of different kinds, such as time series recorded at a fixed or moving observatory (ground station or satellite orbit), time varying latitude/longitude grid (e.g., temperature field in weather forecast), or geolocated images (e.g., mosaic of satellite observations). Storing these objects in relational tables is not efficient because it produces very large indices. A reusable solution is an active storage specifically designed for spatiotemporal data model. It can in parallel read–write, subset, and aggregate chunks of multidimensional data arrays, which are indexed and stored as large binary objects (BLOBs) inside a relational database. In fact, this active storage can be viewed as a specialized parallel file system for large and reasonably diverse scientific data arrays with metadata and a simple web service interface for data mining and visualization tools.

The spatiotemporal data model fits well into the common data model (CDM) designed by UNIDATA [18]. This data model is used for data exchange between

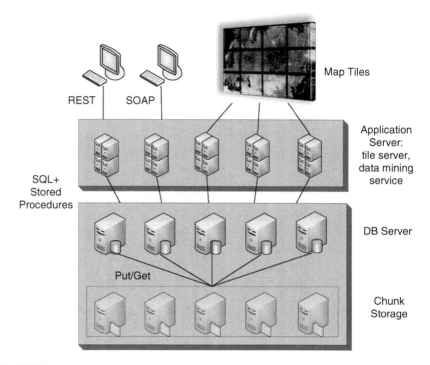

Fig. 15.2 Data cube tiers

the storage, mining, and visualization services in the data cube. To distinguish our solution from other parallel data storage and localized processing methods, we call it "CDM ActiveStorage." The CDM ActiveStorage can be used in numerical modeling, data mining, and interactive visualization of spatiotemporal data. It was successfully used for querying input arrays and storing output of the parallel mesoscale weather model MM5 [19] and space weather model AMIE [20], as well as for the historical climate data mining using environmental scenario search engine (ESSE) [17].

There are two file-based solutions to store CDM, NetCDF-4 [21], and HDF5 [22], but maintaining a large collection of data files can be cumbersome, including problems with remote and multiuser access. Parallel NetCDF [23] combines NetCDF API library with the Luster parallel file system. This parallelization is efficient for the IO operations, but it does not allow parallel execution of data processing functions such as aggregation or convolution.

Data cube adds two more tiers to the "classical" database web application: service layer and visualization farm (Fig. 15.2). The service layer includes (1) SOAP Grid data mining service based on the OGSA-DAI technology [24] and (2) RESTful web service layer for data extraction and image data plots and tile server for visualization.

15.3 Active Storage for Multidimensional Arrays

ActiveStorage is designed to facilitate fast access to multidimensional numeric data, particularly for scientific applications. It consists of a database server and a client library used to access the server from a remote machine. It is horizontally scalable because the data can be distributed between several database servers working in parallel for better performance.

The main idea behind this storage design is data chunking. Multidimensional data array is split into relatively small chunks to avoid addressing the whole array for the query which needs only a small number of chunks. Using a database management system automatically solves the problems with remote and multiuser access. It is true that relational data model is not particularly well suited for scientific data in multidimensional numeric arrays. However, the possibility to write stored procedures in virtually any language makes up for the drawbacks of the data model.

Our design of the active storage is similar to a well-known multidimensional raster server RasDaMan [25]. It also uses a relational database to store large data arrays divided into small blocks. But unlike RasDaMan, which is a middleware server for processing data queries, in ActiveStorage the data processing functions are executed directly by the database server using stored procedures.

On a higher level, the active storage is based on the CDM derived from the NetCDF-4 and HDF5. The actual data model used in ActiveStorage is given in Fig. 15.3:

- A group is a container for other objects. It contains variables, dimensions, attributes, and other groups.

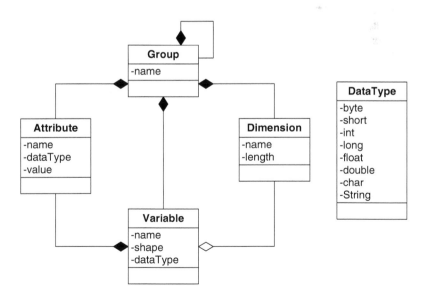

Fig. 15.3 Common data model

- A variable represents a multidimensional data array of specified type indexed by one or more dimensions. A variable may have 0 or more attributes, which contain additional metadata.
- A dimension is a named index used to describe the shape of the data array stored in a variable.
- An attribute is a name–value pair used to store additional metadata for groups and variables.

The classes of the CDM can be directly mapped into relational database tables. The database diagram is given in Fig. 15.4.

Each variable has two additional associated tables, shown in Fig. 15.5. One is a data table; another is an index table. The data are stored in the data table in binary columns. For better performance, each multidimensional array is split into rectangular chunks of arbitrary size and shape. Each chunk occupies one record in a data table. A multidimensional chunk is stored in row-major order (rightmost indices vary faster). Each chunk starts with a small header which contains information about the chunk's boundaries along each dimension. These boundaries are also stored in the index table, allowing for fast selection of chunks inside a given bounding box.

The workflow describing the retrieval of data from the database is shown in Fig. 15.6. Data selection and processing are performed by the database server. The client library joins the pieces of data together and provides an abstraction layer (API) for data access.

Server-side subsetting of data chunks is in itself a time-consuming process. We can speed up data queries by processing chunks simultaneously on several database servers. To do this, we need to create several databases that would contain nonoverlapping subsets of the global data array. As shown in Fig. 15.7, the client library collects subsets of data from multiple databases and merges them as if they came from a single database.

We have tested access performance to the data cube for a large climatological database resulting from the NCEP/NCAR Reanalysis project [26]. It has about 80 different variables (including geopotential height, temperature, relative humidity, cloud cover, etc.), available from 1 January 1948 till present with output every 6 h in coordinate system with 17 height levels or at the surface of the Earth on the regular 2.5 × 2.5° grids. The total size of the NCEP/NCAR Reanalysis database in CDM ActiveStorage is around 500 GB.

Before moving to CDM ActiveStorage, we have used MySQL databases optimized for two different shapes of spatiotemporal queries: time series at a given location or spatial grid at a given time [15]. The results below show performance comparison between the two MySQL databases and the CDM ActiveStorage dataset configured for all-round performance with data chunk shape {1500, 10, 10} (time, lat, lon). The queries produce result sets with the same size but different shape. The time length of the queries changes from 8 to 32,768 in geometric progression, while the length of space dimensions (lat, lon) declines accordingly to keep the size of the result array constant. Test machines configuration was 2× AMD Opteron dual

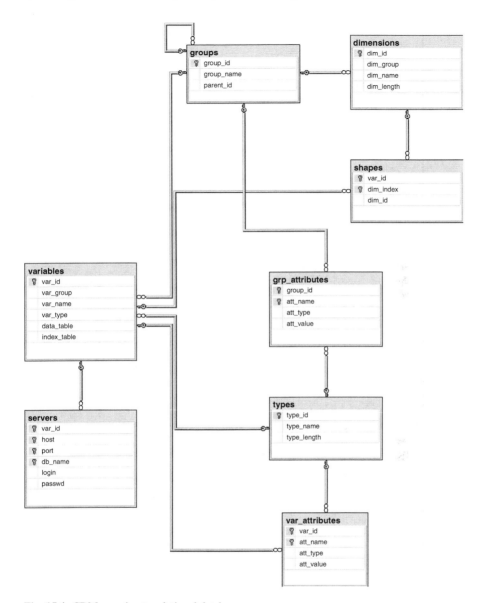

Fig. 15.4 CDM mapping to relational database

core 2.2 GHz with 4 Gb DDR2-667 RAM and a SATA-II 750 Gb hard drive. We also tested two configurations for ActiveStorage: single server and four parallel servers.

Figure 15.8 shows the results for 4D arrays (air temperature on multiple height levels at 6 h time step, 34 GB data array). As expected, space-optimized and time-optimized MySQL databases show excellent performance on their specific queries but their results steadily decline as queries become more diverse. Note that the

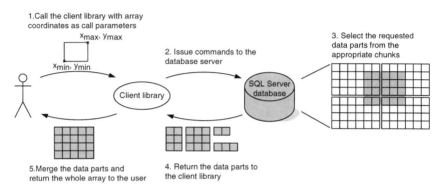

Fig. 15.5 Data chunks and index tables

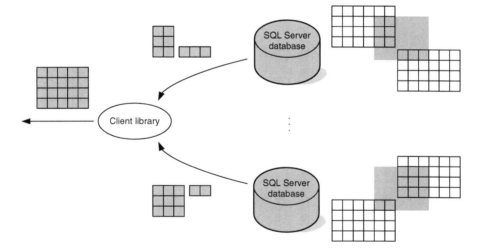

Fig. 15.6 Retrieving data from the database

Fig. 15.7 Distributed queries over multiple database servers

horizontal axis corresponds to logarithmic change of time dimension length, so the curves for space-optimized and time-optimized curves, appearing as exponents, actually correspond to linear change. The CDM ActiveStorage provides better all-round performance, though in certain cases it is not as efficient as the MySQL databases.

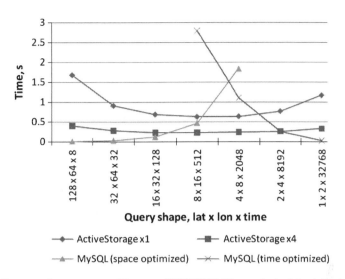

Fig. 15.8 Query performance on a 4D array of NCEP/NCAR reanalysis dataset (air temperature at multiple pressure levels)

15.4 Data Cube API

Transparent Data Cube provides several interface layers for data access (Fig. 15.9). The low-level interface is simply the CDM ActiveStorage client. It provides a rich set of methods for data manipulation, but it cannot be used directly for remote method invocation. The middle level interface is implemented as a SOAP web service, capable of executing remote synchronous and asynchronous data requests. The SOAP web service is used as a wrapper for the active storage client with XML serialization of the client's function calls and the returned data. At this level, several Data Cubes connected via network can be used for distributed data processing. At the top level of the data access interface, there are RESTful web services, which can map basic data extraction and visualization tasks into URLs.

The most straightforward way is to connect to the CDM ActiveStorage using its client library. The client library is available for Java and .NET platforms. The Java client library is integrated with MATLAB. It is compatible with and extends the NetCDF MATLAB library [27] with CDM ActiveStorage functions for parallel data processing, such as aggregation.

15.4.1 SOAP Web Services

Another option is to use the open grid service architecture data access interface (OGSA-DAI) [24] to expose the CDM ActiveStorage to grid-aware applications. The basic abstraction introduced in OGSA-DAI is a virtual data resource capable of performing a standard set of data access and transformation activities using

Fig. 15.9 Data cube API
layers

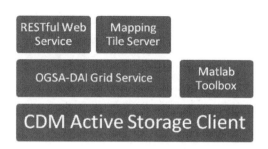

SOAP web service protocol. The SOAP protocol defines the XML format of messages sent between clients and servers and how a message can be interpreted as a remote procedure call consisting of the procedure to be called, the arguments and the results. An OGSA-DAI resource can represent a relational database, an XML database, or a collection of files. Typically, each database is represented as a separate data resource, but the concept is general enough to represent heterogeneous databases. Using these web services, data can be queried, updated, transformed, and aggregated.

OGSA-DAI data resources may differ in a set of activities they are able to perform. For example, a data resource representing a relational database may execute SQL queries while xPath queries may be submitted to a data resource representing an XML database. The advantage of OGSA-DAI is that clients use standard self-describing web service protocols with grid-aware authentication to submit queries and obtain results. OGSA-DAI allows data resources to be federated and accessed via web services on the web or within grids or clouds. The data resources may be orchestrated in such a way that result set from one of it goes as input data directly to another data resource.

The interface to multidimensional data arrays cannot easily and efficiently use standard query languages like SQL or XQuery due to the fact that these languages do not directly support the Common Data Model. Thus, OGSA-DAI had to be extended to implement a CDM data resource with a set of corresponding activities. The data stream from the OGSA-DAI service with the CDM data resource may come to a user in different formats: NetCDF binary format [21], NcML XML format [28], image representation of a time series plot or a gridded dataset map.

The web service wrappers to the data cube API allow distributed execution of the independent data processing steps in the remote data cubes. To test the performance of the distributed data processing, a gigabit network in Moscow was used between the Geophysical Center and the Space Research Institute of the Russian Academy of Sciences. In both institutions, parallel database clusters for CDM ActiveStorage and OGSA-DAI servers were installed.

Consider the following three scenarios for the case when the data volume is equally split between the two data cubes:

1. Each half of the data volume is extracted and parallelly processed locally in the data cube where it is stored. Then, the partial results are streamed from the data cubes for the final merge on the client.

Fig. 15.10 Distributed data processing performance test. Case 1 – distributed processing, Case 2 – distributed select, Case 3 – sequential select and processing

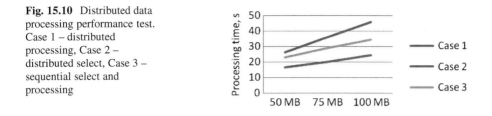

2. Each half of the data volume is extracted and parallelly streamed across the network for the sequential merge and processing on the client.
3. Sequential local data extraction and processing in the same data cube.

A typical data processing request in agroclimatology for the regional climate trends study may be to calculate monthly average temperature (first data cube) and precipitation (second data cube) from the 6−h time step data on the regular latitude–longitude grid with 2.5° step in the latitude range from 40 to 75° North and the longitude range from 25 to 177.5° East (approximate rectangle boundary region for Russia). The size of the time window varies to show how the data processing scales with the size of the input data: 16 years $= 2 \times 25$ MB, 24 years $= 2 \times 38$ MB, 32 years $= 2 \times 50$ MB. In all cases, there is a 120-fold decrease in the data size after the averaging. Wide area network data transfer rate between the two data cubes was 57 Mb s^{-1}. Each data cube was used as a client for sequential data processing and symmetrical data requests for distributed cases to account for the difference in the data cube hardware. Each test was performed three times and the results are presented in Fig. 15.10.

Distributed processing with OGSA-DAI linearly scales with data volume and is significantly faster because of the 120-fold compression of the resulting data size as well as the doubling of the sectional processing time by one server compared to the parallel processing of partial data by two servers. If the size of the resulting data decreases after a processing step (say after averaging and downsampling), then moving of the partial data processing tasks to the remote data cubes (Map) followed by the local merge of the partial results on the client (Reduce) is much faster than the parallel streaming of the extracted data from all remote sources for the local data processing on the client.

15.4.2   RESTful Web Services

At the top of the data cube, there is an API implemented as a RESTful web service. REpresentational State Transfer (REST) is a lightweight HTTP-based protocol for client communication with data resources. It uses the concepts of resources identified by URIs, resources states expressed in XML or JSON, client statelessness and a uniform interface using the standard HTTP request methods GET, PUT, POST, DELETE to retrieve, create, update, delete, and describe resources.

The data cube RESTful web service is implemented as a servlet communicating with the SOAP-based OGSA-DAI web service. The REST common query language (CQL) was developed to simplify the environmental data source queries for the end users. The CQL can be considered as a mapping from parameter–value pairs in the web service query to the OGSA-DAI service request XML message (perform document). The servlet has to parse the query parameters and build a corresponding SOAP request document. The resulting REST service supports synchronous and asynchronous calls, as both are supported by OGSA-DAI.

The lightweight, easy-portable, and having few dependencies RESTful servlet can be easily integrated into any remote Java application or independent Java application server to be a proxy between the client and the heavier SOAP data access engine residing, for example, at a protected network space server. Multiple REST services can be used for the load balancing purposes using its stateless and replaceable nature by having a balancing broker redirecting the user request to a random REST service.

The lightweight REST services are useful for rapidly changing Cloud environments. By registering the newly created service in a metadata repository and periodically checking its state, a developer can have a pool of REST services that are ready to serve the user's data queries. Dynamically adjusting the number of running web service virtual instances according to the current requests load enables the automatic interactive load balancing inside the Cloud cluster.

Using the SOAP services registry and database-specific metadata containing information about the database content, time and spatial coverage allows integrating the location-dependent SOAP data processing service into the global location-independent network of REST services accumulating the data from various distributed data sources.

Currently, the Virtual Observatory metadata management system VxOWare [29] is used as a registry for the running REST data services. It provides the REST service for pushing information about the newly started data service into the metadata registry, such as the service endpoint URL, the list of parameters, current state (enabled or off). Using the VO "outersearch" functionality, an application can receive a server-generated list of selected data services complying with specific requirements, such as data parameters containing a keyword, networking nearest location, specific data topic (Space physics, Earth physics, solid Earth data). This requires the REST data cube services to provide information collected from underlying services layers. The CQL is still in active development. Data query options supported by the current CQL version are listed in Table 15.1.

For example, a GET query string for the RESTful web service call to the NCEP dataset in ActiveStorage can look like:

```
http://<server.name>/servlet/GetData?format=xml&
datefrom=2010-03-01T00:00:00UTC&
dateto=2010-03-03T00:00:00UTC&
dataset=SkinTemperature.Surface@Weather&location=(57.5,37.5)
```

Table 15.1 Data query options supported by common query language

Parameter	Description	Values
Command	Command to be queried at the data service	Get (default) – queries the OGSA-DAI getData activity; provides the main functionality for getting data from service describe – queries the OGSA-DAI resource metadata in XML format and transforms it into an HTML page
Dataset	Specifies parameter name to query in a resource specific format	parameter.vertical_level [.#]]@resource, e.g., SkinTemperature.Surface@NCEP
datefrom	Date–time before or equal to the first data sample	yyyy-mm-ddThh:mm:ssUTC
dateto	Date–time after or equal to the last data sample	yyyy-mm-ddThh:mm:ssUTC
Location	Data query spatial constraints	Point: (39.0,27.5), or station: BOULDER
Format	Data export format	xml, ncml, jpg, png, ...
async	Enables request asynchronous mode	True enables the async requests
asyncorder	ID of the asynchronous data order to check its status or get result	

The web service reply to this call would be an NcML format XML document:

```xml
<?xml version="1.0" encoding="UTF-8"?>
<netcdf xmlns="http://www.unidata.ucar.edu/namespaces
    /netcdf/ncml-2.2">
<dimension name="time" length="20"/>
<dimension name="lat" length="1"/>
<dimension name="lon" length="1"/>
<variable name="time" shape="time" type="double">
<attribute name="units" value="days since 1970-01-01 00:00:00"/>
<attribute name="long_name" value="time"/>
<attribute name="standard_name" value="time"/>
<values>14426.0 14426.25 14426.5 14426.75 14427.0 14427.25
14427.5 14427.75 14428.0 14428.25 14428.5 14428.75 14429.0
14429.25 14429.5 14429.75 14430.0 14430.25 14430.5 14430.75
</values>
</variable>
<variable name="lat" shape="lat" type="float">
<attribute name="units" value="degrees_north"/>
<attribute name="long_name" value="latitude"/>
<attribute name="standard_name" value="latitude"/>
<values>57.5</values>
</variable>
<variable name="lon" shape="lon" type="float">
<attribute name="units" value="degrees_east"/>
<attribute name="long_name" value="longitude"/>
```

```
<attribute name="standard_name" value="longitude"/>
<values>37.5</values>
</variable>
<variable name="TMP_SFC" shape="lat lon time" type="float">
<attribute name="standard_name" value="SkinTemperature"/>
<attribute name="long_name" value="Ground or water surface
Temperature"/>
<attribute name="units" value="C"/>
<values>....</values>
</variable>
</netcdf>
```

Environmental data scientists have developed several RESTful data access protocols which can be compared to the CQL. Open Geospatial Consortium [30] supports a stack of web services for interactive visualization of digital maps in GIS applications. The functionality of OGC web map service (WMS) and web feature services (WFS) is similar to the CQL *get* and *describe* commands. The WMS is used for subsetting and reprojection of digital raster maps. The WFS provides a service for querying and modification of vector features by geographical location (e.g., city population or road length). The main difference between the OGC and the data cube CQL services is in the data model: the OGC model is flat because there is no need in time change for the most of the mapping applications. Thus, the OGC services are not well suited for working with time series data.

OPeNDAP or the open-source project for a network data access protocol, is another example of the HTTP data transport architecture and RESTful web service protocol widely used by environmental scientists [31]. OPeNDAP is based on the same Common Data Model, which is used by active storage and the CQL, and it can perform selection and serialization of the CDM objects from a remote data source. An OPeNDAP client could be a web browser, a graphical desktop program or a web application linked with the OPeNDAP library. The client sends requests to an OPeNDAP server and receives various types of documents or binary data as a response. A dataset description structure (DDS) document describes syntax of a multidimensional dataset. A data attribute structure (DAS) document provides semantics of the dataset variables. A binary subset of data is sent to a client in a DODS type document. Data on the OPeNDAP server can be stored in file collections in HDF, NetCDF, or user-defined format. Compared to ordinary file transfer protocols (e.g., FTP), a major advantage using OPeNDAP is the ability to retrieve subsets of files, and also the ability to aggregate data from several files in one transfer operation.

Although based on the same data model, the CQL and OPeNDAP protocols are different in several aspects. To extract a data object from server, an OPeNDAP client has to call the web service three times: for the DDS, DAS, and DODS documents. In contrast, using the CQL language, a client can call data server once to receive an object in either XML or binary NetCDF self-describing format. The CQL syntax uses data value ranges to subset the data, where the OPeNDAP request for the DODS binary data uses index ranges for the multidimensional data array.

Support of the asynchronous calls by the data cube RESTful web service is important to work with a large data queries, which requires long execution time and cannot be done in highly interactive applications (response time – several seconds). Currently, neither OGC nor OPeNDAP web services support asynchronous calls.

15.5 Data Cube Search Engine

People often use qualitative notions to describe variables such as temperature, pressure, and wind speed. In reality, it is difficult to put a single threshold between what is called "warm" and "hot weather." In this section, we describe the ESSE that allows for parallel evaluating such qualitative queries in several distributed data cubes [15].

Fuzzy set theory serves as a translator from vague linguistic terms into strict mathematical objects. Fuzzy logic was introduced by Lotfi Zadeh [32] in the 1960s as a superset of conventional (Boolean) logic that has been extended to handle the concept of partial truth to model the uncertainty of natural language. Currently, ESSE supports fuzzy linguistic (large, small), numerical (less than, in the range between), and causal (before, after) terms to query for events described as sequences of states of environment at some locations such as grid point or stations, or along a spatial trajectory. Spatial (near, far in space) and temporal (close, far apart in time) reasoning, as well as inverse path of knowledge discovery to learn an event scenario from data are in our plans.

The base data model for ESSE search engine is a vector-valued time series which can be seen as a trajectory in the M-dimensional phase space.

$$\mathbf{X} = \{\mathbf{x}(t_i)\}, i = 1 \ldots N, \mathbf{x}(t_i) = (x_1(t_i), \ldots, x_M(t_i))$$

For example, Fig. 15.11 shows a two-dimensional trajectory in the air pressure–temperature $(P-T)$ space. Using the dualism between set theory and logic, we call the "state" S any subregion of the phase space which can be described by a fuzzy logic expression on predicates describing the parameter values in each dimension in numerical or linguistic terms [33]. In Fig. 15.11, the state $S1$ corresponding to the upper right region can be described by the fuzzy expression:

$$S1 = (Very\ Large\ (P)\mathrm{AND}(Very\ small(T)),$$

where the fuzzy linguistic term $Very\ Large()$ is a predicate, and the operator AND stands for the fuzzy logic conjunction. In the same way, the state $S2$ corresponding to the lower left region is:

$$S2 = (Very\ Small(P))\ \mathrm{AND}(Very\ Small\ (T)).$$

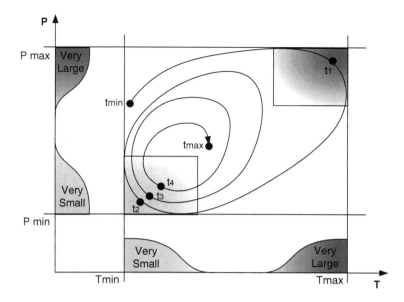

Fig. 15.11 Time series as a trajectory in the two-dimensional phase space (P pressure, T temperature)

Now, combining the fuzzy set descriptions of the states with the "time shift" operator $Shift(dT,)$ to describe transitions between the states, we can write a symbolic expression for the environmental scenario "a day with very low after a day with very high temperature and pressure":

$$Scenario = (Shift(dT = 1 \text{ day}, S1)) \text{AND}(S2).$$

The only pair of observations in Fig. 15.11 which fit the above scenario is the pair $(t1, t2)$.

Simple climatology analysis gives normalization limits $x_{\min} \leq x(t) \leq x_{\max}$ used in calculations of linguistic predicates like "very large." The limits are set to the minimum and maximum parameter values observed within the continuous or seasonal intervals given by the time constraints of the fuzzy search.

The concept of *event duration* $k\Delta t$ for any multiple $k = 1, 2, \ldots$ of the time step Δt of the input is used when searching for events like a "cold day" or a "cold week." We do a moving average of the input parameters with the time window of the event duration before calculation of predicates in the fuzzy logic expression:

$$\bar{x}(t_i) = \frac{1}{k} \sum_{j=i}^{i+k-1} x(t_j), \quad t_i = t_0 + i\,\Delta t.$$

For example, searching for a "very cold day" in the NCEP/NCAR Reanalysis involves smoothing of the air temperature at a given location with the time window of 1 day ($k = 4$), then calculation of the linguistic predicate $Very\ Small(\bar{x}(t_i))$ over the smoothed time series, sort of the fuzzy scores in descending order, and finally selection of the several days with the highest scores as the candidate events.

The ESSE is designed to mine for the phase space transitions like that in very large scientific data cubes. It is implemented as a special data mining activity `fuzzySearchActivity` added to the OGSA-DAI data cube API. The new activity input is a combination of fuzzy conditions on environmental parameters' values localized in space and time; in fact it is an XML-formatted environmental scenario description. The activity output is a time series with values between 0 and 1 for the fuzzy likeliness of the occurrence of the scenario at every moment in time.

`fuzzySearchActivity` is the OGSA-DAI data transformation activity which is not linked to a specific type of data resource. This makes the whole data mining system extremely flexible. One can search the environmental scenario over several parameters stored in a local data cube. This is accomplished by combining several data query and processing steps together with the `fuzzySearchActivity` in a single transaction. In a more advanced scenario, it is possible to combine data mining results from several OGSA-DAI resources. In addition, the ESSE search engine provides a user interface implemented as an interactive web application. In the web application, it is possible:

- To discover data sources by keyword-based metadata search in the Virtual Observatory
- To use predefined weather events (e.g., "magnetic storm" or "heat wave") as well as to define the searching event as a combination of fuzzy conditions on a set of environmental parameters (e.g., "high temperature and low relative humidity") for data mining
- To download the data for selected event in self-describing format (NetCDF or NcML) to the user's workstation
- To visualize the selected event as a time series plot or as a surface map

15.6   Transparent Data Cube

The data cube is called "transparent" because it can be used not only for storage and processing but also for high-resolution interactive visualization of large datasets [34]. In the real-world installation in Moscow, we have combined parallel subsetting of the data array by storage engine with the parallel rendering of the selected data for a 100 Mpixel tiled-display video wall, which can be driven by the same computer cluster or receive visualization streams for the display tiles via network. The visualization API library was provided by the Scalable Adaptive Graphics Environment (SAGE) [35], developed by the Electronic Visualization Lab at UIC. The visualization client for the CDM active storage was developed as a special plugin for the NASA WorldWind 3D virtual globe [36]. To be coherent with the current trend of using 2D tile servers (Google Maps, Bing Maps, or Openstreet Maps [37]) and KML [38] language for visualization of the geospatial data, a tile server service for the data cube was developed, as well as the parallel tile server visualization software for video walls called MultiViewer [39, 40].

A common method of displaying large multi-resolution imagery is building a pyramid of tiles. This method is used in many web-mapping "neogeography" applications such as Google Maps and Bing Maps. In our case, it is convenient for displaying detailed data grids and georeferenced data from CDM ActiveStorage on top of the existing satellite or road maps.

A Tile-server is a web service which produces a requested fragment of the full image (a tile). Tile addressing in Google Maps uses three coordinates: zoom level (Z), row (Y), and column (X). The tile size is fixed at 256×256 pixels. At zoom level 0 (lowest possible resolution), the whole image is covered by a single tile with coordinates (0,0,0). At each subsequent zoom level, the number of tiles is multiplied by 4.

For example, a tile server for mapping the weather forecast is basically a REST service which takes three request parameters: tile coordinates, weather parameter name, and forecast time. The variable part of the tile server URL is given below:

```
GetTile?zxy={Z_X_Y}&layer={paramete}&time={ISO8601_time}
```

A single tile request for the Skin Temperature map might look like:

```
GetTile?zxy=
1_1_0&layer=SkinTemperature.Surface@NCEP&time=2010-08-
20T00:00:00UTC
```

The actual web application uses the tile server output to display a data layer in the Google Maps window as a semi-transparent map overlay (Fig. 15.12).

Our tiled-display visualization system uses the same computer cluster as the CDM Active Storage and OGSA-DAI services and it is based on the SAGE [35]. The network-centered architecture of SAGE allows collaborators to run various applications simultaneously (such as 3D rendering, remote desktop, video streams, and 2D maps) on local or remote clusters, and share them by streaming the pixels of each application window over ultra-high-speed networks to large tiled displays.

The open-source SAGE framework consists of free space manager (FSM), SAGE application interface library (SAIL), SAGE Receiver, and User Interface (UI client) as shown in Fig. 15.13. The FSM gets user commands from UI clients and controls pixel streams between SAIL and SAGE Receivers. SAIL captures output pixels from applications, and streams them to appropriate SAGE Receivers. A SAGE Receiver can get multiple pixel streams from different applications on the network, and displays streamed pixels on multiple tiles. A UI Client sends user commands to control the FSM and receives messages that inform users of the current SAGE status.

The SAIL interface library was developed for OpenGL graphics on the Linux platform. We have bridged the SAGE OpenGL input with the DirectX output from

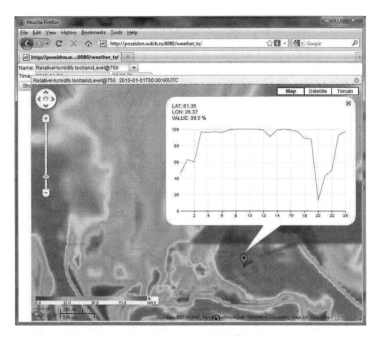

Fig. 15.12 Google maps overlay with CDM ActiveStorage data

the MS Windows applications using a virtual frame buffer. After that a SAGE video wall can be used for 3 D visualization of environmental data with the open-source NASA WorldWind viewer [36]. WorldWind collects images, digital elevation, and other geotagged data from various network sources and creates an interactive zoomable mashup over a 3 D globe.

A 12-display video wall at the Space Research Institute in Russia, in Fig. 15.13, displays a WorldWind application with air temperature color map. The data for the color map are transferred from the CDM ActiveStorage through the OGSA-DAI connector. For the virtual frame buffer size of $1,920 \times 1,280$ pixels, the refresh rate of 15 frames per second was reached on the SAGE video wall.

The total resolution of the video wall is $7,680 \times 3,840$ pixels. This resolution exceeds the maximum frame buffer size available for a WorldWind application. To fully utilize the potential of the video wall, we use 2D Earth surface images provided by Microsoft Virtual Earth as background when displaying the data. The Virtual Earth imagery as well as Google Maps is a hierarchical quad tree of 256×256 pixel tiles. These tiles are requested by HTTP GET operation and stored in the local disk cache. It appears that the throughput of storage system of one computer is not enough for fully interactive navigation including map zooming on the entire video wall. This approach was implemented in the Multiviewer application for videocluster [40].

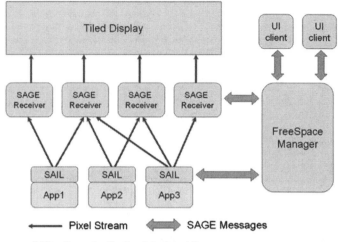

Pixel Stream   SAGE Messages

SAIL : Sage Application Interface Library

Fig. 15.13 SAGE framework architecture and a 12-display video wall running NASA WorldWind with SAGE

In Multiviewer, each video wall computer performs data fetching and basic processing, like a web browser for a small fraction of the total map. When working with ultrahigh-resolution image, each of the six client nodes has its own cache and the total throughput of storage system increases six times (Fig. 15.14). This allows implementing a magnifying glass tool that shows a selected region with higher

Fig. 15.14 Multiviewer
architecture and temperature
color map over Microsoft
Bing maps at the 12-display
video wall

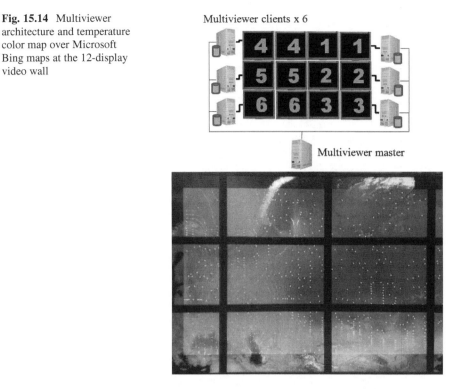

detail. The master computer receives user input and provides shared mouse cursor
for the entire video wall. Additional increase in performance can be achieved using
some kind of distributed cache such as Microsoft Velocity project [41].

15.7 Conclusion

We have designed and tested a scalable and inexpensive transparent data cube
for interactive analysis and high-resolution mapping of environmental and remote
sensing data which we use for comparative study of the climate and vegetation
change. On a modest computer cluster with 6–12 nodes each with several terabytes
of disk space we can deploy parallel active storage for scientific data arrays which
can be used by the same cluster for parallel numerical modeling and visualization.

The question of scalability of that solution to clusters with hundreds or even
thousands of nodes is still open. The answer depends on scalability of the database
cluster (in our case MS SQL or PostgeSQL Server) and on the scalability of the
video wall (above 100 Mpixels).

To make our data storage really "active," we have implemented inside the data
base engine stored procedures for parallel subsampling and aggregation of data

chunks. Spatial convolution, which is frequently used for image processing and data filtering, still needs to be implemented.

Acknowledgements This research was supported by the Russian Foundation for Basic Research Grant "Parallel scalable Grid-center for data mining," Russian-Belorussian "SKIF-Grid" Project, CRDF Grant "Space Physics Interactive Data Resource," and the Microsoft Research Grants "Environmental Scenario Search Engine."

References

1. Riedel, E., Gibson, G., Faloutsos, C.: Active storage for large-scale data mining and multimedia. In: Proceedings of 24th International Conference on Very Large Data Bases (VLDB), pp. 62–73 (1998)
2. Mesnier, M., Ganger, G., Riedel, E.: Object-based storage. IEEE Commun. Mag. **41**, 84–90 (2005)
3. Wang, F., Oral, S., Shipman, G., Drokin, O., Wang, T., Huang, I.: Understanding Lustre File System Internals, Technical Report, National Center for Computational Sciences, ORNL/TM-2009/117 (2009). http://wiki.lustre.org/images/d/da/Understanding_Lustre_Filesystem_Internals.pdf. Accessed 9 Jan 2011
4. Felix, E.J., Fox, K., Regimbal, K., Nieplocha, J.: Active Storage processing in a parallel file system. In: Proceedings of the 6th LCI International Conference on Linux Clusters: The HPC Revolution (2006)
5. Piernas, J., Nieplocha, J., Felix, E.J.: http://sc07.supercomputing.org/schedule/pdf/pap287.pdf (2007). Accessed 9 Jan 2011
6. Ghemawat, S., Gobioff, H., Leung, S.T.: The Google File System, SOSP'03, Bolton Landing. http://labs.google.com/papers/gfs-sosp2003.pdf (2003). Accessed 9 Jan 2011
7. Chang, F., Dean, J., Ghemawat, S., Hsieh, W.C., Wallach, D. A., Burrows, M., Chandra, T., Fikes, A., Gruber, R.E.: Bigtable: A Distributed Storage System for Structured Data, OSDI'06: Seventh Symposium on Operating System Design and Implementation, Seattle (2006). http://labs.google.com/papers/bigtable.html. Accessed 9 Jan 2011
8. Dean, J., Ghemawat, S.: MapReduce: Simplified Data Processing on Large Clusters, OSDI'04: Sixth Symposium on Operating System Design and Implementation, San Francisco. http://labs.google.com/papers/mapreduce.html (2004). Accessed 9 Jan 2011
9. Lam, C.: Hadoop in Action, p. 325, 1st edn. Manning Publications, CT. ISBN 1935182196 (2010)
10. Isard, M., Budiu, M., Yu, Y., Birrell, A., Fetterly, D.: Dryad: Distributed Data-Parallel Programs from Sequential Building Blocks, European Conference on Computer Systems (EuroSys), Lisbon, Portugal. http://research.microsoft.com/research/sv/Dryad/eurosys07.pdf. (2007). Accessed 9 Jan 2011
11. Szalay, A.S., Bell, G., Vandenberg, J., Wonders, A., Burns, R., Fay, D., Heasley, J., Hey, T., Nieto-SantiSteban, M., Thakar, A., van Ingen, C., Wilton, R.: GrayWulf: Scalable Clustered Architecture for Data Intensive Computing. In: Proceedings of 42nd Hawaii International Conference System Sciences, pp. 1–10. http://hssl.cs.jhu.edu/papers/szalay_hicss09.pdf (2009). Accessed 9 Jan 2011
12. Hey, T., Tansley, S., Tolle, K.: The Fourth Paradigm: Data-Intensive Scientific Discovery. Microsoft Research, p. 287 http://research.microsoft.com/en-us/collaboration/fourthparadigm/4th_paradigm_book_complete_lr.pdf (2009). Accessed 9 Jan 2011
13. Kossmann, D., Kraska, T., Loesing, S.: An Evaluation of Alternative Architectures for Transaction Processing in the Cloud, SIGMOD'10, Indianapolis, pp. 579–590. http://systems.ethz.pubzone.org/pages/publications/showPublication.do?pos=0&publicationId=1363428 (2010). Accessed 9 Jan 2011

14. Zhizhin, M.N., Rouland, D., Bonnin, J., Gvishiani, A.D., Burtsev, A.: Rapid estimation of earthquake source parameters from pattern analysis of waveforms recorded at a single three-component broadband station. Bull. Seism. Soc. Am. **96**, 2329–2347 (2006). doi:10.1029/2005SW000199
15. Zhizhin, M., Poyda, A., Mishin, D., Medvedev, D., Kihn, E., Lyutsarev, V.: Grid data mining with environmental scenario search engine (ESSE). In: Dubitsky, W. (ed.) Data Mining Techniques in Grid Computing Environments, pp. 281–306. Wiley, NY (2008)
16. Elvidge, C.D., Ziskin, D., Baugh, K.E., Tuttle, B.T., Ghosh, T., Pack, D.W., Erwin, E.H., Zhizhin, M.: A fifteen year record of global natural gas flaring derived from satellite data. Energies **2**, 595–622 (2009). doi:10.3390/en20300595
17. Zhizhin, ., Kihn, E., Redmon, R., Medvedev, D., Mishin, D.: Space physics interactive data resource – SPIDR. Earth Sci. Informat. **1**, 79–91 (2008). doi: 10.1007/s12145–008–0012–5
18. Common Data Model (CDM) by UNIDATA. http://www.unidata.ucar.edu/software/netcdf/CDM/(2011). Accessed 9 Jan 2011
19. Michalakes, J.: The same-source parallel MM5. Sci. Program. **8**, 5–12 (2000)
20. Kihn, E.A., Zhizhin, M., Kamide, Y.: An analog forecast model for the high-latitude ionospheric potential based on assimilative mapping of ionospheric electrodynamics archives. Space Weather **4**, S05001 (2006)
21. NetCDF file format and API by UNIDATA. http://www.unidata.ucar.edu/software/netcdf/(2011). Accessed 9 Jan 2011
22. National Center for Supercomputing Applications Introduction to HDF5. University of Illinois at Urbana Champaign. http://hdf2.ncsa.uiuc.edu/HDF5/doc/H5.intro.html (1998). Accessed 9 Jan 2011
23. Jianwei, L., Liao, W., Choudhary, A., Ross, R., Thakur, R., Gropp, W., Latham, R., Siegel, A., Gallagher, B., Zingale, M.: Parallel netCDF: A high-performance scientific I/O interface, Supercomputing ACM/IEEE Conference, p. 39 (2003)
24. Antonioletti, M., Atkinson, M.P., Baxter, R., Borley, A., Chue Hong, N.P., Collins, B., Hardman, N., Hume, A., Knox, A., Jackson, M., Krause, A., Laws, S., Magowan, J., Paton, N.W., Pearson, D., Sugden, T., Watson, P., Westhead, M.: The design and implementation of grid database services in OGSA-DAI. Concurrency Comput. Pract. Ex. **17**, 357–376 (2005)
25. http://www.ogsadai.org.uk/(2011). Accessed 9 Jan 2011
26. Baumann, P., Dehmel, A., Furtado, P., Ritsch, R., Widmann, N.: The multidimensional database system RasDaMan. In: Proceedings of ACM SIGMOD International Conference on Management of data, Seattle WA, 575–577. http://www.rasdaman.com(1998). Accessed 9 Jan 2011
27. Kalnay, E., et al.: The NCEP/NCAR 40-year reanalysis project. Bull Am. Meteorol. Soc. **77**, 437–471. http://www.cdc.noaa.gov/cdc/reanalysis/(1996). Accessed 9 Jan 2011
28. Matlab NetCDF Toolbox. http://mexcdf.sourceforge.net/index.php(2011). Accessed 9 Jan 2011
29. NetCDF XML Markaup Langauge. http://www.unidata.ucar.edu/software/netcdf/ncml/(2011). Accessed 9 Jan 2011
30. Weigel, R.S., Zhizhin, M., Mishin, D., Kokovin, D., Kihn, E., Faden, J.: VxOware: Software for managing virtual observatory metadata. Earth Sci. Informat. **3**, 19–28 (2010). doi: 10.1007/s12145–010–0048–1
31. Open Geospatial Consortium standards and specifications for Web Map Services. http://www.opengeospatial.org/standards(2011). Accessed 9 Jan 2011
32. Open-source Project for a Network Data Access Protocol (OPeNDAP). http://www.opendap.org(2011). Accessed 9 Jan 2011
33. Zadeh, L.: Fuzzy sets. Inf. Contr. **8**, 338–353 (1965)
34. Jang, J.S.R., Sun, C.T., Mizutani, E.: Neuro-Fuzzy and Soft Computing. Prentice Hall, NJ (1997)
35. Berezin, S.B., Voitsekhovsky, D.V., Zhizhin, M.N., Mishin, D.Y., Novikov, A.M.: Video walls for Multiresolution Visualization of Natural Environment, Scientific Visualization 1:100–107 (in Russian). http://sv-journal.com/2009--1/04.php?lang=en(2009). Accessed 9 Jan 2011

36. Renambot, L., Rao, A., Singh, R., Byungil, J., Krishnaprasad, N., Vishwanath, V., Chandrasekhar, V., Schwarz, N., Spale, A., Zhang, C., Goldman, G., Leigh, J., Johnson, A.: SAGE: The Scalable Adaptive Graphics Environment. Electronic Visualization Laboratory, Dept. of Computer Science, University of Illinois at Chicago. http://www.optiputer.net/publications/articles/RENAMBOT-WACE2004-SAGE.pdf(2004). Accessed 9 Jan 2011
37. NASA WorldWind virtual 3D globe. http://worldwind.arc.nasa.gov/(2011). Accessed 9 Jan 2011
38. OpenStreetMap tile-server project http://www.openstreetmap.org(2011). Accessed 9 Jan 2011
39. KML documentation. http://code.google.com/apis/kml/documentation/(2011). Accessed 9 Jan 2011
40. Zhizhin, M., Kihn, E., Lyutsarev, V., Berezin, S., Poyda, A., Mishin, D., Medvedev, D., Voitsekhovsky, D.: Environmental scenario search and visualization. In: Proceedings of 15th ACM symposium on advances in geographic information systems (2007)
41. Multiviewer source code. http://www.codeplex.com/multiviewer(2011). Accessed 9 Jan 2011

Chapter 16
Distributed Storage of Large-Scale Multidimensional Electroencephalogram Data Using Hadoop and HBase

Haimonti Dutta, Alex Kamil, Manoj Pooleery, Simha Sethumadhavan, and John Demme

Abstract Huge volumes of data are being accumulated from a variety of sources in engineering and scientific disciplines; this has been referred to as the *"Data Avalanche"*. Cloud computing infrastructures (such as Amazon Elastic Compute Cloud (EC2)) are specifically designed to combine high compute performance with high performance network capability to meet the needs of data-intensive science. Reliable, scalable, and *distributed* computing is used extensively on the cloud. Apache Hadoop is one such open-source project that provides a distributed file system to create multiple replicas of data blocks and distribute them on compute nodes throughout a cluster to enable reliable and rapid computations. Column-oriented databases built on Hadoop (such as HBase) along with MapReduce programming paradigm allows development of large-scale distributed computing applications with ease. In this chapter, benchmarking results on a small in-house Hadoop cluster composed of 29 nodes each with 8-core processors is presented along with a case-study on distributed storage of electroencephalogram (EEG) data. Our results indicate that the Hadoop / HBase projects are still in their nascent stages but provide promising performance characteristics with regard to latency

H. Dutta (✉) · M. Pooleery
Center for Computational Learning Systems (CCLS), Columbia University, NY 10115, USA
e-mail: haimonti@ccls.columbia.edu; manoj@ccls.columbia.edu

A. Kamil
School of General Studies, Columbia University, NY 10027, USA
e-mail: alex.kamil@gmail.com

S. Sethumadhavan
Computer Architecture Laboratory, Department of Computer Science, Columbia University, NY 10115, USA
e-mail: simha@cs.columbia.edu

J. Demme
Department of Computer Science, Columbia University, NY 10115, USA
e-mail: jdd@cs.columbia.edu

S. Fiore and G. Aloisio (eds.), *Grid and Cloud Database Management*,
DOI 10.1007/978-3-642-20045-8_16, © Springer-Verlag Berlin Heidelberg 2011

and throughput. In future work, we will explore the development of novel machine learning algorithms on this infrastructure.

16.1 Introduction

The science of the twenty-first century requires large amounts of computation power, storage capacity, and high speed communication. These requirements are increasing at an exponential rate and scientists are demanding much more than is available today. Several astronomy and physical science projects such as CERN's[1] Large Hadron Collider [21], Sloan Digital Sky Survey [31], The Two Micron All Sky Survey [1], bioinformatics projects including the Human Genome Project [32], gene and protein archives [33, 34], meteorological and environmental surveys [15, 35] are already producing peta- and tera-bytes of data which requires to be stored, analyzed, queried, and transferred to other sites. To work with collaborators at different geographical locations on peta scale data sets, researchers require communication of the order of Gigabits/s. Thus, computing resources are failing to keep up with the challenges they face.

Traditionally, supercomputers [18, 24, 25] are used for highly compute-intensive tasks such as problems involving quantum mechanical physics, weather forecasting, climate research, and molecular modeling. They use custom-made CPUs with innovative designs that allow them to perform many tasks in parallel and hence can gain substantial speed over conventional computers. However, the data has to be stored in a separate repository which has to be brought in each time for computation. This is time consuming and limits interactivity; furthermore, the programs are written at very low level language and rely on a small number of specific packages written by experts. Also, in a super-computer, users submit jobs in a batch mode; the job is done when resources are available – so it does not support flexible programming and runtime environment.

Cloud computing paradigms are being considered for data intensive science in recent years. The concept of the "cloud" has been envisioned to provide a solution to the increasing data demands and offer a shared, distributed computing infrastructure. The *sharing* of distributed computing resources including software, hardware, data, etc. is an important aspect of cloud computing. Sharing can be dynamic depending on the current need, may not be limited to client server architectures, and the same resources can be used in different ways depending on the objective of sharing. Systems such as the Amazon Simple Storage Service [17] and Amazon Elastic Compute Cloud [16] are good examples of the Storage and Compute Clusters which provide cloud computing infrastructures.

In this chapter, we present a mechanism for distributed storage of multidimensional electroencephalogram (EEG) time series obtained from epilepsy patients on a

---

[1]Conseil Europen pour la Recherche Nuclaire – European Organization for Nuclear Research.

cloud computing infrastructure (Hadoop cluster) using a column-oriented database (HBase). It is organized as follows: Section 16.2 presents prior research in storing large time series databases; Section 16.3 describes preliminaries on Hadoop; and Sect. 16.4 discusses the basic structure of a column-oriented database HBase. We present a case study on large-scale storage of intracranial EEG data on HBase in Sect. 16.5. Finally, Sect. 16.7 concludes the chapter and presents directions for future work.

16.2   Related Work

To the best of our knowledge, there is very little work that addresses the problem of time series data *storage* on large distributed environments. The Harvard Time Series Center[2]) is known to be one of the largest data centers hosting approximately a billion time series mainly from the field of astronomy but also expanding to economics, health, and real-estate data.

There is a relatively large body of literature pertaining to analysis and extraction of patterns from time series data [26, 27]; however, very few are known to scale to large datasets [10, 11, 28–30, 36] stored in distributed environments. Reeves et al. [29] describe *Cypress*, a framework to archive and query massive time series streams by sparse (frequency and time domain) representations of the data. The sparsity enables archiving of data in a reduced storage space. Trends such as histograms and correlations can be answered directly from the compressed data. Das et al. [10] consider the problem of distributed eigen monitoring algorithms in petascale astronomy pipelines. They propose an asynchronous algorithm for monitoring principal components of dynamic data streams. Again, the problem of large scale data storage is largely ignored in this work. The Zohmg [2] system is probably closest in spirit to the current chapter, in which they describe a large-scale data store for aggregated time series data. The goal is to model multidimensional time series as data cubes on top of a distributed, column-oriented database (HBase) to reap the scalability benefits of such databases. To import data to Zohmg, the user writes a custom map function and uses Dumbo[3] (a platform that enables usage of streaming Hadoop instances), to execute it on an Apache Hadoop instance. The Zohmg user map function emits triples, which consist of a time-stamp, a hash table of dimensions and their respective values, and a hash table with units and their respective values. A reducer function that is specific to Zohmg is used to perform aggregation on the output from the map phase. The reducer sums the measurements for each point in an n-dimensional space for each unit. The output from the reducer is interpreted by a custom output reader.

Since our case study deals with storage of large-scale EEG data, we discuss related literature in this research area also. A collaboration between the University

---

[2]http://timemachine.iic.harvard.edu/publications/.

[3]http://www.audioscrobbler.net/development/dumbo/.

of Pennsylvania and the Mayo Clinic (which is currently funded by National Institutes of Health) deals with large volumes of intracranial electroencephalogram (iEEG) data from epileptic patients. As part of research done in this collaboration, Brinkman et al. [6] describe a platform for acquisition, compression, encryption, and storage of large-scale EEG data. Continuous, long-term electrophysiological recordings in human subjects undergoing evaluation for epilepsy surgery using intracranial electrodes and clinical macroelectrode arrays generates approximately 3 terabytes of data per day (at 4 bytes per sample). Their work studies real-time data compression techniques enabling random access to data segments of varying sizes. Yet another database that provides invasive EEG recordings for patients suffering from intractable focal epilepsy is hosted at the Epilepsy Center of the University Hospital of Freiburg,[4] Germany [4]. Recordings from 21 patients are acquired using 128 channels, 256 Hz sampling rate, and a 16-bit analogue-to-digital converter. A relatively small EEG data set with sampling rate of 173.61 Hz is hosted at the University of Bonn [3].

16.3   Preliminaries on Hadoop

Apache Hadoop [19] inspired by Google Map-Reduce [12, 13] and Google File System [14] is a framework for supporting data intensive applications on a cluster. It has a free open source MapReduce implementation and was first used on a commercial level for the Nutch [7] search engine project. Since then, both the industry and academia have been working together to develop tools and architectures for supporting Data Intensive Scalable Computing [23] using Hadoop.

MapReduce is a distributed computing framework for large datasets and has two computation phases – map and reduce. In the map phase, a dataset is partitioned into disjoint parts and distributed to workers called mappers. The mappers implement compute-intensive tasks (such as clustering) on local data. The power of MapReduce stems from the fact that many map tasks can run in parallel. The output of the map phase is of the form $\langle key, value \rangle$ pairs which are passed to the second phase of MapReduce called the reduce phase. The workers in reduce phase (called reducers) then partition, process, and sort the $\langle key, value \rangle$ pairs received from the Map phase according to the key value and make final output. For a complex computation task, several MapReduce phase pairs may be involved. The architecture comprises of two main parts: (1) *Data Storage*, using the Hadoop Distributed File System (HDFS). (2) *Computation*, using MapReduce programming paradigm to meet the computation needs of the clustering algorithm. Figure 16.1 shows the above components in our cluster. Specifically, these include:

1. *The HDFS Namespace*: The Namenode maintains the file system namespace and records any changes made to it. It also keeps track of the number of replicas of a file that should be maintained in the HDFS typically called the replication factor.

---

[4]https://epilepsy.uni-freiburg.de/freiburg-seizure-prediction-project/eeg-database.

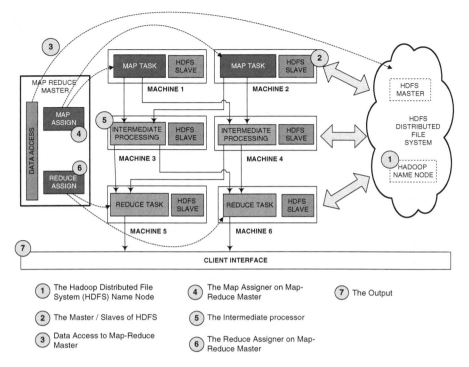

Fig. 16.1 The architecture of a cluster illustrating the Map, Reduce, and Intermediate operations along with the Hadoop distributed file system (HDFS)

2. *The master/slaves of HDFS*: A master server manages the file system namespace and regulates access to files by clients. In addition, there are a number of HDFS Slaves, usually one per node, which manage the data associated with that node. They serve read and write requests from the users and are also responsible for block creation, deletion, and replication upon instruction from the NameNode.

3. *Data access to MapReduce master*: The HDFS file system will be accessed by a MapReduce (MR) master. The input files to the MR Master can be processed in parallel by different machines in cluster.

4. *The Map assigner on MapReduce master*: It stores data structures such as the current state (idle, in-progress or completed) of each map task in the cluster. It is also responsible for pinging the map-workers occasionally. If no response is received from the worker, it assumes that the process has failed and re-schedules the job.

5. *The intermediate processor*: The intermediate $\langle key, value \rangle$ pairs produced by map function are buffered in the local memory of machines. This information is sent to the MR Master which then informs the Reduce Assigner.

Table 16.1 Execution times for read and write operations in the HDFS for a small cluster of 29 nodes

Test	No of files	File Size (MB)	HDFS bytes read or written	No. of Map tasks	Execution time (s)
Write	10	1,000	10485760000	19	219.095
Write	30	1,000	31457280079	53	403.295
Write	60	1,000	62914560082	101	651.326
Write	100	1,000	104857600084	150	1013.916
Read	10	1,000	10446515887	17	157.752
Read	30	1,000	30415428946	51	259.864
Read	60	1,000	60844599982	98	387.055
Read	100	1,000	100933188734	140	506.269

6. *The reduce assigner on MapReduce master*: This takes in the location of the intermediate files produced from a Map operation and assigns reduce jobs to the respective machines.
7. *The Output*: The output of a Reduce function is appended to a final output file. When all the map and reduce tasks are over, the MR Master wakes up the user program.

16.3.1   Read/Write Benchmarks on a Small Scale Cluster

The in-house cluster available to us for experimentation had 29, 8-core processors each with 24 GB RAM, 1 TB RAID connected via a fiber channel. Apache Hadoop (version 0.20.2) was set-up on this cluster and the IO throughput of the distributed file system was tested. Table 16.1 summarizes the results and Figs. 16.2 and 16.3 present the read and write throughputs for varying number of files, keeping the total file size constant at 1,000 MB.

Having described the MapReduce and Hadoop frameworks, we proceed in the next section to provide a brief review of HBase.

16.4   Preliminaries on HBase

HBase [20, 22] is an Apache open source project whose goal is to provide Bigtable-like [8] storage (designed to scale to very large databases) for the HDFS. Applications store rows of data in labeled tables. A data row has a sortable row key and an arbitrary number of columns (illustrated in Table 16.2). Physically, the table is stored sparsely, so that rows in the same table can have widely varying numbers of columns as illustrated in Tables 16.3–16.5. The empty cells shown in the conceptual view are not stored and hence when queried they will return no value.

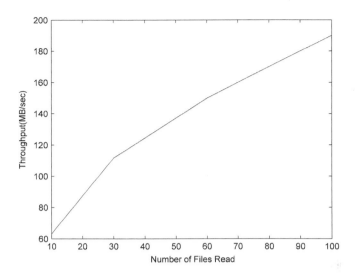

Fig. 16.2   Throughput vs. number of files read on the Hadoop cluster (29 nodes)

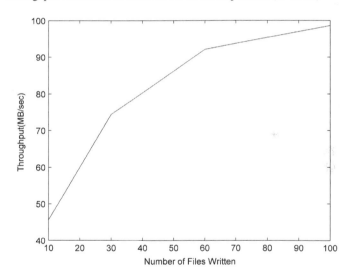

Fig. 16.3   Throughput vs. number of files written on the Hadoop cluster (29 nodes)

The row keys are arbitrary strings of up to 64 KB in size and sorted in lexicographical order. Every read or write under a single row key is atomic, which simplifies the handling of concurrent read/writes. Tables are broken up into row ranges called regions (equivalent Bigtable term is `tablet`). Each row range contains rows from start-key to end-key. A set of regions, sorted appropriately, forms an entire table. As a result, reads of short row ranges are efficient and typically require communication with only a small number of machines. Clients can exploit

Table 16.2 Conceptual view of the entire table

Row key	Time stamp	Column00	Column01	Column02
Row0	t0	Value = String00_t0	Value = String01_t0	Value = String02_t0
Row0	t1	Value = String00_t1		Value = String01_t2
Row0	t3	Value = String00_t3	Value = String01_t3	Value = String02_t3
Row0	t5	Value = String00_t5	Value = String01_t5	

Table 16.3 Physical storage of column00 for Table 16.2

Row key	Time stamp	Column00
Row0	t0	Value = String00_t0
Row0	t1	Value = String00_t1
Row0	t3	Value = String00_t3
Row0	t5	Value = String00_t5

Table 16.4 Physical storage of column01 for Table 16.2

Row key	Time stamp	Column01
Row0	t0	Value = String00_t0
Row0	t3	Value = String00_t3
Row0	t5	Value = String00_t5

Table 16.5 Physical storage of column02 for Table 16.2

Row key	Time stamp	Column02
Row0	t0	Value = String00_t0
Row0	t3	Value = String00_t3

this property by selecting their row keys so that they get good locality for their data accesses.

Column keys are grouped into sets called `column families`, which form the basic unit of access control. A column family must be created before data can be stored under any column key in that family. A column key has the following syntax: ⟨family:qualifier⟩ where ⟨family⟩ and ⟨qualifier⟩ can be byte arrays of arbitrary lengths. HBase stores column families physically close on disk, so that items in a given column family have roughly the same read/write characteristics.

Each cell in a table can contain multiple versions of the same data – these versions are indexed by timestamp. HBase timestamps are 64-bit integers. Applications that need to avoid collisions must generate unique timestamps themselves. Different versions of a cell are stored in decreasing timestamp order, so that the most recent versions can be read first.

16.4.1 Architecture

The three major components of HBase are:

- *The HBase master*: The HBase Master is responsible for assigning regions to HRegion Servers and monitoring their health. The first region to be assigned is the ROOT[5] region which locates all the META[6] regions to be assigned. Each META region maps a number of user regions which comprise the multiple tables that a particular HBase instance serves. In addition, the HBase Master handles table administrative functions such as on/off-lining of tables, changes to the table schema (adding and removing column families), etc.

- *The HRegion server*: The HRegion Server is responsible for handling client read and write requests. It communicates with the HBase Master to get a list of regions to serve and to tell the master that it is alive. When a write request is received, it is first written to a Write-Ahead Log (WAL) called a *HLog* and is stored in an in-memory cache called the Memcache. Reads are handled by first checking the Memcache, and if the requested data are not found, the MapFiles are searched for results. The HRegion Server is also responsible for region splits.
- *The HBase client*: The HBase client is responsible for finding the HRegion Servers that are serving the particular row range of interest. On instantiation, the HBase client communicates with the HBase Master to find the location of the ROOT region. Once located, the client contacts the region server of interest and scans the ROOT region to find the META region that will contain the location of the user region that contains the desired row range.

Figure 16.4 provides the architecture[7] of HBase. In the following section, we present a case study on storing multidimensional time series data obtained from epilepsy patients.

16.5 Case Study: Large-Scale Storage and Indexing of EEG Data

A large-scale EEG database obtained from the University Hospital of Freiburg Germany was used for benchmarking experiments and testing the design of the multidimensional time series index using Apache Hadoop and HBase.

16.5.1 Data Set Description

Epilepsy is a chronic neurological disorder characterized by recurrent, unprovoked seizures that manifest in a variety of ways, including emotional or behavioral

---

[5]The ROOT table is confined to a single region and maps all the regions in the META table. Each row in the ROOT and META tables is approximately 1 KB in size. At the default region size of 256 MB, this means that the ROOT region can map 2.6×10^5 META regions, which in turn map a total $6.9 \times 1,010$ user regions, meaning that approximately $1.8 \times 1,019$ (2^{64}) bytes of user data.

[6]The META table stores information about every user region in HBase such as start and end row keys, whether region is on or off-line and address that is currently serving the region.

[7]This figure is obtained from http://www.larsgeorge.com/2009/10/hbase-architecture-101-storage.html.

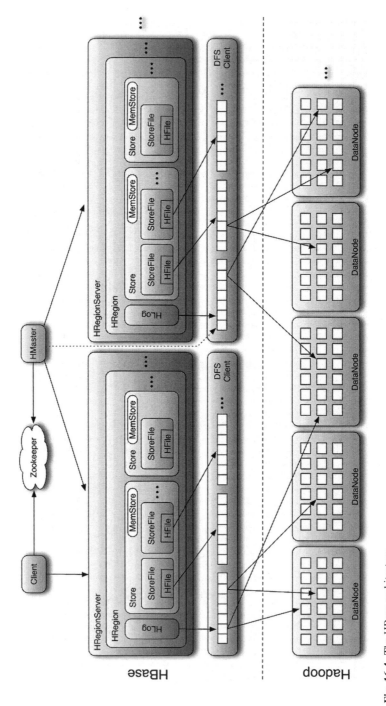

Fig. 16.4 The HBase architecture

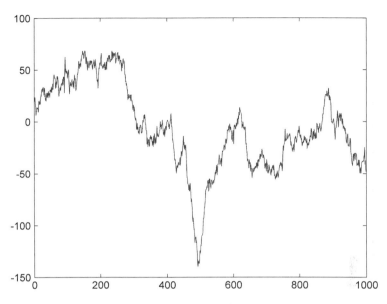

Fig. 16.5 A snippet of EEG time series data. The x-axis plots the time (in seconds) and the y-axis plots the amplitude of the signal

disturbances, convulsive movements, and loss of awareness. EEG has been used in the evaluation of epilepsy and related brain disorders since the 1930s [5]. Interpretation of EEG has traditionally been carried out by visually scanning for recognizable patterns.

The EEG database we use for benchmarking and experimentation contains invasive EEG recordings of 21 patients suffering from medically intractable focal epilepsy. The data were recorded during an invasive presurgical epilepsy monitoring at the Epilepsy Center of the University Hospital of Freiburg, Germany[8] and was recorded using a Neurofile NT digital video EEG system with 128 channels, 256 Hz sampling rate, and a 16-bit analogue-to-digital converter. No prior signal processing (such as application of notch or band pass filters) has been done on the data. For each of the patients, there are datasets called "ictal" and "interictal," the former containing files with epileptic seizures and at least 50 min pre-ictal data and the latter containing approximately 24 h of EEG recordings without seizure activity. Figure 16.5 shows a snippet of EEG time series data obtained from a patient suffering from epilepsy. For each point in time, the amplitude of the signal is recorded. Thus, the conceptual view of the EEG data can be thought of as shown in Table 16.6. For each time point, there are recordings for all the 128 channels. This representation corresponds to a row-oriented database. Using a column-oriented representation, the data can be stored as shown in Table 16.7.

---

[8]http://www.uniklinik-freiburg.de/epilepsie/live/index$_$en.html.

Table 16.6 The conceptual table storing the iEEG data

Row key	Time	Channel 1	Channel 2	$\cdots$	Channel 128
R0	200602190201237000	Value:[123]	Value:[345]	$\cdots$	Value:[0]
R0	200602190201237001	Value:[876]	Value:[123]	$\cdots$	Value:[123]
R0	200602190201237002	Value:[56]	Value:[348]	$\cdots$	Value:[121]

Table 16.7 Physical storage of Channel1 for Table 16.6

Row key	Time stamp	Channel1
Row0	20060219020123700	123
Row0	200602190201237001	876
Row0	200602190201237002	56

16.6 Benchmarking Workloads Using the Yahoo! Cloud Serving Benchmark

The Yahoo! Cloud Serving Benchmark (YCSB) [9] helps to create a standard benchmarking framework to assist in the development of cloud computing systems. The YCSB client is a java program for generating data to be loaded to the database and operations to be made on the workload. The workload executor manages multiple threads; each thread executes a sequential series of operations by which load, read, and write commands are executed on the workload. We use YCSB to test the performance of the EEG dataset stored in HBase running on a HDFS in a small in-house cluster using different workload compositions (read and write access to the database). The Hadoop/HBase combination allows us to scale horizontally by simply adding more nodes; this framework also helps to deal with fault tolerance, data partitioning, and provide an elaborate API, so that the client code can be implemented on a high level of abstraction without the users worrying about low level details such as distributed file system implementation, scheduling, and task coordination. HBase allows random access reads and writes with minimum disk IO overhead. Our in-house cluster is configured in a student laboratory, and there were frequent workstation restarts and shutdowns which resulted in loss of nodes, but Hadoop (HDFS) was able to recover since the data were replicated with replication factor of 3. Thus, there were no outages in HDFS availability.

To LOAD one million records of EEG data into the column-oriented database HBase, the throughput was found to be 1511.46 operations/s with an average latency of 0.6469 ms. The YCSB client allows to benchmark new database systems by implementing the read, insert, update, delete, or scan methods to represent the standard "CRUD" operations (Create, Read, Update, and Delete operations). The workload executor is created and shared among worker nodes. The first workload we tested consisted of 50% read and 50% write operations on 10,000 records in the database of size one million loaded as discussed above. Figures 16.6–16.8 illustrate the statistics obtained over four different runs of the same workload. It is important to note that our usage of HBase and Hadoop will be mostly write once, read many times so the read latency and throughput are more important than write latency and

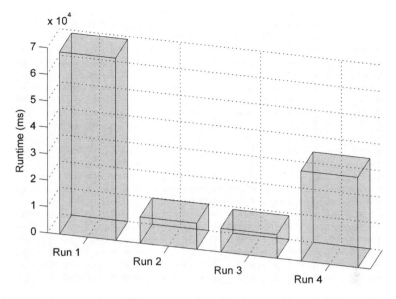

Fig. 16.6 Runtime over four different runs of workload comprising of 50% read and write operations on 10,000 records (6 nodes)

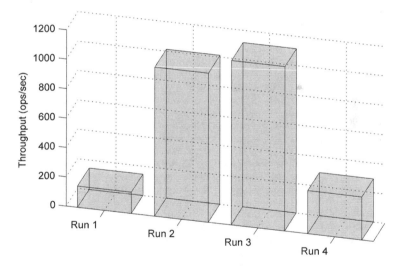

Fig. 16.7 Throughput over four different runs of workload comprising of 50% read and write operations on 10,000 records (6 nodes)

throughput. Read latency was found to be less than a millisecond on average under controlled load which satisfies our initial performance criteria.

The next experiment demonstrated the effect of the number of threads on the runtime, throughput, and average latency of the read and update commands. Table 16.8

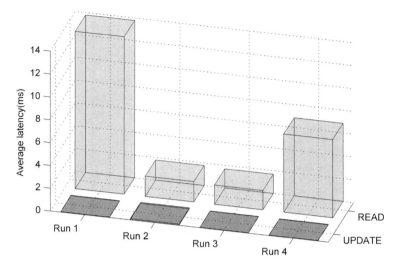

Fig. 16.8 Average latency over four different runs of workload comprising of 50% read and write operations on 10,000 records(6 nodes)

Table 16.8 Average latency with varying number of HBase client threads

No. of threads	Runtime (ms)	Throughput (ops/s)	Avg READ latency (ms)	Avg UPDATE latency (ms)
10	200,310	49.92	7.93	0.372
30	3,071	3256.26	3.11	2.46
50	2,690	3717.47	5.09	4.28
100	2,466	4055.15	11.46	11.17
150	2,217	4510.6	19.24	19.81
200	2,622	3813.88	33.47	28.27

demonstrates the results. If the maximum number of threads is set at 25, for one million records, the throughput achievable is approximately 7,000 operations/s. The throughput generally increases with number of threads used for experimentation; however, this is limited by the number of database accesses made. For example, empirical results revealed that if the number of threads is increased from 25 to 100, performance eventually drops as there is a trade-off between the number of database accesses made and parallelization of jobs run.[9] This indirectly establishes a limit on how many clients the system can serve concurrently. It must also be noted that there are some inconsistencies across runs in terms of throughput since HBase at the time of the experiments was under active development and relatively unstable. Its performance and stability have been significantly improved in

---

[9]We ascertained that this is not a limitation on the client and it was not overloaded in terms of either cpu or bandwidth utilization.

subsequent versions.[10] Finally, the capacity of our system is limited by the hardware and networking configuration of the cluster used. To avoid this dependency, we plan to experiment in future with Amazon AWS (`http://aws.amazon.com/`) as a cloud hosting platform where the inter-cluster network bandwidth is much higher (1-Gbps Ethernet on Amazon EC2 vs. 100 Mbps in our stand-alone setup) than the cluster setup we used for experiments here – large clusters such as Amazon's EC2 also provide more powerful server grade nodes which provide more compute power than workstations typically used in small clusters.

16.7   Conclusion and Future Work

In many scientific domains such as astronomy, social science, and medicine, researchers are faced with a data avalanche. Cloud computing paradigms are being used in these domains for data-intensive science. A popular distributed file system that has been used for large-scale data storage is the Apache Hadoop framework. Column-oriented databases built on the Hadoop, such as HBase, are known to have several advantages over traditional row-oriented databases. MapReduce enables development of large-scale computational tasks on the Hadoop framework. In this chapter, we first provide a brief overview of the Hadoop and HBase infrastructures and then present benchmarking results on a small in-house cluster composed of 29 nodes with 8-core processors each. We present a case study on distributed storage of multidimensional EEG data using Hadoop and HBase and present extensive scalability results using the YCSB. Our results indicate that the Hadoop and HBase ecosystem including the dependencies (services like zookeeper) are still quite immature in terms of stability but promising in terms of the performance characteristics with regard to latency and throughput. The issues pertaining to stability of Hadoop and HBase are being investigated by the project developers in more recent releases. Future work involves design and benchmarking of machine learning algorithms on this infrastructure and pattern matching from large scale EEG data.

Acknowledgements Funding for this work is provided by National Science Foundation award, IIS–0916186. Data for this project were provided by the Freiburg Seizure Prediction EEG Database (FSPEEG). The authors would like to thank administrators of the CLIC Lab Cluster at the Department of Computer Science, Columbia University for help with cluster set-up and experimentation; Dr. David Waltz, Dr. Catherine A Schevon, Dr. Ronald Emerson, Phil Gross, and Shen Wang provided their insightful comments during different phases of the project.

---

[10]HBase 0.20.6 and the more recent development releases which specifically addressed and fixed many bugs which affected performance and stability.

References

1. 2-Micron All Sky Survey. http://pegasus.phast.umass.edu
2. Andersson, P., Mollerstrand, F.: Zohmg – a large scale data store for aggregated time-series-based data. Master's thesis, Chalmers University of Technology (2009)
3. Andrzejak, R.G., Lehnertz, K., Rieke, C., Mormann, F., David, P., Elger, C.E.: Indications of nonlinear deterministic and finite dimensional structures in time series of brain electrical activity: Dependence on recording region and brain state. Phys. Rev. E **64**, 061907 (2001)
4. Aschenbrenner-Scheibe, R., Maiwald, T., Winterhalder, M., Voss, H.U., Timmer, J., Schulze-Bonhage, A.: How well can epileptic seizures be predicted, an evaluation of a nonlinear method. Brain **126**, 2616–2626 (2003)
5. Berger, H.: Uber das elektroencephalogramm des menschen (on the electroencephalogram of man). Archiv fiir Psychiatrie und Nervenkrankheiten **87**, 527–570 (1929)
6. Brinkmann, B.H., Bower, M.R., Stengel, K.A., Worrell, G.A., Stead. M.: Large-scale electro-physiology: Acquisition, compression, encryption, and storage of big data. J. Neurosci. Meth. **180**(1), 185–192 (2009)
7. Cafarella, M., Cutting, D.: Building nutch: Open source search. In: ACM Queue, April 2004
8. Chang, F., Dean, J., Ghemawat, S., Hsieh, W.C., Wallach, D.A., Burrows, M., Chandra, T., Fikes, A., Gruber, R.E.: Bigtable: A distributed storage system for structured data. ACM Trans. Comput. Syst. **26**(2), 1–26 (2008)
9. Cooper, B.F., Silberstein, A., Tam, E., Ramakrishnan, R., Sears, R.: Benchmarking cloud serving systems with ycsb. ACM Symposium on Cloud Computing. ACM, IN, USA (2010)
10. Das, K., Bhaduri, K., Arora, S., Griffin, W., Borne, K., Giannella, C., Kargupta, H.: Scalable distributed change detection from astronomy data streams using local, asynchronous eigen monitoring algorithms. In: Proceedings of the SIAM International Conference on Data Mining, Sparks, Nevada, 2009
11. Dave, R.: Scaling Astronomy. Oreilly Ignite 4, Boston, MA, September 2008. http://timemachine.iic.harvard.edu/publications/#scaling-astronomy
12. Dean, J., Ghemawat, S.: MapReduce: simplified data processing on large clusters. In: Proceedings of the Sixth Symposium on Operating System Design and Implementation, San Francisco, CA, December 2004, pp. 137–150
13. Dean, J., Ghemawat, S.: Mapreduce: A flexible data processing tool. Commun. ACM **53**(1), 72–77 (2010)
14. Ghemawat, S., Gobioff, H., Leung, S.T.: The google file system. In: The 19th ACM Symposium on Operating Systems Principles, lake George, NY (2003)
15. Graves, S.J., Conover, H., Keiser, K., Ramachandran, R., Redman, S., Rushing, J., Tanner, S.: Mining and Modeling in the Linked Environments for Atmospheric Discovery (LEAD). In: Huntsville Simulation Conference, Huntsville, AL, 19 Oct 2004
16. Amazon Elastic Compute Cloud, Amazon EC2. http://aws.amazon.com/ec2/
17. Amazon Simple Storage Service, Amazon S3. http://aws.amazon.com/s3/
18. Grape 6. http://grape.mtk.nao.ac.jp/grape/news/ABC/ABC-cuttingedge000602.html
19. Apache Hadoop. http://hadoop.apache.org/core/
20. Apache Hbase. http://hbase.apache.org/
21. Large Hadron Collider, European Organization for Nuclear Research. http://lhc.web.cern.ch/lhc/
22. Hbase Architecture. http://wiki.apache.org/hadoop/Hbase/HbaseArchitecture
23. Randy Bryant's Home Page. http://www.cs.cmu.edu/~bryant/
24. San Diego Supercomputer Center, SDSC. http://www.sdsc.edu/
25. Hsu, F.H.: Behind Deep Blue: Building the Computer that Defeated the World Chess Champion. Princeton University Press, NJ (2002)
26. Keogh, E.J.: Recent advances in mining time series data. In: PKDD, p. 6 (2005)
27. Keogh, E.J.: A decade of progress in indexing and mining large time series databases. In: VLDB, p. 1268 (2006)

28. Lin, J., Vlachos, M., Keogh, E., Gunopulos, D.: Iterative incremental clustering of time series. In: Proceedings of the IX Conference on Extending Database Technology (2004)
29. Reeves, G., Liu, J., Nath, S., Zhao, F.: Managing massive time series streams with multi-scale compressed trickles. In Proceedings of the 35th Conference on Very Large Data Bases, Lyon, France, 2009
30. Shieh, J., Keogh, E.J.: isax: indexing and mining terabyte sized time series. In: KDD, pp. 623–631 (2008)
31. Sloan Digital Sky Survey. http://www.sdss.org
32. The human genome project. http://www.ornl.gov/sci/techresources/Human$_$Genome/home.shtml
33. The protein data bank (pdb). http://www.rcsb.org/pdb/Welcome.do
34. The swiss-prot protein knowledge base. http://www.expasy.org/sprot/
35. World data center for meterology. http://www.ncdc.noaa.gov/oa/wmo/wdcamet.html
36. Yankov, D., Keogh, E.J., Rebbapragada, U.: Disk aware discord discovery: Finding unusual time series in terabyte sized datasets. In: ICDM, pp. 381–390 (2007)

Index

Algorithm
- cache replacement, 99
- classification, 252
- clustering, 263, 266, 334
- compression, 112
- data mining, 196
- data processing, 238
- decision tree, 207
- development, 172
- hashing, 93
- iterative, 273
- join, 161
- machine learning, 265, 332, 345
- meta-heuristic, 226
- microarray, 262
- PAM, 267
- query processing, 135
- scheduling, 240
- sort, 161
- SPRINT, 209

Authentication
- LDAP-based, 110
- password-based, 110
- RBAC, 110

Cloud
- computing, 23, 194
- database, 109, 133, 181
- environments, 318
- infrastructures, 261, 262, 331
- service, 185, 193

Cloud computing
- IaaS, 24, 109, 170, 196
- PaaS, 24, 109, 170, 197
- Software as a Service (Saas), 109, 170, 193, 197
- XaaS, 109

Cloud offering
- Amazon EC2, 170
- Microsoft Windows Azure, 170

Cloud providers
- Amazon, 170
- Microsoft, 170

Computing utility, 187

Data, 215
- access, 188, 203
- acquisition, 280
- aggregation, 176
- caching, 162
- calibration, 296
- collections, 261
- consistency, 180, 183
- cube, 308
- curation, 228
- delivery, 280
- distribution, 269
- environmental, 320
- federation, 204
- formats, 203
- geospatial, 284
- integration, 203, 280
- locality, 242, 274
- meteorological, 299
- mining, 193, 307
- multidimensional, 312
- ownership, 188
- partitioning, 140, 160, 342
- postprocessing, 299
- preprocessing, 299
- processing, 261, 280, 317
- provenance, 215